Springer Series on Polymer and Materials

CW00516372

Series Editor

Susheel Kalia, Army Cadet College Wing, Indian Military Academy, Dehradun, India

The "Springer Series on Polymer and Composite Materials" publishes monographs and edited works in the areas of Polymer Science and Composite Materials. These compound classes form the basis for the development of many new materials for various applications. The series covers biomaterials, nanomaterials, polymeric nanofibers, and electrospun materials, polymer hybrids, composite materials from macro- to nano-scale, and many more; from fundamentals, over the synthesis and development of the new materials, to their applications. The authored or edited books in this series address researchers and professionals, academic and industrial chemists involved in the areas of Polymer Science and the development of new Materials. They cover aspects such as the chemistry, physics, characterization, and material science of Polymers, and Polymer and Composite Materials. The books in this series can serve a growing demand for concise and comprehensive treatments of specific topics in this rapidly growing field. The series will be interesting for researchers working in this field and cover the latest advances in polymers and composite materials. Potential topics include, but are not limited to:

Fibers and Polymers:

- Lignocellulosic biomass and natural fibers
- Polymer nanofibers
- Polysaccharides and their derivatives
- Conducting polymers
- Surface functionalization of polymers
- Bio-inspired and stimuli-responsive polymers
- Shape-memory and self-healing polymers
- Hydrogels
- Rubber
- Polymeric foams
- Biodegradation and recycling of polymers

Bio- and Nano- Composites:-

- Fiber-reinforced composites including both long and short fibers
- Wood-based composites
- Polymer blends
- Hybrid materials (organic-inorganic)
- Nanocomposite hydrogels
- Mechanical behavior of composites
- The Interface and Interphase in polymer composites
- Biodegradation and recycling of polymer composites
- Applications of composite materials

More information about this series at http://www.springer.com/series/13173

Santosh Kumar Tiwari · Kashma Sharma ·
Vishal Sharma · Vijay Kumar
Editors

Electrospun Nanofibers

Fabrication, Functionalisation
and Applications

 Springer

Editors
Santosh Kumar Tiwari (iD)
School of Resources, Environment
and Materials
Guangxi University
Nanning, China

Vishal Sharma
Institute of Forensic Science
and Criminology
Panjab University
Chandigarh, India

Kashma Sharma
Department of Chemistry
DAV College, Sector-10
Chandigarh, India

Vijay Kumar
Department of Physics
National Institue of Technology Srinagar
Hazratbal, Jammu and Kashmir, India

Department of Physics
University of the Free State
Bloemfontein, South Africa

ISSN 2364-1878   ISSN 2364-1886 (electronic)
Springer Series on Polymer and Composite Materials
ISBN 978-3-030-79981-6   ISBN 978-3-030-79979-3 (eBook)
https://doi.org/10.1007/978-3-030-79979-3

This Springer imprint is published by the registered company Springer Nature Switzerland AG
The registered company address is: Gewerbestrasse 11, 6330 Cham, Switzerland

Preface

Nanofibers have stimulated a new class of material which employed for a considerable range of applications from medical to consumer products such as, filtration, wound dressing, adsorbent, energy storage, protein separation, immobilization, drug delivery, and composites. Currently, the electrospinning of nanofibers is emerging as a specialized processing technique for the formation of sub-micron fibers, with high specific surface areas. This procedure does not need the use of high temperature or the coagulation process to yield solid threads of the materials from the solution. This makes the electrospinning technique, particularly suitable to produce the desired kind of fibers using a complex molecular system. Owing to the scalable production rate and simplicity of the experimental setup turn out electrospinning device exceedingly attractive to both industry and academia. The main advantage of this technique is that the researchers could tailor and produce desired types of nanofibers using a variety of materials with required functional groups onto the surface of the fiber by tuning the assemblies of the electrospinning device. This technique offers an adorable route to incorporate nanoparticles into the polymer fibers. Therefore, in this book, we would like to deliberate and explore in deep about the current advances in research activities related to the electrospun nanofiber, selective functionalization of nanofiber, incorporation of 1D, 2D, or 3D nanomaterials into nanofiber with the help of the electrospinning technique, and applications of these nanomaterials in various research areas. Moreover, in this book, special attention has been paid to the growing application of electrospun nanofibers in energy as well as environmental research. The book contains 13 chapters and each chapter provides the basic knowledge and deep ongoing research in recent days. The present book assesses the delineation of various techniques of fabrication and functionalization of electrospun nanofibers and their advanced applications in different areas, which will be an asset to beginners. The book will provide the execution of fabrication and functionalization of electrospun nanofibers into practical devices via the knowledge of various materials. However, our main aim with this book is to inspire and develop young minds towards electrospun nanofibers research for future prospects. We are grateful to the authors of the chapters for their exceptional assistance in the completion of the book. We would also like to acknowledge the admirable control at Springer Nature for their proficiency and commitment to publishing the book. We hope the present book will help the readers to comprehend Electrospun Nanofibers: Fabrication, Functionalization, and Application at their end.

Nanning, China   Santosh Kumar Tiwari
Chandigarh, India   Kashma Sharma
Chandigarh, India   Vishal Sharma
Hazratbal, India   Vijay Kumar

Contents

About the Editors

Dr. Santosh Kumar Tiwari Scientist (NAWA), Laboratory of Nanomaterials Physics and Chemistry,
Department of Chemistry, Warsaw University, 1 Pasteur Str., 02-093 Warsaw, Poland
Email. ismgraphene@gmail.com, nanochem@chem.uw.edu.pl

He received his Ph.D. from IIT Dhanbad, India, in Graphene-based Polymer nanocomposites and then worked with HSCL, Hanyang University, Seoul, South Korea for 1.5 years. In 2019, he joined the University of Guangxi, Nanning, China as a tenured faculty and then moved to Poland where he got a prestigious NAWA research grant from the Govt. of Poland and presently working as a scientist in the Department of Chemistry, University of Warsaw. In addition, He is a visiting faculty in the Department of Chemical Science & Engineering, Kathmandu University, Nepal. He published more than 45 research articles in reputed scientific journals and presented his works as an invited speaker in different countries. To date, he has edited four books on graphene-based nanomaterials for Springer, Nature. He got several awards and qualified for the top national exams organized in India for research and admission in PhDs including IIT Dhanbad Ph.D. entrance, NET, UGC/CSIR-JRF, and GATE organized by HRD. His research interest includes 2D Carbon Nanomaterials, Biomass-Based Nanostructures, Mechanical Properties, and Smart Materials.

Dr. Kashma Sharma Assistant Professor (Chemistry)
DAV College, Sector -10, Chandigarh—160011 (INDIA)
Email: kashma@davchd.ac.in; shama2788@gmail.com

She has received her Ph.D. from Shoolini University of Biotechnology and Management Sciences, Solan (H.P.) in Collaboration with the National Institute of Technology (NIT) Jalandhar. She has done her master's degree in Chemistry from Punjab University, Chandigarh. In September 2014, she joined the material science group at the University of the Free State, South Africa, as a postdoctoral research fellow. She was awarded UGC Postdoctoral Fellowship to Women in 2017 and carried out the research work in the Institute of Forensic Science & Criminology, Panjab University Chandigarh. She started her independent academic career as an Assistant Professor in the Department of Chemistry, DAV College, Chandigarh, at the end of April 2021. Her research experience, as well as research interests, lies in the following fields: synthesis, processing, and characterization of bio-based/biodegradable polymers and composite, drug delivery devices, tissue engineering, wastewater treatment, polymer nanocomposites, and ion solid interaction. She also worked in Grant-in-aid for the joint center project titled "Joint Center for generating tissue-engineered organs and controlling cell behavior" under Indo-US Joint R&D Networked Center. So far, she has been published more than 25 research papers in many of the reputed international journals on polymeric materials, which attracted more than 620 citations with an h-index of 14 and an i10-index of 17. She is editing three books and published five chapters.

Dr. Vishal Sharma Professor (Assistant) (Physical Sciences)
Institute of Forensic Science & Criminology
(1st Floor Old Biomedical Sciences-BMS Block)
Panjab University, Sector-14, Chandigarh-160 014 (India)
Email: vsharma@pu.ac.in, vishalsharma.pu@gmail.com

He is presently working as Professor (Assistant), Institute of Forensic Science and Criminology (IFSC), Panjab University, Chandigarh (INDIA). He has obtained his Ph.D. degree in Physics discipline from Kurukshetra University, Kurukshetra, India, and Inter-University Accelerator Centre (IUAC—an autonomous center of UGC, Govt. of India), New Delhi (INDIA), in the year 2007. He is the recipient of the DAE Young scientist research award in the year 2011. He is leading a material science research group at the IFSC, Panjab University, Chandigarh. His current research interest is in the field of material science—Functional materials, Ion beam analysis, Sensors, Nanophosphors, Hydrogels; Analytical Chemistry; IR Spectroscopy; Forensic Chemistry; Chemometrics. He is the author of over 90 peer-reviewed scientific papers, 6 book chapters, and 1 book as editor from Springer Nature. He is on the editorial board of two International journals of Elsevier. He has delivered Keynote, invited talk, session chair, and presented his work at various national and International Conferences. He has been invited for expert lectures abroad by various prestigious societies and visited countries like Spain, South Africa, Canada, Poland, and Thailand.

Dr. Vijay Kumar Assistant Professor, Department of Physics
National Institute of Technology Srinagar,
Jammu and Kashmir—190006, India
Mob No.: +91-6005495506
Email: vijaykumar@nitsri.ac.in; vj.physics@gmail.com

He is presently working as Assistant Professor, Department of Physics, National Institute of Technology Srinagar (J&K), India. He received his Ph.D. in Physics at the beginning of 2013 from the SLIET Longowal. From 2013 to 2015, he was a post-doctoral research fellow in the Phosphor Research Group, Department of Physics, University of the Free State, South Africa. He has received the Young Scientist

Award under the fast track scheme of the Department of Science and Technology (Ministry of Science and Technology, Government of India), New Delhi. He also worked in Grant-in-aid for the joint center project titled "Joint Center for generating tissue-engineered organs and controlling cell behavior" under Indo-US Joint R&D Networked Center. He has been a nominated member of the Scientific Advisory Committee for Initiative for Research and Innovation in Science (IRIS) by DST. He has also received the "Teacher with Best Research Contribution Award" by Hon'ble Chancellor, Chandigarh University on Teachers Day. He is currently engaged in the research of functional materials, solid-state luminescent materials, ion beam analysis, biodegradable composites, smart polymers, and biomedical applications. He has more than 70 research papers in international peer-reviewed journals, 10 peer-reviewed conference proceedings, 6 book chapters, and edited 4 books (authored and co-authored). He has more than 3220 citations with an h-index of 33 and an i10-index of 66. He has edited the Virtual Special Issue of VACUUM and Materials Today: Proceedings and currently editing a special issue "Nano Biocomposites for Future Bioeconomy" of Crystals (MDPI Journal). He has organized various short-term courses and conferences/workshops.

Production and Application of Biodegradable Nanofibers Using Electrospinning Techniques

Tomasz Blachowicz and Andrea Ehrmann

Abstract Electrospinning is a versatile method to produce nanofibers or nanofiber mats from diverse polymers or polymer blends. Including ceramic or metallic nanoparticles can even be used to create purely inorganic nanofibers for diverse applications. On the other hand, biocompatible and biodegradable polymers are of high interest especially for biomedical applications. Biodegradable nanofiber mats as scaffolds can be used in tissue engineering, especially when degradation times are in the same order of magnitude as cell proliferation on these substrates. Biodegradation, however, involves more aspects than the pure time profile. Especially for utilization in vitro and in vivo, byproducts of degradation processes may lead to undesired reactions with the surrounding tissue, and vice versa. Here, we give an overview of the production techniques of biodegradable nanofibers and nanofiber mats by different electrospinning techniques. In addition, we report on biotechnological and biomedical applications of such fully or partly biodegradable nanofibers and show the chances and challenges in interaction with living tissue and organisms.

Keywords Electrospinning · Biodegradable nanofibers · Degradation processes · Biocompatibility · Biodegradability

1 Introduction

Electrospinning can be used to prepare fibers with diameters in a typical range of some ten to some hundred nanometers, sometimes up to the range of a few micrometers [1–3]. Due to their small diameter and the corresponding large surface-to-volume ratio, there are diverse applications of such nanofibers or nanofiber mats,

T. Blachowicz
Institute of Physics—Center for Science and Education, Silesian University of Technology, ul. Konarskiego 22B, 44-100 Gliwice, Poland
e-mail: tomasz.blachowicz@polsl.pl

A. Ehrmann (✉)
Faculty of Engineering and Mathematics, Bielefeld University of Applied Sciences, Interaction 1, 33619 Bielefeld, Germany
e-mail: andrea.ehrmann@fh-bielefeld.de

© The Author(s), under exclusive license to Springer Nature Switzerland AG 2021
S. K. Tiwari et al. (eds.), *Electrospun Nanofibers*, Springer Series on Polymer and Composite Materials, https://doi.org/10.1007/978-3-030-79979-3_1

for example, in the biomedical or biotechnological area [4–6], in filters [7–9], batteries, solar cells, and supercapacitors [10–12].

While nanofiber mats are often spun from polymers like polyacrylonitrile (PAN) and other petrochemical polymers [13–15], there are also diverse biopolymers which can be electrospun, e.g. proteins like gelatin [16], collagen [17], etc., polysaccharides like chitosan [18], cellulose [19], dextrose [20], etc., and diverse composites of two or more biopolymers as well as biopolymers blended with petrochemical polymers. While biopolymers generally stem from non-oil-based resources, here we have a deeper look into biodegradable polymers, i.e. polymers that are degraded by microorganisms or enzymes on time scales between hours and years [21]. Degrading means a high-molecular weight polymer is degraded into lower molecular weight fractions, in addition to modifications of CO_2 and oxygen content, finally resulting in a full collapse of the structure and the corresponding loss of the mechanical properties [22]. Especially in biomedical applications, biodegradable nanofibers are of high interest since they can be used for degradable implants, making surgical implant removal unnecessary; they are used in tissue engineering and similar life science applications [23].

This chapter is organized as follows: The next sub-chapter gives a short overview of biodegradation mechanism, followed by a sub-chapter describing possibilities to use typical biodegradable polymers for electrospinning, either solely or combined with a spinning agent which can be biodegradable or long-term stable. It should be mentioned that different biodegradable polymers show strongly different physical and chemical properties, which make the corresponding nanofibers highly interesting for different possible applications, of which several examples are given.

2 Biodegradation

Far more than 100 million tons of synthetic polymers worldwide are produced yearly, resulting in large amounts of household and industrial waste [24–26]. The idea of using biodegradable polymers is thus related not only to biomedical applications but also a reasonable method to reduce especially agricultural polymer waste [27]. Biodegradation means that microorganisms degrade a polymer, typically in the form of bacteria, fungi, and algae [28], by oxidation and hydrolysis to produce carbon dioxide, methane, and residual biomass as well as carbon in case of typical synthetic polymers, which is also converted into carbon dioxide [25–27]. While aerobic biodegradation, in the presence of oxygen, results in carbon dioxide production, anaerobic degradation mostly leads to methane production [29, 30].

Biodegradation of most synthetic polymers is complicated, but often nevertheless possible under well-suited conditions [31, 32]. Typical approaches to prepare such polymers are using synthetic polymers with special groups which are prone to hydrolytic microbial attack, biopolyesters which can be derived from bacterial sources, and mixing synthetic polymers with natural ones which can be easily degraded by microorganisms, such as starch [31–33].

It should be mentioned that biodegradation does not only mean that the final step, resulting in carbon dioxide, water, and some other byproducts but also that several smaller molecules are usually formed along the way [34]. For poly(lactic acid (PLA) and poly(glycolides) (PGA), e.g. the small molecule lactic acid and glycolic acid are formed [35], which must be taken into account if these materials are used as biodegradable screws fixing broken bones since biodegradation will make the surrounding of the treated bone area more acidic and can result in inflammations [36]. Neutralizing or at least reducing this effect belongs to the important topics of recent research on PLA implants [37–39].

Another important point to mention is related to the time scales and environmental requirements of biodegradation—some materials which are claimed to be biodegradable may necessitate years or longer for at least partial biodegradation, and in many cases highly specialized environments are required which cannot be reached, e.g. in the common compost [40–42].

3 Electrospinning

A large amount of natural and synthetic polymers has been made available for electrospinning, including many biodegradable and biocompatible polymers which are typically used for biomedical and biotechnological applications [43]. The electrospinning process generally uses a strong electric field to drag polymer droplets from a polymer solution or melt from an electrode to a counter electrode. The most common setup is based on a syringe which constantly ejects the polymer solution or melt through a needle into the electric field [44]. At the tip of the needle, a so-called Taylor cone is formed. If the surface tension is overcome by the applied electric field, the polymer is ejected from the tip, stretched and accelerated until it reaches the collector where polymeric nanofibers are deposited [45]. This short description already suggests that many parameters will influence fiber formation, including conductivity, surface tension, molecular weight, and viscosity of the spinning solution, dimensions of the electrospinning equipment and the applied electric field, as well as environmental conditions such as temperature and humidity in the spinning chamber [46]. Besides the needle-based electrospinning process, electrospinning can also be performed using wires (Fig. 1), rotating cylinders, or other shapes as ejecting electrodes [47–49]. Similarly, the substrates may be composed of different materials and show different shapes, including fast rotating electrodes which can be used to align the nanofibers [50–52].

3.1 Solvents

An important issue in electrospinning biodegradable polymers is related to the solvent used to prepare the spinning solution. Several polymers, such as poly(ethylene oxide)

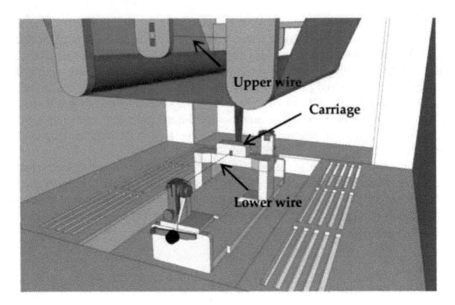

Fig. 1 Sketch of the Nanospider Lab, a commercially available equipment for wire-based electrospinning. From [47], originally published under a CC-BY license

(PEO) or gelatin, can be dissolved in water [53–55]. The disadvantage that the corresponding nanofiber mats can again be dissolved in water can be overcome by blending water-soluble and water-resistant polymers [56, 57] or by crosslinking the nanofiber mats after electrospinning [58–61].

On the other hand, some biodegradable polymers such as PLA need much more sophisticated solvents or solvent combinations to be electrospun. Septiyanti et al. reported recently on stereocomplex PLA, formed by solution blending poly(L-lactic acid) (PLLA) and poly(D-lactic acid) (PDLA) via electrospinning. While PLA is generally dissolvable in chloroform, they added N,N-dimethylformamide (DMF) to decrease the viscosity and hexafluoroisopropanol (HFIP) as a polar solvent with low surface tension to prepare thin fibers by needle-based electrospinning [62].

On the other hand, combining chloroform with a non-solvent of PLA such as dimethyl sulfoxide (DMSO) results in electrospinning of nanofibers with internal porosity, while surface porosity can be reached by adding ethanol to chloroform [63]. Blending PLA with ethylene vinyl acetate (EVA) was enabled using a solvent mixture from acetone and dichloromethane, resulting in electrospun nanofibers with good strength and flexibility [64]. Ghafari et al. tested solvent mixtures for electrospinning nanofiber mats from PLA, PEO, and cellulose and found that a chloroform/acetone/ethanol mixture showed a good dispersion of cellulose nanofibers and good electrospinnability of the polymer solutions. The resulting nanofiber mats also showed strong deviations, depending on the solvent mixtures, for example, of the mechanical properties and the water uptake [65]. While poly(vinyl alcohol) (PVA) can be electrospun from an aqueous solution, poly [(R)-3-hydroxybutyrate-co-(R)-3-hydroxyhexanoate] (PHBH)/PVA nanofibers—which are interesting due to their high water uptake ability—can be electrospun using hexafluoroisopropanol (HFIP) or HFIP/water as solvent [66]. Hyaluronic acid is water-soluble, but belongs to the biopolymers that are usually claimed to be not electrospinnable solely due to the high electrical conductivity of the polymer solution, resulting in possible short-circuits between both electrodes [67]. Gelatin or another spinning agent was suggested to prepare corresponding blend fibers [68]. Nevertheless, some groups found possibilities not only to spin hyaluronic acid from sophisticated solution mixtures, such as distilled water/formic acid/DMF [69] or DMF/distilled water alkali solutions [70], but also by mixing water with the low-toxic dimethyl sulfoxide (DMSO) [71]. For a recent review on electrospinning hyaluronic acid, the reader is referred to [67]. A broad overview of possible solvent or solvent mixtures for electrospinning blends of natural and natural polymers is given in [72].

4 Biodegradable Natural and Man-Made Polymers

As mentioned before, biodegradable polymers are found among natural as well as man-made polymers. A brief overview of some of the most often used ones is given in Fig. 2 [72]. While biodegradation of the natural polymers occurs via the enzymatic route, by fungi, bacteria, etc. [73], hydrolytic biodegradation occurs in some of the synthetic polymers. Mechanical properties are on the average higher for the synthetic polymers, with exceptions such as the highly water-soluble PEO on the one side and the relatively strong silk fibroin on the other side. While natural polymers are usually hydrophilic, the hydrophobicity of the synthetic polymers depends on the material and can in some cases (e.g. polylactic-co-glycolic acid, PLGA) even be modified by chemical after-treatments. Electrospinning of these and some other interesting biodegradable polymers will be discussed in the next sub-chapters.

5 Electrospinning Biodegradable Polymers

5.1 Electrospinning Collagen

As the main protein of the extracellular matrix, collagen is of high interest for biomedical applications such as tissue engineering and drug delivery [74, 75]. To overcome the aforementioned weak mechanical properties of collagen nanofiber mats (Fig. 2), collagen is often blended with other natural or man-made polymers [76].

Most recently, Fahmi et al. prepared collagen/cellulose acetate (CA) electrospun nanofibers with embedded $MnFe_2O_4$ magnetic nanoparticles, enabling controlling the release of NAP under magnetic induction. They blended CA in acetone with

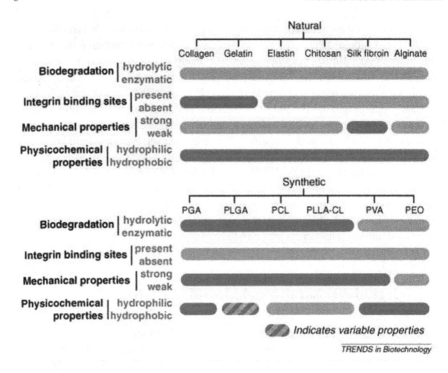

Fig. 2 Often used natural and synthetic biodegradable polymers with some physical, biological, and chemical properties. From [72], reprinted with permission from Elsevier

collagen dissolved in water to prepare a homogenous spinning solution in which the magnetic nanoparticles, dispersed in chloroform, as well as naproxen as a model drug, were added. This solution was electrospun using needle-based equipment and a cylinder collector, as depicted in Fig. 3, after optimizing the flow rate which was found to be critical for stable nanofiber formation [77].

Labbafzadeh et al. prepared polyvinyl alcohol (PVA)/collagen electrospun nanofibers with Fe_3O_4 (magnetite) nanoparticles for bovine serum albumin (BSA) release and also found a strong impact of an external magnetic field on the BSA release, enabling approximately one order of magnitude higher BSA release than without a magnetic field [17]. Combining collagen with polycaprolactone (PCL), Rather et al. prepared a drug-delivery scaffold for early osteogenic differentiation by needle-based electrospinning. Two supporting drugs were encapsulated in the electrospun nanofibers and slowly released for 4 weeks. While these drugs slightly influenced the fiber diameters, the general diameter distributions and morphologies showed high-quality fibers [78].

Blending Zein/PCL dissolved in chloroform/ethanol with collagen dissolved in ethanol resulted in spinning solutions with different amounts of collagen. Needle-based electrospinning resulted in different fiber morphologies, depending on the amount of collagen. These nanofibers could additionally be loaded with aloe vera and

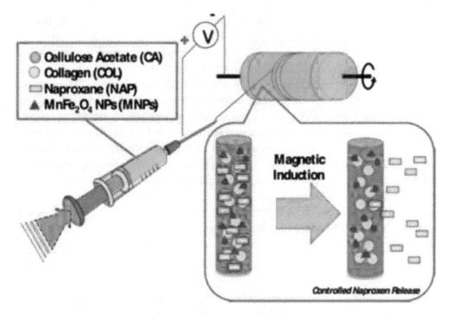

Fig. 3 Needle-based electrospinning of collagen/CA solution with additional magnetic nanoparticles and a model drug. From [77], reprinted with permission from Elsevier

ZnO nanoparticles for wound healing applications [79]. Gao et al. blended collagen type I in different weight ratios with cellulose diacetate-graft-poly(ethylene terephthalate), dissolved in HFIP, for needle-based electrospinning. The resulting nanofiber mats were crosslinked in glutaraldehyde vapor. While the mean fiber diameters and the water contact angles decreased with increasing collagen content, no significant differences were found in the mechanical properties. On the other hand, proliferation of bone marrow mesenchymal stem cells was supported by larger amounts of collagen [80]. Besides these few most recent studies on electrospinning collagen blends, only very few reports exist on electrospinning pure collagen. Most recently, Berechet et al. reported on electrospinning collagen hydrolysate, loaded with essential oils, with a needle-based technology [81]. Several other possibilities exist to combine nanofibrous structures with collagen, for example, by grafting collagen onto a nanofiber mat [82]. Here, however, these attempts are not further described.

5.2 Electrospinning Gelatin

Opposite to collagen, gelatin is more often electrospun purely. Santos de Oliveira et al. used photographic-grade gelatin for electrospinning in a wire-based setup from acetic acid aqueous solution (50 vol%) and crosslinked the nanofiber mats in formaldehyde vapor for different durations. These nanofiber mats were investigated by scanning electron microscopy (SEM) and phase-contrast X-ray computed tomography at the nanoscale (nano-CT) (Fig. 4), showing the strong influence of the crosslinking duration on the nanofiber mat structure [83].

A gelatin inner layer, loaded with menthol, was embedded between Balangu seed gum outer layers in a fully electrospun sandwich structure. Without crosslinking, the menthol release was prolonged, as compared to the pure gelatin nanofiber mat. Nevertheless, it should be mentioned that here release occurred on an order of magnitude of 2 min, not during hours or days, as would be expected for crosslinked nanofiber mats [84]. Zhang et al. added gum Arabic to a gelatin electrospinning solution and found not only better electrospinning characteristics but also high thermal decomposition stability upon heating up to 250 °C due to electrostatic interactions and new hydrogen bonds between these materials [85]. Core–shell fibers with poly(lactic-co-glycolic acid) (PLGA) core and gelatin shell were prepared by co-axial electrospinning. Combining two such electrospun layers with an inner pure PLGA nanofiber

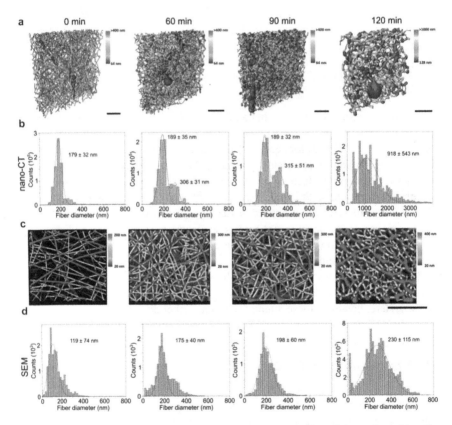

Fig. 4 Fiber characterization by **a** Nano-CT with colors encoding fiber thickness; **b** resulting fiber diameter distributions; **c** scanning electron microscopy images with color-coded fiber thickness; and **d** resulting fiber thickness. From [83], reprinted with permission from Elsevier

layer resulted in good mechanical strength, suture strength, and biocompatibility of this electrospun sandwich, making this approach useful for tissue engineering applications [86].

Besides such studies on pure and blended gelatin nanofibers, there are also attempts to use gelatin as a coating on nanofiber mats of different materials, for example, to improve the biocompatibility of the resulting structure. Du et al. produced calcium silicate nanofibers by electrospinning, followed by calcination at high temperatures between 800 °C and 1200 °C, and coated the resulting nanofiber mats with gelatin to improve their mechanical stability. By implanting these scaffolds in rat calvarial defects, new bone formation was shown [87]. As this brief overview of the most recent investigations in electrospinning gelatin shows, this material offers a broad variety of possible utilizations in electrospinning purely, in blends, sandwiches, or core–shell fibers.

5.3 Electrospinning Elastin

Elastin is only scarcely used in electrospinning. Since it belongs, together with collagen, to the main components of the extracellular matrix of the abdominal wall [88], it is nevertheless interesting for hernia repair and similar applications [89, 90]. On the other hand, elastin is insoluble since it contains crosslinked molecules between amino acid chains, making it necessary to hydrolyze the material before removing these crosslinked molecules [91]. Cao et al. dissolved different ratios of chitosan and elastin in a HFIP/acetic acid mixture for needle-based electrospinning. They found different fiber diameters, depending on the chitosan content, and increasing numbers of beads along the fibers for higher elastin content [92]. Adding elastin to PLGA electrospinning solution resulted in higher mechanical properties of the resulting nanofibers which supported the regeneration of epithelial organs [93]. Electrospun silk-elastin nanofiber mats showed good cytocompatibility and improved cell proliferation [94]. For dermal tissue engineering, human elastin/collagen composite scaffolds were electrospun, increasing cell migration and proliferation [95, 96]. Besides these interesting applications, studies on electrospinning elastin are rare.

5.4 Electrospinning Chitosan

In contrast to elastin, chitosan is often used purely or in blends with other materials. Nikbakth et al. loaded chitosan/PEO blend electrospun nanofibers with aloe vera for biomedical applications. They found the often occurring burst release within the first 5 hours, followed by a stain release over 30 h, and good biocompatibility [97]. For a completely different application, Surgutskaia et al. also used chitosan/PEO nanofibers. They modified chitosan with diethylenetriaminepentaacetic acid (DTPA), using different DTPA contents. DTPA is a chelating agent, forming stable complexes with several metals, in this way supporting sorption properties of chitosan-based materials. As depicted in Fig. 5, the electrospun chitosan-DTPA/PEO was able to adsorb different metal ions after crosslinking in glutaraldehyde vapor [98].

Blending chitosan with PVA and encapsulating the antioxidant peptide ML11, Sannasimuthu et al. found increased wound healing activity in NIH-3T3 mouse embryonic fibroblast cells [99]. Mojaveri et al. also applied chitosan/PVA hybrid electrospun nanofibers to load them with the probiotic Bifidobacterium animalis subsp. lactis Bb12 and the prebiotic inulin. They found a strongly increased surviv-ability of the cells in gastric and intestinal fluids, showing that such nanofiber mats can be used for protecting living probiotics in functional food [100]. Core–shell fibers composed of curcumin loaded cyclodextrin-graphene oxide core and gallic acid loaded chitosan shell were coaxially electrospun for controlled release of both drugs, which showed higher anti-cancer, antioxidant, and antimicrobial activity as well as anti-inflammatory properties than fibers loaded with one of the drugs [101]. Chitosan/PEO/berberine blend nanofibers were prepared by electrospinning, resulting in uniform, bead-free biocompatible fibers with drug release properties which could effectively support wound healing [102].

Pure chitosan nanofibers were prepared by modification with the negatively charged surfactant sodium dodecyl sulfate and after-treated with hemoglobin protein, in this way preparing a biosensor for electrocatalytic monitoring of hydrogen peroxide [103]. Electrospinning chitosan fibers loaded with simvastatin, a restenosis prevention drug, on stents showed constant drug delivery and will be tested in vivo in a future study [104]. While this brief overview already shows the broad range of possible material blends and applications of electrospun chitosan and chitosan blend

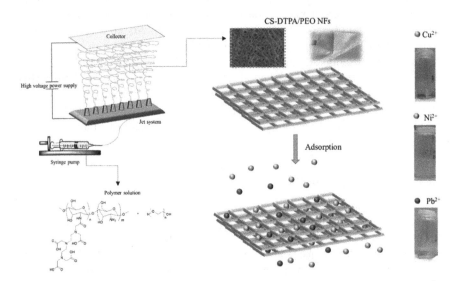

Fig. 5 Scheme of preparation and adsorption of chitosan-DTPA/PEO nanofiber mats. From [98], reprinted with permission from Elsevier

nanofibers, a comprehensive review of chitosan-based nanofiber mats with bioactive and therapeutic agents for would healing and skin regeneration can be found in [105].

5.5   Electrospinning Silk Fibroin

Natural silk has a fibrillary structure, with high elastic strength stemming from the silk fibroin protein [106]. Silk fibroin microparticles were electrospun, e.g. with poly(methylmethacrylate) (PMMA) and hyperbranched poly(ethyleneimine) (PEI) from DMF in different ratios. Mechanical tests showed a significant increase of the tensile stress of PMMA/PEI/silk fibroin nanofibers, as compared to pure PMMA nanofibers [107]. Silk fibroin was also electrospun blended with PCL and poly(glycerol sebacate) (PGS) from acidic solutions, resulting in good fibroblast attachment and growth of the resulting nanofiber mats [108]. To produce conductive silk fibroin nanofiber mats, Liu et al. embedded graphene into the spinning solution and produced highly aligned nanofibers by electrospinning to support cell adhesion and directional growth. They found an optimum balance between electrochemical and mechanical properties for a graphene content of 3%, resulting in enhanced neurite elongation on the corresponding nanofiber mats, making them possible candidates for electrically active scaffolds for neural regeneration [109]. Pure silk fibroin was electrospun, using silk from the mulberry silkworm *Bombyx mori*, by degumming the materials and dissolving it. By using high concentrations of 40–50 wt%, straight nanofibers could be spun from the pure silk fibroin, showing full biocompatibility [110]. Besides the aforementioned applications, silk fibroin nanofibers can also be used for drug delivery. A recent review on this possible application of silk fibroin electrospun nanofibers is given in [111].

5.6   Electrospinning Alginate

Alginate also belongs to the biopolymers which are often used in electrospinning. Alginate nanofibers can, for example, be used to encapsulate probiotic bacteria, similar to the aforementioned chitosan. As depicted in Fig. 6, *Lactobacillus paracasei* KS-199 was encapsulated in electrospun alginate nanofibers. The aqueous spinning solution was prepared from PVA/sodium alginate since pure alginate has a high conductivity and surface tension and is thus poorly electrospinnable. Electrospinning was performed using a needle-based system, followed by drying the nanofiber mats. While the pure spinning solution resulted in bead-free, smooth nanofibers, beaded nanofibers were created by adding the probiotic bacteria in which the viability of the cells was significantly increased under gastrointestinal conditions, as compared to free cells [112].

Similarly, Aloma et al. used PVA as a spinning agent for alginate, also resulting in nanofiber mats with good tensile strength and elongation properties, making

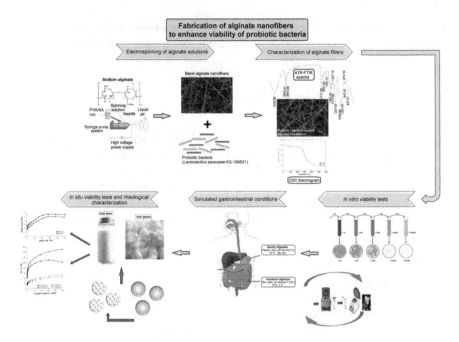

Fig. 6 Preparation of alginate nanofibers mats, loading them with probiotic bacteria and determining their effects to increase viability. From [112], reprinted with permission from Elsevier

them useful for wound dressings [113]. Najafiasl et al. added dexpanthenol to the PVA/sodium alginate core of electrospun nanofibers, while different shells enabled drug release control [114]. Another possible spinning agent for alginate is PEO. Gutierrez-Gonzales et al. loaded sodium alginate/PEO electrospun composites with curcumin and crosslinked the nanofiber mats with trifluoroacetic acid (TFA). Due to the resulting good mechanical properties, such nanofiber mats could be used for filters, tissue engineering, or food packaging [115]. Cesur et al. used PLA as spinning agent to prepare nanofibers for biomedical applications [116], while Amjadi et al. prepared zein/sodium alginate nanofibers for food packaging [117]. Rashtchian et al. applied alginate/PCL blends for electrospinning in which additional cellulose nanocrystals were embedded to increase the bio-mechanical properties [118]. As these few most recent examples show, alginate can be electrospun in blends with diverse other polymers, allowing for using the material in a broad range of applications.

5.7 Electrospinning Poly(Glycolic Acid) (PGA)

Opposite to the previously mentioned polymers, PGA is a synthetic biodegradable polymer. Being the simplest linear polyester, it is a semi-crystalline thermoplastic.

Although widely used in tissue engineering and regenerative medicine, PGA is only scarcely electrospun. In a blend with the aforementioned silk fibroin, Kim et al. used PGA as scaffolds for regeneration of rabbit calvarial defects, similar to the aforementioned procedure using gelatin-based scaffolds. Here, attachment and proliferation of pre-osteoblasts on a PGA and PGA/silk fibroin scaffolds were examined, showing better regeneration support of the hybrid scaffolds, making this material combination promising for guided bone regeneration and tissue regeneration [119]. A triple-blend of PGA with collagen and bioglass was used to prepare nerve guidance channels by electrospinning. This composite structure showed better mechanical, chemical, biocompatible, and biodegradable properties than pure PGA or PGA/collagen guidance channels and is thus promising for nerve regeneration [120]. Another possible blend partner for PGA is PEO. Electrospun PEO microparticles were used to increase the porosity of electrospun PGA and other biodegradable polymers, resulting in increased cell proliferation and human fibroblast infiltration, as compared to pure electrospun nanofiber mats, while the mechanical properties remained nearly unaltered [121]. PGA can also be blended with PCL for electrospinning. Loading these blended nanofibers with curcumin and polyhexamethylene biguanide (PHMB) as bactericides resulted in a strong drug release dependence on the PGA/PCL ratio, while generally showing a bactericide effect in hydrophilic and hydrophobic environments [122]. Finally, it should be mentioned that adding TiO_2 to electrospun PGA nanofibers was used as a degradation retardant for low-molecular-weight PGA. Interestingly, TiO_2 in the hygroscopic anatase modification was found to be a more efficient hydrolytic degrader than TiO_2 in the rutile modification [123].

5.8   Electrospinning Poly(Lactic-Co-Glycolic Acid) (PLGA)

Shen et al. used the copolymer of PLA and PGA, PLGA, as the core of electrospun nanofibers, surrounded by a shell of chitosan with acid-neutralizing capability to avoid the aforementioned acidic degradation products after implantation and found indeed that the pH decrease during degradation of the PLGA core was hindered [124]. PLGA was blended with gelatin to prepare nanofibrous tubular scaffolds with inner and outer layers of fibers with PLGA-core and gelatin shell, separated by a PLGA nanofiber layer, mimicking native vascular structures for cardiovascular tissue engineering. By this sandwich structure, mechanical strength, suture strength, and biocompatibility against human umbilical vain endothelial cells could be improved [125]. Blending PLGA with PCL was used to produce electrospun nanofibers as stent coatings, in order to block the flow toward the aneurysm cavity while allowing nutritional support to the vessel. This nanofiber coating resulted in improved physicochemical and mechanical properties, while it was bioabsorbable [126]. As described before for PGA, PLGA can also be used for guided bone regeneration. Dos Santos et al. used bilayer membranes of a dense PLGA layer doped with hydroxyapatite (HAp) on which PLGA blended with different amounts of HAp:beta-tricalcium phosphate was electrospun, resulting in a dense layer pore size of ~4 µm and a high degree of porosity of the electrospun layer, preventing fibroblast infiltration, but enabling osteoblast migration and nutrient permeation. The combination with calcium phosphates resulted in improved osteoblast attachment, proliferation, and migration, making this system promising for bone reconstruction [127]. A completely different application was chosen by Zheng et al. who blended the piezoelectric PLA with the non-piezoelectric, but faster degradable PLGA in different ratios. By adding additional magnetic nanoparticles, the possibility was examined to control piezo-electricity magnetically. In this way, a potential magneto-electric nanocomposite for biomedical applications was prepared [128].

5.9   Electrospinning Polycaprolactone (PCL)

PCL is an often used polymer in biomedical applications. Rostami et al. used PCL/graphene oxide (GO) nanocomposites loaded with osteogenic drugs as scaffolds to increase the osteogenic differentiation of mesenchymal stem cells. Adding GO and osteogenic drugs resulted in improved hydrophilicity, cell viability and osteogenic differentiation, as compared to pure PCL scaffolds, making these nanocompos-ites promising for bone tissue engineering applications [129]. Bone regeneration was also the target of a study by Sruthi et al., using coaxial electrospinning to produce PCL/polyvinylpyrrolidone (PVP) fibers with embedded chitosan nanopar-ticles, loaded with veratric acid. These nanofibers showed not only biocompatibility with mouse mesenchymal stem cells but also promotion toward osteoblast differ-entiation [130]. PCL electrospun nanofibers loaded with polyaniline (PAni) coated TiO_2 nanoparticles and the aforementioned restenosis prevention drug simvastatin enabled drug release control by the TiO_2/PAni concentration, in this way stabi-lizing cell proliferation and attachment, as compared to pure PCL nanofibers [131]. One of the problems of PCL electrospun nanofiber mats for cartilage repair is their small pore size, combined with hydrophobicity, which prevents cell attachment and proliferation. Blending PCL with gelatin introduces favorable biological properties. Additionally electrosprayed PEO particles, as described before, introduce increased pore sizes when these sacrificial particles are removed after electrospinning. In this way, cell attachment and proliferation could significantly be enhanced [16]. Another application of PCL/gelatin nanofiber mats can be found in wound healing. Jafari et al. found that double-layer electrospun nanofiber mats, including amoxicillin as a model drug in the upper layer and ZnO nanoparticles for increased wound healing in the bottom layer, showed sustained release of the model drug during 144 h, combined with hindered bacterial growth and accelerated cell proliferation. In vivo tests additionally revealed accelerated wound contraction and reduced scar formation, making this system also interesting for wound healing [132]. Similarly, zein/PCL/collagen nanofiber mats were found to support wound healing [79]. Besides these few mentioned examples, many more studies report on PCL as a typical mate-rial for wound healing, bone repair, soft tissue engineering, and other biomedical applications.

5.10 Electrospinning Poly(L-Lactide-E-Caprolacton) (PLLA-CL)

Only very few studies report on electrospinning PLLA-CL, a polymer prepared using the ring-opening polymerization of L-lactide and ε-caprolactone as monomers. Yin et al. used electrospun silk fibroin/PLLA-CL vascular grafts, loaded with growth factor, to grow smooth muscle cells. The growth factor resulted in deeper infil-tration of the cells into the graft, as compared to the pure nanofiber mat, making this system promising for tissue-engineered blood vessels [133]. Block-copolymer PLLA-CL was tested for controlled drug release from electrospun nanofibers, embed-ding protein molecules in the core and PLLA-CL as well as PCL and PLLA as cores. Depending on shell material and concentration, different mechanical prop-erties and drug release kinetics were found, showing a burst release followed by a controlled, constant release [134]. Huang et al. used electrospinning to prepare collagen/PLLA-CL nanofiber mats with sufficient mechanical strength and flexi-bility due to the PLLA-CL, while the collagen supported the biocompatibility of the composite scaffolds, indicating this system's potential for vessel repair [135].

5.11 Electrospinning PVA

PVA can, for example, be used for drug delivery. Li et al. loaded PVA nanofiber mats with finasteride, a drug for prostatic hyperplasia treatment, and stabilized the elec-trospun nanofiber mats by a heat treatment, resulting in increased crystallinity and thus improved water stability. These nanofiber mats showed good cytocompatibility and higher embolization efficacy, resulting in a stronger prostate volume reduction as compared to pure crosslinked PVA nanofibrous particles by combining phys-ical embolization with localized medical therapy [136]. To improve the mechanical properties of PVA nanofiber mats, Choy et al. blended it with α-chitin, in this way increasing the stiffness by nearly a factor of 20 and the extensibility by nearly a factor of 4. At the same time, the thermal stability was increased, suggesting such molec-ular design approaches to improve the thermomechanical performance of electrospun nanofiber mats [137].

A completely different material mix was investigated by Chenari et al. who added CuO to PVA nanofibers and found varying physicochemical properties, depending on the heat treatment after electrospinning [138]. Adding propolis to PVA nanofi-brous wound dressings, Alberti et al. investigated wound healing potential in vitro and in vivo, applying murine NIH-3T3 fibroblasts as model cells. While pure PVA scaffolds showed good fiber morphology and no cytotoxicity to fibroblasts, adding propolis furthermore increased the wound closure rate after 7 days significantly, making this material blend suitable for tissue regeneration [139].

While many applications of PVA are related to biomedicine, Elhami and Habibi used PVA/montmorillonite electrospun nanofiber mats as UV protection. By measuring the degradation of methylene blue dye, shielded by these nanofiber mats, the amount of UV protection was estimated [140]. A more basic investigation on the PVA nanofiber alignment reached by electrospinning on parallel electrode collectors was performed by Icoglu et al. They found increased fiber alignment for decreased tip-to-collector distance and decreased charge density [141]. As another application, air filtration should be mentioned. Li et al. produced PVA/zein nanofiber mats by electrospinning and found an effect of the zein content as well as the alcoholysis degree on the air filtration efficiency, allowing optimizing the filtration effect by tailoring these material properties [142].

5.12   Electrospinning PEO

As the last material in this chapter, PEO is briefly presented. PEO mostly serves as a spinning agent for diverse polymers which are not or hardly electrospinnable solely, e.g. for chitosan [97, 98, 102], alginate [115], and PGA [121]. Nevertheless, PEO can also be used, for example, to prepare nanofiber mats serving as solvent-free electrolytes including lithium salts for the use in Li-ion batteries [12, 143]. Combined with PAni, conductive PEO/PAni nanofibers were prepared using rotating drum electrospinning to prepare supercapacitor electrodes [144]. Finally, PEO can be used as sacrificial material, for example, by electrospraying to allow for preparing pores by diluting it out of the fibers [16, 121] or by electrospinning it together with metallic or semiconducting nanoparticles, such as SnO_2, followed by calcinating the fibers to obtain the pure metal or semiconductor nanofibers [145]. Table 1 summarizes advantages and disadvantages of the aforementioned biodegradable polymers.

6   Conclusion

Many different biodegradable polymers can be used for electrospinning, either purely or in combination with a spinning agent. Water resistance is problematic for some of them, often necessitating an additional crosslinking step before the electrospun nanofiber mats can be used for the desired applications. On the other hand, blending such polymers with water-resistant ones or using them only for fiber formation, followed by a calcination step, are other possibilities to deal with the water solubility of some of them. As the short excerpt of the recent literature, given in this chapter, already shows, a broad range of physical, chemical, and biological properties can be found in biodegradable polymers, often supportive for biomedical applications such as wound healing and tissue engineering, but sometimes also related to quite different fields of research, e.g. using magnetic fields to control properties of composites with magnetic nanoparticles. We hope that this overview will give the readers new ideas

Table 1 Advantages and disadvantages of different biodegradable polymers

Material	Advantages	Disadvantages
Collagen	Main protein of the extracellular matrix Can be used for drug delivery	Weak mechanical properties, thus has to be blended with other polymers
Gelatin	Can be electrospun solely	Low thermal decomposition stability Necessitates crosslinking
Elastin	One of the main components of the extracellular matrix of the abdominal wall	Insoluble, thus needs hydrolyzation before electrospinning
Chitosan	Can be electrospun solely Useful for wound healing, drug release, etc.	
Silk fibroin	High elastic strength Electrospinnable solely	
Alginate	Usable for encapsulation of bacteria, wound dressing, etc.	Poorly electrospinnable due to high conductivity and surface tension
Poly(glycolic acid)	Combined with silk fibroin usable for bone and tissue regeneration	Not much literature available on electrospinning the material
Poly(lactic-co-glycolic acid)	Usable for guided bone regeneration Stent coating to increase physicochemical and mechanical properties	
Polycaprolactone	Blended with diverse materials useful for drug delivery, bone tissue engineering, wound healing, etc.	Small pore size and hydrophobicity of pure material prevent cell attachment
PVA	Electrospinnable purely Usable for drug delivery, wound healing, UV protection, air filtration, etc.	
PEO	Spinning agent, for example, for chitosan, alginate or PGA Sacrificial material to prepare pores	

about possible solutions for their recent applications and stimulate further research on well-known and emerging fields of interest.

References

1. Li D, Xia Y (2004) Electrospinning of Nanofibers: Reinventing the Wheel? Adv Mater 16:1151–1170
2. Subbiah T, Bhat GS, Tock RW et al (2005) Electrospinning of Nanofibers. J Appl Polym Sci 96:557–569
3. Greiner A, Wendorff JH (2007) Electrospinning: a fascinating method for the preparation of ultrathin fibers. Angew Chem Int Edit 46:5670–5703
4. Ashammakhi N, Ndreu A, Yang Y et al (2012) Nanofiber-based scaffolds for tissue engineering. Eur J Plast Surg 35:135–149
5. Klinkhammer K, Seiler N, Grafahrend D et al (2009) Deposition of electrospun fibers on reactive substrates for in vitro investigations. Tissue Eng Part C 15:77–85
6. Großerhode C, Wehlage D, Grothe T et al (2017) Investigation of microalgae growth on electrospun nanofiber mats. AIMS Bioeng 4:376–385
7. Lemma SM, Esposito A, Mason M et al (2015) Removal of bacteria and yeast in water and beer by nylon nanofibrous membranes. J Food Eng 157:1–6
8. Yalcinkaya F, Boyraz E, Maryska J, Kucerova K (2020) A review on membrane technology and chemical surface modification for the oily wastewater treatment. Materials 13:493
9. Boyraz E, Yalcinkaya F, Hruza J, Maryska J (2019) Surface-modified nanofibrous PVDF membranes for liquid separation technology. Materials 12:2702
10. Garcia-Mateos FJ, Ruiz-Rosas R, Rosas JM et al (2020) Activation of electrospun lignin-based carbon fibers and their performance as self-standing supercapacitor electrodes. Separ Purific Technol 241:116724.
11. Kohn S, Wehlage D, Juhász Junger I, Ehrmann A (2019) Electrospinning a dye-sensitized solar cell. Catalysts 9:975
12. Banitaba SN, Semnani D, Heydari-Soureshjani E et al (2020) Electrospun core-shell nanofibers based on polyethylene oxide reinforced by multiwalled carbon nanotube and silicon dioxide nanofillers: a novel and effective solvent-free electrolyte for lithium ion batteries. Inter J Energy Res 44:7000–7014
13. Yördem OS, Papila M, Menceloglu YZ (2008) Effects of electrospinning parameters on poly-acrylonitrile nanofiber diameter: an investigation by response surface methodology. Mater Des 29:34–44
14. Matulevicius J, Cliucininkas L, Martuzevicius D et al (2014) Design and characteriza-tion of electrospun polyamide nanofiber media for air filtration applications. J Nanomater 2014:859656
15. Wortmann M, Frese N, Sabantina L et al (2019) New polymers for needleless electrospinning from low-toxic solvent. Nanomater 9:52
16. Semitela A, Girao AF, Fernandes C et al (2020) Electrospinning of bioactive polycaprolactone-gelatin nanofibres with increased pore size for cartilage tissue engineering applications. J Biomater Applications 35:471–484. https://doi.org/10.1177/0885328220940194
17. Labbafzadeh MR, Vakili MH (2020) Application of magnetic electrospun polyvinyl alcohol/collagen nanofibres for drug delivery systems. Molecul Simul. Early access https://doi.org/10.1080/08927022.2020.1783462
18. Grimmelsmann N, Homburg SV, Ehrmann A (2017) Needleless electrospinning of pure and blended chitosan. IOP Conf Series Mater Sci Eng 225:012098
19. Zahran SME, Abdel-Halim AH, Nassar K, Nada AA (2020) Fabrication of nanofiltration membrane based on non-biofouling PVP/lecithin nanofibers reinforced with microcrystalline cellulose via needle and needle-less electrospinning techniques. International J Biological Macromol 157:530–543
20. Kutzli I, Beljo D, Gibis M et al (2020) Effect of maltodextrin dextrose equivalent on elec-trospinnability and glycation reaction of blends with pea protein isolate. Food Biophys 15:206–215
21. Karlsson S, Albertsson AC (1998) Biodegradable polymers and environmental interaction. Polym Eng Sci 35:1251–1253

22. Albertsson AC, Karlsson S (1988) The three stages in degradation of polymers—polyethylene as a model substance. J Appl Polym 35:1289–1302
23. Kia D, Liow SS, Loh XJ (2014) Biodegradable polymers for electrospinning: towards biomedical applications. Mater Sci Eng C 45:659–670
24. Shimao M (2001) Biodegradation of plastics. Curr Opin Biotechnol 12:242–247
25. Jayasekara R, Harding I, Bowater I, Lonergan G (2005) Biodegradability of selected range of polymers and polymer blends and standard methods for assessment of biodegradation. J Polymer Environ 13:231–251
26. Leja K, Lewandowicz G (2010) Polymer biodegradation and biodegradable polymers—a review. Polish J Environ Stud 19:255–266
27. Kyrikou J, Briassoulis D (2007) Biodegradation of agricultural plastic films: a critical review. J Polym Environ 15:125–150
28. Gautam R, Bassi AS, Yanful EK (2007) A review of biodegradation of synthetic plastic and foams. Appl Biochem Biotechnol 141:85–108
29. Grima S, Bellon-Maurel V, Feuilloley P, Silvestre F (2002) Aerobic biodegradation of polymers in solid-state conditions: a review of environmental and physicochemical parameter settings in laboratory simulation. J Polymer Environ 8:183–195
30. Swift G (1998) Requirements for biodegradable water-soluble polymers. Polymer Degrad Stabil 59:19–24
31. Zhang C, Bennett GN (2005) Biodegradation of xenobiotics by anaerobic bacteria. Appl Microbiol Biotechnol 67:600–618
32. Leahy JG, Colwell RR (1990) Microbial degradation of hydrocarbons in the environment. Microbiol Rev 54:305–315
33. Liao HT, Wu CS (2009) Preparation and characterization of ternary blends composed of polylactide, poly(ε-caprolactone) and starch. Mater Sci Eng 515:207–214
34. Albertsson AC, Barenstedt C, Karlsson S (1992) Susceptibility of enhanced environmentally degradable polyethylene to thermal and photo-oxidation. Polym Degrad Stab 37:163–171
35. Hakkarainen M, Albertsson AC, Karlsson S (1996) Weight losses and molecular weight changes correlated with the evolution of hydroxyacids in simulated in vivo degradation of homo- and copolymers of PLA and PGA. Polym Degrad Stab 52:283–291
36. Li LH, Ding S, Zhou CR (2003) Preparation and degradation of PLA/chitosan composite materials. J Appl Polymer Sci 91:274–277
37. Hakkarainen M, Höglund A, Odelius K, Albertsson AC (2007) Tuning the release rate of acidic degradation products through macromolecular design of caprolactone-based copolymers. J Am Chem Soc 129:6308–6312
38. Zhao CL, Wu HL, Ni JH et al (2017) Development of PLA/Mg composite for orthopedic implant: tunable degradation and enhanced mineralization. Comp Sci Technol 147:8–15
39. Elsawy MA, Kim KH, Park JW, Deep A (2017) Hydrolytic degradation of polylactic acid (PLA) and its composites. Renew Sustain Energy Rev 79:1346–1352
40. Gómez E, Michel FC Jr (2013) Biodegradability of conventional and bio-based plastics and natural fiber composites during composting, anaerobic digestion and long-term soil incubation. Polym Degrad Stab 98:2583–2591
41. Orhan Y, Hrenovic J, Büyükgüngör H (2004) Biodegradation of plastic compost bags under controlled soil conditions. Acta Chim Slov 51:579–588
42. Starnecker A, Menner M (1996) Assessment of biodegradability of plastics under simulated composting conditions in a laboratory test system. Int Biodeterior Biodegrad 37:85–92
43. Agarwal S, Wendorff JH, Greiner A (2009) Progress in the field of electrospinning for tissue engineering applications. Adv Mater 21:3343–3351
44. Katti DS, Robinson KW, Ko FK, Laurencin CT (2004) Bioresorbable nanofiber-based systems for wound healing and drug delivery: optimization of fabrication parameters. J Biomed Mater Res B Appl Biomater 70:286–296
45. Deitzel JM, Kleinmeyer JD, Hirvonen JK, Beck Tan NC (2001) Controlled deposition of electrospun poly(ethylene oxide) fibers. Polymer 42:8163–8170

46. Grothe T, Wehlage D, Böhm T et al (2017) Needleless electrospinning of PAN nanofibre mats. Tekstilec 60:290–295
47. Döpke C, Grothe T, Steblinski P et al (2019) Magnetic nanofiber mats for data storage and transfer. Nanomater 9:92
48. Fang Y, Xu L (2019) Four self-made free surface electrospinning devices for high-throughput preparation of high-quality nanofibers. Beilstein J Nanotech 10:2261–2274
49. Jiang GJ, Qin XH (2014) An improved free surface electrospinning for high throughput manufacturing of core-shell nanofibers. Mater Lett 128:259–262
50. Ubaid S, Kanwal S, Aslam MS, Islam AS (2020) Strategy for the development of stretchable highly aligned electrospun polyacrylamide (PAM) nanofibers. Optoelectron Adv Mater Rapid Comm 14:184–188
51. Katta P, Alessandro M, Ramsier RD et al (2004) Continuous electrospinning of aligned polymer nanofibers onto a wire drum collector. Nano Lett 4:2215–2218
52. Edmondson D, Cooper A, Jana S et al (2012) Centrifugal electrospinning of highly aligned polymer nanofibers over a large area. J Mater Chem 22:18646–18652
53. Grothe T, Großerhode C, Hauser T et al (2017) Needleless electrospinning of PEO nanofiber mats. Adv Eng Res 102:54–58
54. Blomberg T, Borgmeier N, Kramer LT et al (2018) Influence of salts on the spinnability of poly(ethylene glycol). Appl Mech Mater 878:313–317
55. Banner J, Dautzenberg M, Feldhans T et al (2018) Water resistance and morphology of electrospun gelatine blended with citric acid and coconut oil. Tekstilec 61:129–135
56. Wehlage D, Blattner H, Sabantina L et al (2019) Sterilization of PAN/gelatin nanofibrous mats for cell growth. Tekstilec 62:78–88
57. Sabantina L, Wehlage D, Klöcker M et al (2018) Stabilization of electrospun PAN/gelatin nanofiber mats for carbonization. J Nanomater 2018:6131085
58. Zhang YZ, Venugopal J, Huang ZM et al (2006) Crosslinking of the electrospun gelatin nanofibers. Polymer 47:2911–2917
59. Panzavolta S, Gioffrè M, Focarete ML et al (2011) Electrospun gelatin nanofibers: optimization of genipin cross-linking to preserve fiber morphology after exposure to water. Acta Biomater 7:1702–1709
60. Huang CH, Chi CY, Chen YS et al (2012) Evaluation of proanthocyanidin-crosslinked electrospun gelatin nanofibers for drug delivering system. Mater Sci Eng C 32:2476–2483
61. Dias JR, Baptista-Silva S, de Oliveira CMT et al (2017) In situ crosslinked electrospun gelatin nanofibers for skin regeneration. Eur Polym J 95:161–173
62. Septiyanti M, Septevani AA, Ghozali M et al (2020) Effect of solvent combination on electrospun stereocomplex polylactic acid nanofiber properties. Macromol Symp 391:1900134
63. Huang C, Thomas NL (2018) Fabricating porous poly(lactic acid) fibres via electrospinning. European Polym J 99:464–476
64. Cikova E, Kulicek J, Janigova I et al (2018) Electrospinning of Ethylene Vinyl Acetate/Poly(Lactic Acid) Blends on a Water Surface. Materials 11:1737
65. Ghafari R, Scaffaro R, Maio A et al (2020) Processing-structure-property relationships of electrospun PLA-PEO membranes reinforced with enzymatic cellulose nanofibers. Polym Test 81:106182
66. Rebia RA, Rozet S, Tamada Y, Tanaka T (2018) Biodegradable PHBH/PVA blend nanofibers: fabrication, characterization, in vitro degradation, and in vitro biocompatibility. Polym Degrad Stab 154:124–136
67. Snetkov P, Morozkina S, Uspenskaya M, Olekhnovich R (2019) Hyaluronan-based nanofibers: fabrication. Characterization and application. Polymers 11:2036
68. Li J, He A, Han CC et al (2006) Electrospinning of hyaluronic acid (HA) and HA/gelatin blends. Macromol Rapid Commun 27:114–120
69. Liu Y, Ma G, Fang D et al (2011) Effects of solution properties and electric field on the electrospinning of hyaluronic acid. Carbohydr Polym 83:1011–1015
70. Brenner EK, Schiffman JD, Thompson EA et al (2012) Electrospinning of hyaluronic acid nanofibers from aqueous ammonium solutions. Carbohyd Polym 87:926–929

71. Snetkov PP, Uspenskaia TE, Uspenskaya MV, Rzametov KS (2019) Effect of technological parameters on electrospinnability of water-organic solutions of hyaluronic acid. In: Proceedings of the 19th international multidisciplinary scientific GeoConference SGEM-2019, surveying geology and mining ecology management, Albena, Bulgaria, vol 19, 175–182
72. Gunn J, Zhang MQ (2010) Polyblend nanofibers for biomedical applications: perspectives and challenges. Trends Biotech 28:189–197
73. Singh SK (2021) Biological treatment of plant biomass and factors affecting bioactivity. J Clean Product 279:123546
74. Tungprapa S, Jangchud I, Supaphol P (2007) Release characteristics of four model drugs from drug-loaded electrospun cellulose acetate fiber mats. Polymer 48:5030–5041
75. Armedya TP, Dzikri MF, Sakti SCW et al (2019) Kinetical release study of copper ferrite nanoparticle incorporated on PCL/collagen nanofiber for naproxen delivery. BioNanoScience 9:274–284
76. Dotti F, Varesano A, Montarsolo A et al (2007) Electrospun porous mats for high efficiency filtration. J Ind Textil 37:151–162
77. Fahmi MZ, Prasetya RA, Dzikri MF et al (2020) $MnFe_2O_4$ nanoparticles/cellulose acetate composite nanofiber for controllable release of naproxen. Mater Chem Phys 250:123055
78. Rather HA, Patel R, Yadav UCS, Vasita R (2020) Dual drug-delivering polycaprolactone-collagen scaffold to induce early osteogenic differentiation and coupled angiogenesis. Biomed Mater 15:045008
79. Ghorbani M, Nezhad-Mikhtari P, Ramazani S (2020) Aloe vera-loaded nanofibrous scaffold based on Zein/polycaprolactone/collagen for wound healing. Int J Biologic Macromol 153:921–930
80. Gao FF, Jiang ML, Liang WC et al (2020) Co-electrospun cellulose diacetate-graft-poly(ethylene terephthalate) and collagen composite nanofibrous mats for cells culture. J Appl Polymer Sci 137:49350. https://doi.org/10.1002/app.49350
81. Berechet MD, Gaidau C, Miletic A et al (2020) Bioactive Properties of Nanofibres Based on Concentrated Collagen Hydrolysate Loaded with Thyme and Oregano Essential Oils. Materials 13:1618
82. Sadeghi-Avalshahr AR, Nokhasteh S, Molavi AM et al (2020) Tailored PCL scaffolds as skin substitutes using sacrificial PVP fibers and collagen/chitosan blends. Int J Molecular Sci 21:2311
83. Santos de Oliveira C, Trompetero González A, Hedtke T et al (2020) Direct three-dimensional imaging for morphological analysis of electrospun fibers with laboratory-based Zernike X-ray phase-contrast computed tomography. Mater Sci Eng C 115:111045
84. Rezaeinia H, Ghoroni B, Emadzadeh B, Mohebbi M (2020) Prolonged-release of menthol through a superhydrophilic multilayered structure of balangu (*Lallemantia royleana*)-gelatin nanofibers. Mater Sci Eng C 115:111115
85. Zhang C, Li Y, Wang P et al (2020) Core-shell nanofibers electrospun from O/W emulsions stabilized by the mixed monolayer of gelatin-gum Arabic complexes. Food Hydrocolloids 107:105980
86. Wu C, Zhang HG, Hu QX, Ramalingam M (2020) Designing biomimetic triple-layered nanofibrous vascular grafts via combinatorial electrospinning approach. J Nanosci Nanotech 20:6396–6405
87. Du ZY, Zhao ZD, Liu HH et al (2020) Macroporous scaffolds developed from CaSiO3 nanofibers regulating bone regeneration via controlled calcination. Mater Sci Eng C 113:111005
88. Franz MG (2006) The biology of hernias and the abdominal wall. Hernia 10:462–471
89. Minardi S, Taraballi F, Wang X et al (2017) Biomimetic collagen/elastin meshes for ventral hernia repair in a rat model. Acta Biomater 50:165–177
90. Daamen WF, Nillesen ST, Wismans RG et al (2008) A biomaterial composed of collagen and solubilized elastin enhances angiogenesis and elastic fiber formation without calcification. Tissue Eng A 14:349–360

91. Yildiz A, Kara AA, Acartürk F (2020) Peptide-protein based nanofibers in pharmaceutical and biomedical applications. Int J Biological Macromol 148:1084–1097
92. Cao GX, Wang CY, Fan YB, Li XM (2020) Biomimetic SIS-based biocomposites with improved biodegradability, antibacterial activity and angiogenesis for abdominal wall repair. Mater Sci Eng C 109:110538
93. Foraida ZI, Kamaldinov T, Nelson DA et al (2017) Elastin-PLGA hybrid electrospun nanofiber scaffolds for salivary epithelial cell self-organization and polarization. Acta Biomater 62:116–127
94. Machado R, da Costa A, Sencadas V et al (2013) Electrospun silk-elastin-like fibre mats for tissue engineering applications. Biomed Mater 8:065009
95. Rnjak-Kovacina J, Wise SG, Li Z et al (2012) Electrospun synthetic human elastin:collagen composite scaffolds for dermal tissue engineering. Acta Biomater 8:3714–3722
96. Chong C, Wang YW, Fathi A et al (2019) Skin wound repair: results of a pre-clinical study to evaluate electropsun collagen-elastin-PCL scaffolds as dermal substitutes. Burns 45:1639–1648
97. Nikbakht M, Salehi M, Rezayat SM et al (2020) Various parameters in the preparation of chitosan/polyethylene oxide electrospun nanofibers containing Aloe vera extract for medical applications. Nanomed J 7:21–28
98. Surgutskaia NS, di Martino A, Zednik J et al (2020) Efficient Cu^{2+}, Pb^{2+} and Ni^{2+} ion removal from wastewater using electrospun DTPA-modified chitosan/polyethylene oxide nanofibers. Separ Purific Technol 247:116914
99. Sannasimthu A, Ramani M, Paray BA et al (2020) Arthrospira platensis transglutaminase derived antioxidant peptide-packed electrospun chitosan/ poly (vinyl alcohol) nanofibrous mat accelerates wound healing, in vitro, via inducing mouse embryonic fibroblast proliferation. Colloids Surf B Biointerf 193:111124
100. Mojaveri SJ, Hosseini SF, Gharsallaoui A (2020) Viability improvement of *Bifidobacterium animalis* Bb12 by encapsulation in chitosan/poly(vinyl alcohol) hybrid electrospun fiber mats. Carbohydrate Polymers 241:116278
101. Sattari S, Tehrani AD, Adeli M et al (2020) Fabrication of new generation of co-delivery systems based on graphene-g-cyclodextrin/chitosan nanofiber. Int J Biological Macromol 156:1126–1134
102. Tabaei SJS, Rahimi M, Akbaribazm M et al (2020) Chitosan-based nano-scaffolds as antileishmanial wound dressing in BALB/c mice treatment: characterization and design of tissue regeneration. Iranian J Basic Med Sci 23:788–799
103. Kholosi F, Afkhami A, Hashemi P et al (2020) Bioelectrocatalysis and direct determination of H_2O_2 using the high-performance platform: chitosan nanofibers modified with SDS and hemoglobin. J Iranian Chem Soc 17:1401–1409
104. Kersani D, Mougin J, Lopez M et al (2020) Stent coating by electrospinning with chitosan/poly-cyclodextrin based nanofibers loaded with simvastatin for restenosis prevention. Eur J Pharmaceutics Biopharmaceutics 150:156–167
105. Augustine R, Rehman SRU, Ahmed R et al (2020) Electrospun chitosan membranes containing bioactive and therapeutic agents for enhanced wound healing. Int J Biological Macromol 156:153–170
106. Ebrahimi D, Tokareva O, Rim NG et al (2015) Silk-its mysteries, how it is made, and how it is used. ACS Biomater Sci Eng 1:864–876
107. Karatepe UY, Ozdemir T (2020) Improving mechanical and antibacterial properties of PMMA via polyblend electrospinning with silk fibroin and polyethyleneimine towards dental applications. Bioactive Mater 5:510–515
108. Keirouz A, Zakharova M, Kwon J et al (2020) High-throughput production of silk fibroin-based electrospun fibers as biomaterial for skin tissue engineering applications. Mater Sci Eng C 112:110939
109. Liu HF, Wang YQ, Yang Y et al (2020) Aligned graphene/silk fibroin conductive fibrous scaffolds for guiding neurite outgrowth in rat spinal cord neurons. J Biomed Mater Res A. Early access https://doi.org/10.1002/jbm.a.37031

110. Kopp A, Smeets R, Gosau M et al (2020) Effect of process parameters on additive-free electrospinning of regenerated silk fibroin nonwovens. Bioactive Mater 5:241–252

111. Farokhi M, Mottaghitalab F, Reis RL et al (2020) Functionalized silk fibroin nanofibers as drug carriers: advantages and challenges. J Controlled Release 321:324–347

112. Yilmaz MT, Taylan O, Karakas CY, Dertly E (2020) An alternative way to encapsulate probiotics within electrospun alginate nanofibers as monitored under simulated gastrointestinal conditions and in kefir. Carbohydrate Polymers 244:116447

113. Aloma KK, Sukaryo S, Fahlawati NI et al (2020) Synthesis of nanofibers from alginate-polyvinyl alcohol using electrospinning methods. Macromol Symp 391:1900199

114. Najafiasl M, Osfouri S, Azin R et al (2020) Alginate-based electrospun core/shell nanofibers containing dexpanthenol: a good candidate for wound dressing. J Drug Delivery Sci Technol 57:101708

115. Gutierrez-Gonzalez J, Carcia-Cela E, Magan N et al (2020) Electrospinning alginate/polyethylene oxide and curcumin composite nanofibers. Mater Lett 270:127662

116. Cesur S, Oktar FN, Ekren N et al (2020) Preparation and characterization of electrospun polylactic acid/sodium alginate/orange oyster shell composite nanofiber for biomedical application. J Australian Ceramic Soc 56:533–543

117. Amjadi S, Almasi H, Ghorbani M et al (2020) Preparation and characterization of TiO_2 NPs and betanin loaded zein/sodium alginate nanofibers. Food Pack Shelf Life 24:100504

118. Rashtchian M, Hivechi A, Bahrami SH et al (2020) Fabricating alginate/poly(caprolactone) nanofibers with enhanced bio-mechanical properties via cellulose nanocrystal incorporation. Carbohydrate Polym 233:115873

119. Kim BN, Ko Y-G, Yeo TG et al (2019) Guided regeneration of rabbit calvarial defects using silk fibroin nanofiber-poly(glycolic acid) hybrid scaffolds. ACS Biomater Sci Eng 5:5266–5272

120. Dehnavi N, Parivar K, Goodarzi V et al (2019) Systematically engineered electrospun conduit based on PGA/collagen/bioglass nanocomposites: the evaluation of morphological, mechanical, and bio-properties. Polym Adv Technol 30:2192–2206

121. Hodge J, Quint C (2019) The improvement of cell infiltration in an electrospun scaffold with multiple synthetic biodegradable polymers using sacrificial PEO microparticles. J Biomed Mater Res A 107:1954–1964

122. Keridou I, Franco L, Turon P (2018) Scaffolds with tunable properties constituted by electrospun nanofibers of polyglycolide and poly(epsilon-caprolactone). Macromol Mater Eng 303:1800100

123. Silva de la Cruz LI, Medellin Rodriguez F, Velasco-Santos C et al (2016) Hydrolytic degradation and morphological characterization of electrospun poly(glycolic acid) [PGA] thin films of different molecular weights containing TiO_2 nanoparticles. J Polym Res 23:113

124. Shen YB, Tu T, Yi BC et al (2019) Electrospun acid-neutralizing fibers for the amelioration of inflammatory response. Acta Biomater 97:200–215

125. Wu C, Zhang HG, Hu QX et al (2020) Designing biomimetic triple-layered nanofibrous vascular grafts via combinatorial electrospinning approach. J Nanosci Nanotechnol 20:6396–6405

126. Hwang TI, Lee SY, Lee DH et al (2020) Fabrication of bioabsorbable polylactic-co-glycolic acid/polycaprolactone nanofiber coated stent and investigation of biodegradability in porcine animal model. J Nanosci Nanotechnol 20:5360–5264

127. dos Santos VI, Merlini C, Aragones A et al (2020) In vitro evaluation of bilayer membranes of PLGA/hydroxyapatite/beta-tricalcium phosphate for guided bone regeneration. Mater Sci Eng C 112:110849

128. Zheng T, Yue ZL, Wallace G et al (2020) Nanoscale piezoelectric effect of biodegradable PLA-based composite fibers by piezoresponse force microscopy. Nanotechnol 31:375708

129. Rostami F, Tamjid E, Behmanesh M (2020) Drug-eluting PCL/graphene oxide nanocomposite scaffolds for enhanced osteogenic differentiation of mesenchymal stem cells. Mater Sci Eng C 115:111102

130. Sruthi R, Balagangadharan K, Selvamurugan N (2020) Polycaprolactone/polyvinylpyrrolidone coaxial electrospun fibers containing veratric acid-loaded chitosan nanoparticles for bone regeneration. Colloids Surf B 193:111110

131. Rezk AI, Bhattarai DP, Park J et al (2020) Polyaniline-coated titanium oxide nanoparticles and simvastatin-loaded poly(epsilon-caprolactone) composite nanofibers scaffold for bone tissue regeneration application. Colloids Surf B 192:111007
132. Jafari A, Amirsadeghi A, Hassanajili S et al (2020) Bioactive antibacterial bilayer PCL/gelatin nanofibrous scaffold promotes full-thickness wound healing. Int J Pharmaceutics 583:119413
133. Yin AL, Bowlin GL, Luo RF et al (2016) Electrospun silk fibroin/poly (L-lactide-epsilon-caplacton) graft with platelet-rich growth factor for inducing smooth muscle cell growth and infiltration. Regenerat Biomater 3:239–245
134. Zaini AF, Ranganathan B, Sundarrajan S et al (2014) Coaxial electrospun nanofibers as pharmaceutical nanoformulation for controlled drug release. In: IEEE 14th international conference on nanotechnology, 531–534
135. Huang C, Morsi Y, Chen R et al (2010) Fabrication and characterization of Collagen/PLLA-CL composite vascular grafts. In: Proceedings of international forum on biomedical textile materials, 384–388
136. Li XH, Li BS, Ullah MW et al (2020) Water-stable and finasteride-loaded polyvinyl alcohol nanofibrous particles with sustained drug release for improved prostatic artery embolization— In vitro and in vivo evaluation. Mater Sci Eng C 115:111107
137. Choy SW, Moon HW, Park YJ et al (2020) Mechanical properties and thermal stability of intermolecular-fitted poly (vinyl alcohol)/alpha-chitin nanofibrous mat. Carbohydrate Polymers 244:116476
138. Chenari HM, Mottaghian F Electrospun CuO/PVA fibers: effects of heat treatment on the structural, surface morphology, optical and magnetic properties. Mater Sci Semicond Process 115:105121
139. Alberti TB, Coelho DS, de Pra M et al (2020) Electrospun PVA nanoscaffolds associated with propolis nanoparticles with wound healing activity. J Mater Sci 55:9712–9727
140. Elhami M, Habibi S (2020) A study on UV-protection property of poly(vinyl alcohol)-montmorillonite composite nanofibers. J Vinyl Additive Technol. Early access https://doi.org/10.1002/vnl.21786
141. Icoglu HI, Ceylan S, Yildirim B et al (2020) Production of aligned electrospun polyvinyl alcohol nanofibers via parallel electrode method. J Text Inst. Early access https://doi.org/10.1080/00405000.2020.1789274
142. Li K, Li CM, Tian HF et al (2020) Multifunctional and efficient air filtration: a natural nanofilter prepared with Zein and Polyvinyl alcohol. Macromol Mater Eng 305:2000239. https://doi.org/10.1002/mame.202000239
143. Banitaba SN, Semnani D, Fakhrali A et al (2020) Electrospun PEO nanofibrous membrane enable by LiCl, LiClO$_4$, and LiTFSI salts: a versatile solvent-free electrolyte for lithium-ion battery application. Ionics 26:3249–3260
144. Bhattacharya S, Roy I, Tice A et al (2020) High-conductivity and high-capacitance electrospun fibers for supercapacitor applications. ACS Appl Mater Interf 12:19369–19376
145. Mehrabi P, Hui J, Janfaza S et al (2020) Fabrication of SnO$_2$ composite nanofiber-based gas sensor using the electrospinning method for tetrahydrocannabinol (THC) detection. Micromach 11:190

Electrospun Nanofibers for Energy and Environment Protection

Shashikant Shivaji Vhatkar, Ashwini Kumari, Prabhat Kumar, Gurucharan Sahoo, and Ramesh Oraon

Abstract Presently, the continuous depletion of non-renewable resources and their attendant environmental impacts has stimulated tremendous research attention to ensure sustainable energy and environment. To cope with such issues, research trend witnessed a paradigm shift toward designing and fabrication of nanofibrous materials as an effective alternative solution using appropriate processing techniques. Lately, a fascinating fiber fabrication technique, popularly known as electrospinning, has been found to broaden its application reach owing to intriguing physicochemical properties of materials like a huge surface area to volume ratio, complex porous structure, lightweight, and stability. Rising concerns over conventional energy resources, nanoscale featured—an electrospun technique not only promotes the production of continuous fibers (with diameters down to few nanometers) but also utilizes various organic (polymers)–inorganic (metal oxides) materials toward energy storage while resolving many other problems associated with ecological growth. The utilization of nanofibers can efficiently enhance the electrocatalytic activity toward higher electrochemical conversion efficiency. Similarly, nanofiber membrane because of its designed, complex structure with aligned patterning has shown prominence for filtration and detection of heavy metal ion, toxic chemical compounds desirable for environmental protection. Hence, innovative nanoscale manipulation of electrospun nanofibers through different morphologies, patterns, and functionality could pave the way to excel in the need for ever-rising demand of energy and sustainable growth of modern society.

Keywords Electrospun nanofibers · Organic–Inorganic material · Electrocatalytic activity · Filtration

S. S. Vhatkar · A. Kumari · P. Kumar · R. Oraon (✉)
Department of Nanoscience and Technology, Central University of Jharkhand, Jharkhand, India
e-mail: ramesh.oraon@cuj.ac.in

G. Sahoo
Department of Chemistry, Birsa College Khunti, Jharkhand, India

© The Author(s), under exclusive license to Springer Nature Switzerland AG 2021
S. K. Tiwari et al. (eds.), *Electrospun Nanofibers*, Springer Series on Polymer and Composite Materials, https://doi.org/10.1007/978-3-030-79979-3_2

1   Introduction

Due to the ever-rising demand for energy with continuous depletion of fossil fuels, efforts have been made toward environmental pollution and inefficient waste management [1]. So far, various technologies have been employed with acceptable efficiencies to mitigate the common issue of pollution to promote sustainable growth [2]. With the emergence of nanotechnology, reasonable enhancement is found in nanomaterials' unique properties and their relative abundance in the environment. The scientific community has also witnessed a paradigm shift from both the designing and fabrication of nanomaterials to cater to energy demand while reducing severe environmental pollution [3]. With the availability of variable choices for nanoscale material fabrication, Electrospun Nanofibers (ESN) by electrospinning has been most preferred academically and industrially worldwide [4–6].

Electrospinning is a technique that fabricates three-dimensional nano-scaled woven fibers by projecting a solution through injection over a collector, all in the presence of a strong electric field. Its assembly consists of an injection pump, a needle, a high-voltage power source, and a collector plate. In practice, the polymer solution is plunged in controlled mode through the needle-like nozzle in the presence of an electric field, resulting in polymer drop. Further, the polymer drop undergoes a shape change from spherical to cone, known as Taylor cone (Taylor cone is the formation of a cone-like shape which originates at the convex sides and traces up till the converging tip), on account of the loss of surface tension over charge accumulation due to electrostatic interaction. This formation then distorts to electrically charged jet-like ejection and gets collected as solid fibers rolled or deposited over the collector plate. The resulting ultrathin ESN diameters range from 20 nm to several micrometres (approx. 10,000 nm) through polymer solution optimization techniques. Furthermore, they acquire in situ properties like large surface area to volume ratio, high porosity, flexibility in surface functionality, increased number of channels, etc. during electrospinning itself and are viable for multifunctional applications including environmental applications such as water filtration and its treatment, as well as various energy storage applications including solar, batteries, and so on. Remarkably, the intriguing properties of ESN regards among diverse composite materials (like composite polymers, conducting polymers, and metal oxides), as a model entrant for supercapacitors (SC) [7–10]. There are three important considerations, viz. electrospinning variables, solution properties (physical, chemical), and environmental parameters which significantly affect evenness and blob-free ESN. The ultimate challenge is to obtain a smooth and even ESN. This is majorly governed by the method of electrospinning, the solution, and other constraints like the distance between needle and collector, needle dimensions, and so on. Among various factors that affect a decent ESN, many of these can be categorized into physical, solvent solution, and other factors. Physical factors involve magnitude of the electric field, collector–needle gap, needle dimensions, flow rate, etc. With the solvent solution, conductivity, concentration (with respect to polymer), and viscosity, determine the nature of ESN. Notably, polymers bring up several advantages like low density, affordable production costs, along with ease of processibility. Electrospinning further allows to tailor several properties for desired ESN through optimization. This includes material parameters (like a solvent, fiber, and dielectric constant of the material), as well as operating/process parameters (such as voltage, separating distance, and collector plate geometry) in order to achieve desired morphology (such as random and aligned) based on the end application. Additionally, the ESN can also be implemented with composites and in the fabrication of sensors (like optical sensors and pH sensors). In practice though, electrospinning is also affected by factors that include environmental constraints like relative humidity, and ambient temperature that affects the evaporation of the solvent (such as molecular weight, boiling point, and dipole moment), which also counts the final nature of obtained ESN. Also, these are also easily influenced geographically over climatic conditions. Hence, it is imperative to gain a better understanding of the electrospinning technique, all governing parameters toward the fabrication of ESN [11]. Feasibility of obtaining ESN offers versatile advantages, viz. (a) adequate mechanical property, (b) in production of high surface area porous scaffold fabrication conducive to adhesion, (c) growth, and (d) proliferation to a wide range of high-tech energy applications to environmental remediation such as filtration, membrane technology, drug delivery, tissue engineering, and the textile industry [12–15] (Fig. 1).

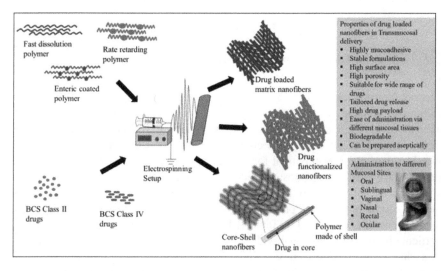

Fig. 1 Schematic diagram of electrospinning and multifunctional application of electrospun nanofibers in diverse area (Reproduced from Ref. [16] with permission from Elsevier)

2   Electrospun Nanofibers for Energy Applications

Global concerns about limited resources of fossil fuels, concomitant energy storage, and conversion have attracted a significant share of research attention of modern era [17]. Further, more than 90% handheld electronic gadgets depend on portable and mobile power source. As a result, energy storage devices too have seen a remarkable development in terms of both charge storage as well as in size. Furthermore, with advances in nanotechnology, new material fabrication techniques have seen a lot of improvements. Among many, electrospinning has been an attractive fabricating technique interest owing to its simplicity, both in terms of material selection and the ability to operate at ambient temperatures. ESN has been extensively employed as a constituent part of the energy and drastically accelerated efforts toward design and improvement of efficient and clean energy storage devices. Due to high surface area to volume ratio, good crystalline structure, mechanical stability, and superior kinetic property, ESN has been frequently used in various devices. Besides, the electrospinning techniques have also been already deployed in various fields like those powering sensors, enhancing charge storage of batteries, supercapacitors, and fuel cells. Although ESN has been deployed over a variety of applications (like aerospace, automotive, and biomedical), however, in this chapter, we have focused on applications pertaining specifically to energy and environmental protective ones.

2.1   Charge Storage Mechanism in Electrospun Nanofibers

Most of the research performed for the advancement of energy storage/conversion technology is mainly focused on material selection and their optimization. Remarkably, the intriguing properties of Electrospun Nanofibers (ESN) are regarded among diverse composite materials (like composite polymers, conducting polymers, and metal oxides) as a model entrant for supercapacitors (SC). In ESN, energy storage performance exploits electrostatic interaction and their transport of interacting species from bulk electrolyte to electrode interface. Further, these are accompanied by certain redox-based processes. Recent advances of ESN in different energy applications has been listed in Table 1. Further, charge storage in energy devices, particularly the ones fabricated with electrospinning, involves electrostatic adsorption which is governed by Coulomb's law shown in Fig. 2. This additionally facilitates ion exchange, thereby releasing free electrons for electricity. In addition to this, electrostatic charge adsorption is a morphological phenomenon, which is also observed in energy devices such as battery, supercapacitors, solar cells, and so on. Additionally, isotherms enable in analyzing adsorption action quantitatively. Moreover, no covalent bonds are formed in the process. ESN, by virtue, provides enhanced porous as well as layered framework, which contributes to storage and effective delivery of charge across the load and back to the energy device. Electrostatic adsorption is also governed by isotherms such as Langmuir.

Table 1 Comparison of recent advances of ESN in different energy applications

Nanofiber base	Modifier	Deposition	End application	References
ZnO	–	SnO_2	Batteries	
Carbon	Carboxyl groups	MnO_2 particles	Supercapacitors	[18]
Carbon	Cobalt oxide nanohairs	Silver nanoparticles	Supercapacitors	[19]
poly(3,4-ethylenedioxythiophene)	Polystyrene sulfonate–sulfur	Polyacrylonitrile	Batteries	[20]
Carbon	–	SnSe	Batteries	[21]
Activated carbon	Polyacrylonitrile (PAN)	–	DSSC	[22]
(Co/CoP@NC)	Nitrogen-doped CNF	–	Fuel cell	[23]

Fig. 2 Working principle of DSSC (Reproduced from Ref. [25] with permission from Elsevier)

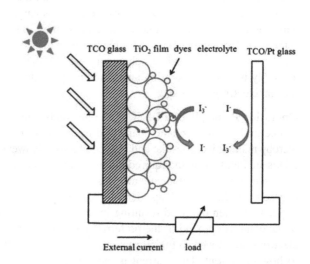

3 Dye-Sensitized Solar Cells

ESN has been an attractive alternative for Dye-Sensitized Solar Cell (DSSC) both environmentally and economically. Interconnected network structure, high contact surface area for greater absorption of solar energy, and resultant greater efficiency in adverse environmental condition make ESN a novel electrode material for application in DSSCs [24].

The past few decades have seen depletion of fossil fuel, increased crude oil prices and consumption, which has led to look out for alternate, affordable, and sustainable energy forms like solar. Among these, thin-film solar cells, which are third generation solar cells, also popularly known as DSSC or Grätzel cells (named after the Swiss chemist Michael Grätzel), have seen a growing interest among researchers and academicians globally. Apart from these, DSSC present several advantages including low cost, ease of fabrication, mechanical robustness, efficiency (in comparison to silicon-based solar cells), and its ability to be deployed in both indoor as well as outdoor lighting conditions including artificial lighting which makes it an attractive alternative over other solar cells. Additionally, they even operate at ambient temperature conditions. Also, current DSSCs have also been subjected to a variety of materials in order to reduce the operating cost and increase overall stability.

3.1 DSSC Construction, Assembly, and Operating Principle

DSSC consists of a glass plate which is usually coated with a transparent conductive oxide (TCO) film such as ITO, F-doped Tin Oxide by various methods. This is followed by the application of a thin layer of titanium dioxide (TiO_2) owing to its semiconducting nature, as well as its mesoporous nature. The anode is then soaked with photosensitizers (organic metal compounds), which can bond to TiO_2. The cathode is also a glass plate with a thin Pt film which serves as a catalyst. In practice, typically an iodide/triiodide solution is used as the electrolyte. Later, a compact assembly is made by pressing anode and cathode electrodes together and closed to prevent electrolyte leakage.

Operating Principle of a DSSC: The photoactive material, in its ground state, has its energy level near valence band and hence does not show any electron activity, thereby making it diode-like and non-conductive. However, when solar energy radiations are incident (with required intensity) on the surface, these cause molecules of photosensitizers to promote themselves to a higher energy state, and as a result cross over semiconductor bandgap, thereby becoming conductive at this stage. This excited molecule is then oxidized resulting in releasing an electron which is then inserted in the conduction band of the preferred semiconductor (such as TiO_2). As a result, electrons are moved to the current collector through diffusion of electrons leading to flow of current. This current is then connected across the load, thus powering the end device. Additionally, this current is also affected by cell voltage or potential (current decreases with increasing potential) and intensity of the incident solar radiation (which increases with increase in intensity). This further can be analyzed by Current-potential curves (also called I-V curves) which illustrates the relationship. IV curves also assist in DSSC characterizations. Meanwhile, the oxidized photosensitized molecule is regenerated by available electron from iodide in the electrolyte. Consequently, iodide is also regenerated by reducing triiodide present at the cathode.

Important parameter definition:

Short-circuit current (Isc): It is the maximum current that can be obtained from a solar cell. The cell voltage or potential is zero. Hence the power output is also zero. Further, it is directly proportional to light intensity.

Open-Circuit Potential (Voc): It is the maximum voltage of a solar cell for incident light intensity. The current flow through a solar cell at Voc is zero. It is unaffected by shunts as current flow is minimal or zero.

Power (P): It is the generated power by the DSSC and is calculated empirically. The power curve is plotted against the cell potential, which translates the maximum power (P_{max}) the cell can deliver. Maximum power delivery is also affected by shunts appearing at contact interfaces.

Fill Factor: The fill factor (**FF**) parameter specifies the overall capacities of a DSSC and qualitatively describes the DSSC. Mathematically, the FF is the ratio of maximum generated power P_{max} to ideal power maximum P_{ideal} of a DSSC. FF can also be interpreted from IV curves. FF is unity when V_{oc} and I_{sc} are maximum. Similar to efficiency, FF is never unity in real world, due to imperfections, and impurities present in DSSC components.

Efficiency of DSSC: Mathematically, it is a ratio of maximum power delivered to incident solar power.

While in practise, DSSCs fabricated through electrospinning have been found to possess exceptional flexibility, tunable surface functionality, and lighter weight of ESN, these thus acquire certain advantages over first- and second-generation solar cells including reduced manufacturing cost. Fabrication of electrode material via electrolyte state modification for DSSC using electrospinning technique has been reported successfully with comparable electrochemical activities [13]. Conventional materials offer limited surface area at the bulk level, whereas nanoscale ESN opens up with new pores, provides higher surface area to volume, and smaller fiber-diameter dimensions. Subsequently, ESN successfully addresses the inherent problem of electrolyte evaporation and lower efficiency of quasi-type solid-state electrolytes in DSSCs owing to the better electrode–electrolyte interface than the conventional counterpart. As a result, ESN provides room for enhancement in electrochemical activity toward optimum surface utilization of electrode material and thus better efficiency. Recently, electrospun Polyacrylonitrile (PAN) nanofiber membrane matrix activated by electrolyte had been used to fabricate gel-type electrolyte. DSSCs that comprised optimized nanofiber membrane thickness showed an efficiency of 5.44% with an open-circuit voltage of 735.0 mV, short-circuit current density of 14.49 mA cm^{-2}, and fill factor of 51.06% [22]. For improved catalytic activity, nanocomposites were also extensively studied for developing counter electrodes for DSSCs. Aboagye et al. also reported on electrospun carbon nanofibers with Pt nanoparticles with an efficiency of 7.6% [26]. Carbon nanofibers decorated with Pt–Ni_2P nanoparticles exhibited efficiency as high as 9.11% as counter electrode for dye-sensitized solar cells as compared to native Co_3S_4/CNs with an efficiency of 9.23% on account of electrospinning as reported by Zhao et al. [6]. Gnanasekar et al. 2019 fabricated a DSSC based on carbon wrapped VO_2(M) nanofiber synthesized by a sol–gel based simple and versatile electrospinning and post-carbonization technique [27]. The as-fabricated counter electrode for DSSC showed high power conversion efficiency of 6.53%.

Table 2 Comparison of various electrospun photoelectrode materials used in DSSC

Sr. no.	Photoelectrode material	Counter electrode material	Current density (J_{sc}) [mA cm^{-2}]	Open-circuit voltage (V_{oc}) [mV]	Fill factor (FF)%	Photo-conversion efficiency (PCE) %	References
1	TiO$_2$	Pt/GQD	11.72	680	78	4.81	[32]
2	TiO$_2$	PVDF-HFP	11.8	710	65.3	5.3	[33]
2a	Acetylated TiO$_2$	Pt/FTO	8.8	780	58	4.0	[34]
2b	De-acetylated TiO$_2$	Pt/FTO	8.6	780	55	3.7	[32]
3	5% Ag–ZnO nanofibers	Pt/FTO	5.64	724	78.1	3.19	[35]
6	TiO$_2$	Pt/FTO	9.73	650	62	3.92	[36]
7	TiO$_2$	WSe$_2$/MoS$_2$	16.89	690	72.4	8.44	[37]
8	TiO$_2$	Pt/FTO	17.84	710	68.9	8.73	[35]

In another approach, 15 nm graphene quantum dots (GQD) decorated with TiO$_2$, electrospun using carbon nanofibers, enhanced photovoltaic activity by 6.22% in comparison to native TiO$_2$ nanofibers. Among various gel electrolytes, electrospun polymer gel electrolyte presented a 5.52% photoelectric conversion efficiency with long-term stability [28]. Electrospun Co–TiC has also been found as a stable counter electrode along with enhanced electrochemical activity for methanol fuel cells as well as DSSC [29]. Similarly, Zhang et al. 2019 demonstrated carbon-based materials (viz. carbon nanotubes, graphite, conductive carbon black, and graphene) as a counter electrode with a cell efficiency of 6.29% for DSSC. A power conversion efficiency of 4.32% was also achieved by DSSC using optimum carbon material mixture [30]. A new type of eco-friendly, cost-effective stainless steel mesh-based flexible quasi-solid dye-sensitized solar cell (DSSC) using electrospun, a high surface to volume ratio with highly porous ZnO nanofibers as photoelectrode was also reported [31]. The as-prepared electrospun ZnO exhibited a solar cell conversion efficiency of 0.13% with a short-circuit photocurrent, open-circuit voltage, and fill factor of 28 μA, 0.321 mV, and 32.77%, respectively. Comparison of different DSSCs are listed in Table 2.

4   Lithium (Li)-Ion Batteries

As an electrochemical device, Li-ion battery has shown great promise as an appealing power source of the current era. Among various characteristics of Li-ion battery, the most favored ones are long cycle life, high energy density, high operational voltage, low self-discharge rate, no memory effect, finds potential applications in, viz. portable electronic devices, energy storage systems, and electric vehicles, etc. [38]. In general, Li-ion batteries are mainly comprised of anode, cathode, separators, and electrolyte. Charge–discharge occurs by a redox process where Li-ions transfer between electrodes and their charge capacity depends on the accumulation of Li-ion into electrode material [39]. Further, carbon allotropes (like Graphite, Graphene) are most preferred ones owing to their layered structure.

4.1 Assembly of Lithium Ion and Charge Storage Mechanism

The electrochemical series places Lithium at the top for its ability to get oxidized at maximum. Lithium-ion batteries complete a full cycle by a full charge followed by complete discharge. This is accomplished by providing separate electron as well as ion pathways followed by the flow of the electron back to the device. Thus, the former is the consequence of the very first charge whereas the latter is the result of electricity being produced across the load through the circuit. As for the first charge, when the device is connected across a voltage, the positive polarity promotes electron ejection from the lithium atoms from the metal oxide resulting in an electron and a lithium ion. Thus, a series of electrons leads to build up of electron flow. Further, the electrons then drift through the external circuit and reach to intercalate within the graphite/graphene layer. Unlike other components in the lithium-ion battery, carbon allotropes, viz. graphite or graphene, are most preferred for charge storage owing to their layered structure. These loose layers are weakly bonded through Van der Wall forces, which in turn facilitate easy intercalation of lithium ions thereby allowing stable storage space for lithium atoms. Also, the electrolyte between the graphite and the metal oxide prevents the passage of electrons and only allows lithium ions to pass through. Meanwhile, the positively charged lithium ions pass through the electrolyte and move on toward the negative terminal. Finally, when all the lithium atoms reach the graphite sheet, the cell is then fully charged, ready for discharge. This completes the first half-cycle which involves the formation of lithium ions, followed by the release of an electron from metal oxide and its intercalation between graphite layers. For the discharging part, the lithium ions move through the electrolyte to reunite with its metal oxide and electrons across the load once connected. Thus, an electric current is obtained through the load. This completes the next half of the cycle. While graphite acts as a storage medium, it does not contribute any part in chemical reactions of the lithium-ion cells. In the assembly of the lithium cell, another important component—separator, which acts as insulator, is positioned between the electrodes. The function of the separator is to ensure safe working of the cell. This is accomplished by preventing a short circuit in case of abnormal functioning of the cell. This undesired condition may arise on account of interconnectivity between cathode and anode. To prevent such unwanted situation, an insulating layer called the separator is placed between the electrodes. The separator is semi-permeable and non-conducting for the lithium ions because of its microporosity and can be fabricated by the electrospinning technique. So far, graphite and metal oxides are poor

Fig. 3 Working principle of
Li-ion battery (Reproduced
from Ref. [40] with
permission from Elsevier)

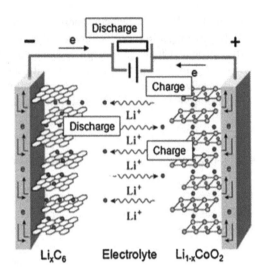

at collecting or delivering the electrons. Thus, a conductive metal layer is inserted next to the graphite, and metal oxides. These layers are called collectors. The number of collectors in a cell may vary depending on the number of unit cells present. In practice, some percentage of lithium always remains in anode, cathode, and electrolyte despite the battery being fully charged or discharged, respectively, owing to the intrinsic imperfections or defects. In commercialization, the whole assembly is made compact by coating electrolyte over separator membrane in the form of sheets, which are then wrapped around a supportive core. Similarly, aluminum sheets are also coated by graphite or any other layered element and metal oxide layer which further sandwiches the separator sheet is preferred, thereby achieving a mobile, portable, and ready-to-use lithium-based energy device (Fig. 3).

As of now, various materials have been developed for electrode material, however, the need for further advances in Li-ion technology toward designing and fabrication of nanoscale composite materials still exists. ESN fabrication incorporates pores which act as locking sites for the charge, which further promotes charge storage. This has been demonstrated in various reports, such as by Li et al. and Xu Tian et al., through optimization [41, 42]. Apart from chemical properties, ESN technique have also been found to improve interesting physical properties like tensile strength, Young's modulus, and electrical properties like ionic conductivity too. A report evident to this has found that electrospun cellulose/PEO fabricated as a solid polymer electrolyte for Li-ion batteries increased not only tensile strength by twofold but also a fivefold increase in Young's modulus, along with an ionic conductivity of $104 \, S \, cm^{-1}$ [43]. Aravindan et al. also reported on high electrochemical performance of 1D NiO nanofibers as anode material. Half-cell assembly is used to evaluate the Li-uptake properties and found maximum reversible capacity around $784 \, mA \, h \, g^{-1}$ at a current density of $80 \, mA \, g^{-1}$ with an operating potential of 1.27 V versus Li [44]. A free-standing, binder-free Na-ion and Li-ion were also fabricated by utilizing electrospun Polyacrylonitrile (PAN) carbon nanofibers (CNF) as anode material. Better cycling performance of batteries was achieved due to unique 3D interconnected nanofiber web morphology of PAN-CNF [45]. Xiang et al. also used flexible nanofiber membrane containing dispersed MoO_2 nanoparticles for the binder-free self-standing anode in Li-ion battery with reversible capacity of 811 mA h g^{-1} after 100 cycles at a current density of 100 mA h g^{-1}, higher compared to most of MoO_2-based anodes [46]. Qui et al. fabricated carboxymethyl cellulose acetate butyrate (CMCAB) nanofiber through electrospinning for high rate Li-ion battery with increased specific capacity from 147.6 mA h g^{-1} to 160.8 mA h g^{-1} for first discharge at the rate of 2 C. [47]. Wu et al. also reported on the superior electrochemical performance of Li-ion battery anode with 94% of initial specific capacity (720 mA h g^{-1} maintained even after 55 cycles) [48].

Yanilmaz et al. fabricated SiO_2/nylon-6,6 nanofiber membrane and demonstrated that better thermal stability, mechanical strength with high porous structure along with enhanced electrochemical property were achieved on account of the electrospinning technique. This nanofiber membrane had larger liquid electrolyte uptake for 77% porosity, higher electrochemical oxidation limit, and lower interfacial resistance with Li [49]. SnO_2–ZnO nanofibers were fabricated as anode material for Li-ion battery not only exhibited superior Li storage capacity but also enhanced cyclic performance along with rate property over native SnO_2 nanofibers at different current densities 500 and 1000 mA g^{-1} discharge capacities were 200.1 and 125.7 mA h g^{-1}, respectively, [50]. Another report suggests electrospun fabricated rGO–TiO_2 nanocomposite enhanced charge storage to 200 mA h g^{-1} [51]. ESN based-Li-ion batteries are listed in Table 3.

Table 3 Comparison of various materials, method of fabrication used for Li-ion batteries

Composite material	Electrospinning solution	Electrochemical performance	References
Free-Standing CNF	PAN/DMF	460 mA h g^{-1} @ 0.1 A g^{-1}	[52]
WS$_2$/carbon nanofibers (WS$_2$/CNFs)	–	Discharge/charge Capacity: 761/604 mA h g^{-1} First Coulombic Efficiency: 79.4%	[53]
Bi-embedded one-dimensional (1D) carbon nanofibers (Bi/C nanofibers)	–	Capacitance—316 mA h g^{-1} at 500 cycle versus Capacitance—190 mA h g^{-1} at 90 cycle	[54]
Tin oxide–carbon	–	Discharge capacity—722 mA h g^{-1} after 100 cycles at the rate of 1 C	[55]
NFHP-SnO$_2$	PAN/DMF	609 mA h g^{-1} @ 3A g^{-1}	[56]

5 Supercapacitors (SCs)

Supercapacitors (SC) now popularly known have been reported earlier as electro-chemical capacitors, which discharge more comparatively with respect to time. Supercapacitors (also known as an electrochemical capacitor) were first intro-duced in late 1950 which demonstrated activated charcoal plates. Among several electrochemical-based energy devices, supercapacitors are the ones that hold them-selves amid batteries and capacitors with respect to charge–discharge cycles. Unlike capacitors, SCs are free from dielectric media, and instead house an electrolyte. Further, their composition is almost similar to capacitors which include an anode where electrons and protons are formed and an electrolytic membrane which acts as a channel for proton transport between electrolyte and cathode. The electrolyte is further sandwiched between two plates using a separator. The separator is an extremely thin (<1 nm or angstroms) insulator material (such as carbon allotropes, cellulose or polymer) and can be fabricated using electrospinning. Furthermore, to promote electron transport, viz. from anode to load, an external circuit is devised which ultimately converges at the cathode where all protons, electrons combine up to from some product. Additionally, they sport several thousands of capacitances in comparison to electrolytic capacitors. SCs present several advantages including higher charge–discharge cycles, minimal or negligible internal resistance, and better efficiency than batteries making them an irresistible choice for energy storage device. They store energy electrostatically in extremely large quantity in comparison to capacitors. Depending on the electrode design, they are classified into three cate-gories, viz. Double layered, Pseudocapacitors, and Hybrid capacitors. Double-layer capacitors are based on carbon or their derivative with respect to electrode material, Pseudocapacitors, on the other hand, have metal oxides or conducting polymer as electrode material and finally, Hybrid capacitors have asymmetric electrodes that feature both double-layer capacitance as well as pseudocapacitance.

5.1 Construction, Assembly, and Working of SC

An SC consists of two metal plates coated with porous material. Further, they are immersed into an electrolyte. A charged-up SC plate has an array of opposite charged ions on account of electrostatic attraction at the interface (Helmholtz double layer) of metal plate and electrolyte, thus the name, Electric Double Layered Capacitor (EDLC). This arrangement of charge storage enables higher capacitance compared to the normal capacitors, which holds a thick insulating dielectric medium for charge storage. Another way of charge storage, which takes place on account of redox reactions (faradic), is known as Pseudocapacitance. Also, the capacitance of the SC depends upon surface area (capacitance increases with increase in area) of the plate and distance between the plates (capacitance increases with decrease in distance between the plates) (Fig. 4).

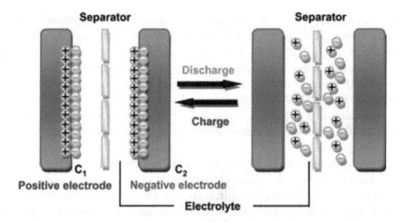

Fig. 4 Charge storage mechanism in Supercapacitor (Reproduced from Ref. [57] with permission from Elsevier)

Few terminologies in supercapacitors:

Specific Power Density: It is the quantity of power transferred per unit of mass (Wh kg^{-1}).

Current density: It is the quantity of charge transferred per unit time per unit area (mA cm^{-2}).

Due to their higher power density, extremely longer cycle life, low maintenance, and safe operation cost, SCs are considered to be highly complementary to batteries [58]. Primary components for any SC are capacitor plate, electrolyte, and a thin insulator between two plates. Based on charge storage mechanism, SCs can be classified into (a) Electric Double Layer Capacitors (EDLC) which store the charge through electrostatic accumulation by utilizing high surface area in carbon materials, viz. graphene, carbon nanotubes, and activated carbon and (b) Pseudocapacitors, which store charges through Faradaic reaction using active redox materials.

In practice, factors that govern the overall performance of SCs are pore size, surface area, functional groups, the electrical conductivity of material, and type of electrolyte used. Remarkably, the intriguing properties of ESN are regarded among diverse composite materials (like composite polymers, conducting polymers, metal oxides) as a model entrant for SC. Numerous works have been done on the use of ESN for SC application. ESN of 200–300 nm were synthesized by carbonization of MoS_2/PVA to give $MoS_2@C$ for SC applications. It was found that as-synthesized ESN exhibited maximum specific capacitance of 335.6 F g^{-1} in 6 M KOH electrolytic solution with superior capacitance retention of 93% up to 2000 cycles [59]. Recent reporting also suggests bio-mimic in regard to electrospun fabrication. One of them, Tian et al., 2018, fabricated a worm-like NiMo4 coaxially decorated through ESN as a binder-free electrode that exhibited a better specific capacitance of 1088 Fg^{-1} at 1 Ag^{-1} with 73% capacitance retention even after 5000 cycles.

ESN-based CNF–CNT-coated polyaniline composite electrode material showed specific capacitance of 278 Fg^{-1} which retained 98% of its capacitance for continuous 2000 cycles at 10 Ag^{-1} [60]. Di Tian et al. highlighted on improved conductivity and active surface area of electrospun CNF by coating polyaniline thorns (Ni-G-CNFs@PANI) through carbonization and in situ polymerization to develop binder-free electrode for practical application in the energy storage field. This electrode had a specific capacitance of 318 F g^{-1} at 0.5 A g^{-1}. Capacitive retention remained at 85.8% after 1000 cycles of the charge–discharge process at a high current density of 10 A g^{-1} [61].

Another electrospinning technique, known as Coaxial electrospinning technique, has also been reported with various enhancements in specific capacitance. One of these, Yu et al., fabricated decorated Iron Oxide particle, hollow carbon nanofibers through Coaxial electrospinning technique which had a high specific capacitance of 121 F g^{-1} at 0.5 A g^{-1} with 90% retention rate after 1000 cycles [62]. On another approach, Pech et al. found electrospun carbon Manganese oxide composite nanofibers C/MnOx exhibited not only higher specific capacitance of 213.7 F g^{-1} at 0.5 A g^{-1} but also energy density of 30 mW h g^{-1} at a power density of 30 mW h g^{-1} along with 97% retention after 1000 cycles [63]. Similarly, high porosity on account of electrospinning, along with the incorporation of Boron and MnO_2 enriched electrochemical performance, was detailed by [64]. SCs, fabricated through ESN techniques, have shown to improve performance, as evidenced by recent publications. One of them have reported an increase in specific capacitance of 208 F g^{-1}, energy density of 25.66–18.54 Wh kg^{-1}, and power density 400–10,000 W kg^{-1} and 1 mA cm^{-2}. Furthermore, 90% retention of initial capacitance till 3000 cycles was also accounted. At another instance, an increase in surface area was reported to 718 m^2 g^{-1} as a consequence of the electrospinning technique. X. Zhao et al. also reported a device with a power density of 320 mW^{-2}, which was accounted to better interconnectivity, as a result of electrospinning Zhao et al. [65]. Tyagi et al. further extended their approach to obtain even higher density symmetric supercapacitor devices by doping 2% Ta in TiO_2 through electrospinning [66]. Z. Dai et al. fabricated GNS-N-S co-doped ACNF with various GNs addition derived from renewable plane resources by electrospinning, thermostabilization and carbonization process [67]. Fabrication of COOH@MnO_2 device through ESN technique demonstrated a better specific capacitance of 415 F g^{-1} at 5 mV s^{-1} along with a retention of 94% in a 10,000-cycle test for asymmetric capacitor [18]. Comparison of different ESN materials for SC application is also tabulated in Table 4.

6 Fuel Cell

Unlike conventional combustion technologies, fuel cells endure chemical processes to convert hydrogen-rich fuel into some form of energy, say electricity. Additionally, unlike batteries, they are free from regular recharge schedules and go on producing electricity as long as a fuel source is provided [73].

Table 4 Comparison of various materials through electrospinning along with electrochemical parameters used for supercapacitors

Composite electrode	Fabrication technique	Electrochemical parameters	References
Binder-free mesoporous $Mn_{0.56}V_{0.42}O@C$	Electrospinning	Specific capacitance—668.5 F g^{-1} at 1 A g^{-1} Power density—900 W kg^{-1}	[68]
Poly(acrylonitrile-co-vinylimidazole) (PAV)	Electrospinning	Specific capacitance—114 F g^{-1} Energy Density—70.1 Wh/kg at 1 A g^{-1} Power Density—9.5 W kg^{-1} at 6 A g^{-1}	[69]
Polyacrylamide-graft-polyaniline (PAAm-g-PANI)	Electrospinning	Capacitance—102 F g^{-1} at 0.3 A g^{-1}	[70]
CeMO$_3$ (M = Co, Ni, Cu)	Electrospinning	Specific capacitance (F g^{-1})— CeCoO$_3$–128, CeNiO$_3$–189, CeCuO$_3$–117 Current Density—0.5 A g^{-1}	[71]
PAni–MnO$_2$	Electrodeposition	Specific capacitance— 396.89 F/g	[72]

6.1 Fuel Cell Assembly and Their Charge Storage Mechanism

A typical fuel cell assembly consists of two electrodes along with an electrolyte. Further, the electrode is porous in nature providing large surface to volume ratio, and can be fabricated by electrospinning. Electrodes are also made from different materials (like Platinum, porous PVC (Poly vinyl chloride), Nickel boride, Teflon, and so on) are coated with silver to promote efficient electron transport between electrode and current collector. Fuel and an oxidant (oxidant may be air, oxygen, hydrogen, H_2O_2, HNO_3, and so on) are supplied continuously and individually to the two electrodes of the cell. Additionally, fuel for fuel cells can be liquid fuels such as methanol, ethanol, hydrazine, and formaldehyde. Further, gaseous fuels such as hydrogen, alkane, and carbon monoxide have also been used. In practice, fuel cell mechanism primarily involves conversion of chemical energy into heat, followed by its conversion into mechanical energy (based on end application) and finally into electrical energy. The degree into which heat is converted into mechanical energy is limited, however in a fuel cell, chemical energy is almost converted into electrical energy. Fuel cells are classified based on temperature, viz. Low temperature (<100 °C), Medium temperature (100–250 °C), and High Temperature (>500 °C). Also, fuel cells where microorganisms disintegrate organic compounds are known as Biochemical cells.

Polymer electrolyte membrane fuel cell or PEMFC is an electrochemical device which converts hydrogen and oxygen into electricity. Both the cathode and anode catalysts are typically platinum nanoparticles which are deposited on a carbon substrate. Catalysts facilitate splitting of electrode material into protons and electrons, from which protons make through the electrolyte barrier, whereas electrons travel across circuit and then get together with ion that passed through the electrolyte membrane. Reaction advances with two hydrogen molecules splitting into four protons and four electrons, which can be used through an external load. Also, at the center of the PEMFC is a polymer called nafion, which is a proton conducting ionomer. Nafion consists of sulphuric acid groups forced into Teflon backbone. Four protons move through nafion onto cathode where oxygen is reduced to two water molecules. While the theoretical output of a single cell is typically about 1.23 V, the optimal performance in practice is around 0.7 V.

Another type of fuel cell, Solid Oxide Fuel Cell (SOFC), is an electrochemical device which converts chemical energy to electrical energy. It typically operates at temperatures from 600 to 950 °C; at these temperatures, the electrolyte, cathode, and anode act as oxide ion conductors. Like PEMFC, this fuel cell too needs a continuous fuel supply such as hydrogen or methane. This fuel is then directed to the anode where it reacts with oxide ions from electrolyte lattice. Released electrons are transferred to cathode through an external load. In the cathode, the oxygen molecules absorb and dissociate the oxygen atom in the presence of electrons that absorb oxygen atom will reduce the oxide ion. The oxide ions then move through the electrolyte to the anode and react with hydrogen from a fuel. In SOFC, electrolyte is an ion conductive ceramic solid oxide membrane. Electrolyte material should be dense and not have electronic conductivity as well as impermeability so as to prevent gas flow through it. One-unit cell of SOFC contains two current collectors. One of the advantages of Solid Oxide fuel cell is that no platinum is needed and in addition to hydrogen, it is able to utilize natural gases like methane or other carbonaceous fuel. Additionally, Electrical efficiency of solid oxide fuel cell is approximately up to 70% which is almost twice the classical efficiency of fossil fuel-based engines.

Although in practice, efficiency of fuel cell still needs to be improved, various materials have been tried at anode, cathode, and electrolyte to overcome most of the shortcomings faced by traditional fabrication methods. Some of them like decreasing fiber diameter to the nanoscale, increasing pore size, and engineering channels in electrolyte have proved to increase overall device efficiency during the last decade [74]. Table 5 consolidates most of the recent composite materials, along with the fabrication process and their effect on efficiencies used in fuel cells to provide a better understanding in this regard.

Table 5 Parameter-wise comparison for various electrospun polymer composites in fuel cell

Polymer composite	Fabrication technique	Component	Parameter	References
Optimized CNF	Electrospinning	Membrane	In-plane electrical conductivity: -30 S cm^{-1}	[82]
Antimony-doped Tin oxide (ATO) versus Pt/C	Electrospinning	Pt catalyst support material versus Pt/C	Electrochemical Surface Area (ECSA) 33 m2 g^{-1} versus 55 m2 g^{-1}	[83]
TiO$_2$-CNF	Electrospinning	Electrocatalyst—DMFC	Current density at 345.64 mA mg$_{catalyst}$$^{-1}$	[84]
SiO$_2$@C	Coaxial Electrospinning	Microporous layer—PEMFC	Power density: 687.1 mW cm^{-2}, Relative Humidity: 15%, Temperature (°C): Anode/Cell/Cathode 50/50/50	[85]
La$_{0.8}$Sr$_{0.2}$MnO$_3$ (LSM)	Electrospinning	Cathode	Power density: 0.85 W cm^{-2} at 700 °C	[86]

7 Electrospinning Techniques

Multiaxial Electrospinning technique: This technique is preferred whenever hollow fiber morphology is desired. To achieve this, a multiaxial setup is established. Unlike conventional setup, coaxial electrospinning uses dual feed, likewise triaxial three, tetra-axial four, and so on. This facilitates injection of the solution above one another over single spinneret. Further, the sheath, which acts as an inner hollow housing, is drawn at the Taylor cone. Different morphologies can be obtained by varying the solvent miscibility. For instance, immiscible solutions lead to a core–shell structure, whereas miscible solutions lead to a highly porous structure. This is on account of the separation of phases on solidification of fibers.

Emulsion electrospinning technique: This technique is favored if desired morphology involves composite materials. As a result, the technique involves multiple variables which directly affect spinnability, thereby making the technique even more challenging in comparison to coaxial technique. In order to control these variables, emulsion should be carefully produced by accounting electrospinning parameters. The emulsion can be produced using an immiscible solvent along with a water phase in the presence of a stabilizer or emulgent. Additionally, as long as the immiscible phase interfaces are stabilized, a variety of agents can be deployed (for instance, surfactants, nano particles, and so on). In practice, electrospinning involves stretching of emulsion droplets result in coalesced nanofibers. This technique also facilitates the blending of a variety of polymer solutions given their immiscibility with each other.

Furthermore, they can stabilize and also the phase segregates even without emulgents. Additionally, this technique can even be further tailored if a miscible solvent is used.

Melt electrospinning: Human exposure to volatile solvents has already been proven hazardous. As a result, this technique features the very advantage of excluding such volatile solvents. Although the above-mentioned techniques, viz. Multiaxial Electrospinning technique and Emulsion electrospinning techniques, favor crystalline polymers, they are incapable of dealing with semi-crystalline polymers. Thus, melt electrospinning overcomes this limitation and allows employing such solutions for spinning. Melt electrospinning features similar construction as that of traditional electrospinning. To generate the melt, the polymer is heated by several methods such as resistance heating, lasers, and so on. As a consequence of the heat, resulting fibers are thick (>250 nm) in comparison to those obtained from conventional electrospinning. The effect of the melt on spinnability tends to attain thermal equilibrium after attaining stable flow rates. Also, unlike other techniques, the whipping phenomena are not generally observed on account of high viscosity and lower conductivity values of the melt. Apart from these factors, the fiber diameter is also affected by several factors, including feed rate, the molecular weight of the polymer, and spinneret diameter.

In line with existing reports, Park et al. found a significant 46% higher power output in case of electrospun Nafion/PVDF membrane compared to native Nafion [75]. Jindal et al. achieved a higher power density of 3.43 mW cm^{-2} and a current density of 9.79 mA cm^{-2} using electrospun carbon nitride nanofibers as a cathode in a microfluidic fuel cell (MFC). This electrochemical performance was better in comparison to MFC that used Au (2.72 mW cm^{-2}, 6.4 mA cm^{-2}) as anode and Pt (3.09 mW cm^{-2}, 6.18 mA m^{-2}) as an anode [76]. Yang et al. used electrospinning and thermal treatments to plant Co and CoP nanograins in porous Nitrogen-doped CNF (Co/CoP@NC). This resulted in a robust hydrogen evolution reaction (HER) with 117 mV in acidic and 180 mV in alkaline media which delivered a current density of 10 mA cm^{-2}. These were accounted to a strong synergistic relationship between Co and CoP nanoparticles, N-doped carbon configuration, and the interconnected 3D porous network [23]. Salahuddin et al. demonstrated superhydrophobic PAN nanofibers through electrospinning for gas diffusion layers as proton exchange membrane fuel cell (PEMFC) for cathodic water management [77]. Furthermore, improved poor oxygen reduction reaction (ORR) of supercapacitor and microbial fuel cell by electrospun carbon nanofiber composite as cathode material was also reported by them. High ORR catalytic activity with low internal resistance (0.18 Ωcm^{-2}) and high exchange current density (13.68 Am^{-2}) was contributed to the interconnectivity of fiber aggregates and thorn-like structure of carbon nanofiber composite electrode, a maximum power density of 320 mW was obtained, which was 140% higher than of the Pt/C [65].

Ahn et al. fabricated composite nanofiber electrode through the electrospinning technique which showed least polarization resistance of 0.0310 Ωcm^{-2} and power density of 0.7 Wcm^2 at 650 °C with 39.2% and 12.5% improvement to catalyst-free composite nanofiber electrode, respectively [78]. Powers et al. used electro-spinning for H$_2$/air fuel cell. This tri-layer morphology contained a 20 μm thick membrane which exhibited an area specific resistance of 0.023 Ω cm^{-2} and a 5% in plane water swelling, with a tensile strength of 26 MPa and H$_2$/air fuel cell, performance comparable to Nafion 212 [79]. NAM Barkat et al. synthesized NiSn nanoparticle-incorporated carbon nanofiber via electrospinning mats of Nickel acetate. Current density increased to 175 mA cm^{-2} at 55% for urea oxidation reaction [80]. Ekrami-Kakhki et al. synthesized Pt/PVA CuO-Co3O4/Chitosan via single nozzle electrospun technique. As a result, higher electrocatalytic activity for methanol electro-oxidation was observed [81].

8   Electrospun Nanofibers for Environmental Protection

Due to rapid industrialization, the quality of the environment (soil, water, and air system) has deteriorated due to the release of many pollutants in the atmosphere [15]. Due to the ongoing threats to the world's collective energy security and environment, there is an urgent need to develop technology and devices from different standpoints [87]. Rising concerns over conventional energy resources, nanoscale featured—an electrospun technique promotes not only the production of continuous fibers (with diameters down to few nanometers) but also utilizes various organic (polymers)-inorganic (metal oxides) materials toward energy storage while resolving many other problems associated to ecological growth. Furthermore, the development of low cost, eco-friendly material which can be expanded industrially and academically could ensure sustainable energy and environment. One of these applications includes deployments of filtration membranes fabricated by electrospinning techniques in water or air treatment.

Membranes offer better possibilities, especially when labored into environmental applications. Recently, membranes attained through ESN (ESN membranes) could remove not only pollutants but also lower operational costs. These already have been demonstrated by Akduman et al. [88], recently, for example, their better surface area to volume ratio, and tunability of surface functionality as adsorbate in wastewater treatment plants [88]. ESN membranes have also been reported to enhance mechanical, thermal, elastic, and stability features. One of them was demonstrated to have better performance in dye absorption, and was achieved by supporting polyacrylonitrile and polyamide nanofibers with Halloysite nanotube (HNT) through the electrospinning technique [89]. Bai et al. used plasma-etched ESN membrane for efficient removal of cationic methylene blue (MB) dye which occurred due to enhancement in surface area of nanofibers composed of reduced diameter of nanofiber and increased area of generated holes [90]. Qureshi et al. fabricated Zein nanofiber through electrospinning as a green and recyclable adsorbent for the removal of reactive black 5 from aqueous phase toward water purification [91]. Electrospun nanofibers of anionic chiral polysaccharide, hyaluronic acid/transparent polymer, and polyethylene oxides were effectively utilized for improved polarization properties of fluorescent dyes [92]. Apart from the excellent dye absorbent role of ESN, they have also been broadly explored for removal of toxic heavy metal ions and water purification. Hamad et al. employed cellulose acetate nanofiber incorporated with hydroxyapatite for removal of heavy metals with separation efficiency of about 99.7% for 35 min and 95.46% for 40 min [93]. Zhang et al. reported the fabrication of Poly(vinyl) alcohol/poly(acrylic) acid (PVA/PAA) nanofiber membrane using electrospinning technique for separation of lead Pb (II) from water in a continuous fixed bed column under variable conditions. Another evidence is where Zhang et al. [94], found the arrangement as efficient with lower feed concentration and low flow rate [94]. Also, the removal of Cu (II) ions from aqueous solution had already been reported by Guan et al., where modified Hordein electrospun nanofibers were used [95]. Currently, detection of Atrazine, an agricultural herbicide, has been a concern and several reports exist for its remediation. Among those, P. Supraja emphasized its detection in the range of 10–21 g/mL by Mn_2O_3 nanofibers through the electrospinning technique [96]. Mehrani et al. synthesized modified ESN of poly(m-phenylene diamine/carbon nanotube) (Pm PDA/CNT), a nanosorbent for thin film microextraction (TFME) of trace levels of copper from food and water sources [97]. Du et al. demonstrated the electrospun support layer for conductive CNT/nanofiber composite hollow fiber membrane for water purification. The merits of the above system derive from continuous interlaced nanofiber scaffold structure of electrospun nanofiber support layer and cross-linked CNT separation layer with low tortuosity membrane pore and mechanical strength [98]. W. Zhang et al. found strong intermolecular hydrogen bonds between hydroxyl groups of PVA and Lignin in electrospun lignin/poly (vinyl alcohol) (lignin PVA) composite. This membrane as an adsorbent for safranine T (ST) for water purification has found to increase with pH and temperature [99]. Wu et al. incorporated Gold nanocages into electrospun nanofibers of poly(vinylidene fluoride) for efficient water evaporation via photothermal heating with evaporation efficiencies 97% and 79.8% at wt. percentage of 0.05% and 0.10%, respectively [100]. At another instance, electrospun polyamide-6 hybrid membrane has been successfully implemented for wastewater treatment by Medeiros et al. [101]. Given the nano-structural build up by ESN, Bortolassi et al. [102] a composite of Silver/polyacrylonitrile (Ag/PAN) reported efficient pollutant-based nanoparticle removal and significant anti-bacterial activity [102]. On a similar approach, Zhu et al. also emphasized the principle and air purification by ESN-fabricated membranes [103].

In line, ESN of Polyurethane was successfully prepared by rotating bead spinneret electrospinning and utilized in air filtration for high-efficiency $PM_{2.5}$ capture. ESN-fabricated air filter was noticed to be up to 99.654% with good optical transparency of 60%. It also enhanced contact angles, ventilation rates at 128.5°, and 3480 mm/s, respectively [104]. Similarly, poly (Ɛ-caprolactone/polyethylene oxide (PCL/PEO)-based air filter were successfully fabricated through electrospinning technique, and solvent annealing methods as prepared ESN endowed wrinkled fiber surface and observed to enhance $PM_{2.5}$ capacity of protective masks which can effectively filter

Table 6 Comparison of various electrospun nanofiber composites used for different environmental remediation

Electrospun polymer–composite nanofiber	Remediating technique	Hazardous agent	References
Polyhydroxyalkanoate film	Adsorption	Malachite Green	[106]
Fibroin/Cellulose Acetate (80/20)	Adsorption	Cu (II)	[106]
Chitosan PMMA	Filtration	Chromium (II)	[107]
ZnO/0.5 Wt% CuO	Photocatalytic reduction	MB dye	[108]
PAN-CNT/TiO$_2$-NH2	Photocatalytic reduction	Chromium	[109]
Plasma–etched PLLA	Adsorption	MB dye	[90]
Graphene oxide/PAN	Filtration	Improved protein rejection rate	[107]
PVA/Lignin	Adsorption	Safranine T (ST)	[99]
Ag/PAN	Filtration	E. Coli	[102]
(PVA/PAA)	Adsorption	Pb (II)	[110]

with a removal efficiency of 80.01% [105]. All such comprehensive reviews demonstrate a simple approach toward the fabrication of ESN with excellent degradation characteristics and wide potential application in eco-friendly environmental remediation toward sustainable energy and development [104]. Recent advances of ESN-based filter, membrane, etc. are tabulated in Table 6.

9 Conclusion

This chapter summarizes current progressions over several modes of fabricated nanostructured electrospun nanofiber (ESN) for energy and environmental applications. Electrospinning is a simple way of generating multiple active components in single nanofiber with large surface area to volume ratio, surface tuneable functionality, high porosity, hierarchical nanostructure, etc. This 1D hierarchical nanostructure can afford the possibility of fabricating flexible devices for energy storage coupled with environmental remediation. Despite such intriguing properties of ESN, selection of material and their optimization (viz. concentration, solvent, viscosity, and voltage) still need to be addressed for its multifunctional application. For commercial industrialization, the promise of good structural, thermal, and mechanical integrity is still a huge challenge for large-scale and downstream processing of ESN toward future application.

References

1. Manea LR, Bertea A, Popa A, Bertea AP (2018) Electrospun membranes for environmental protection. IOP Conf Ser Mater Sci Eng 374:12081. https://doi.org/10.1088/1757-899X/374/1/012081
2. Gür TM (2018) Review of electrical energy storage technologies, materials and systems: challenges and prospects for large-scale grid storage. Energy Environ Sci 11:2696–2767. https://doi.org/10.1039/C8EE01419A
3. Vitale G, Mosna D, Bottani E et al (2018) Environmental impact of a new industrial process for the recovery and valorisation of packaging materials derived from packaged food waste. Sustain Prod Consum 14:105–121. https://doi.org/10.1016/j.spc.2018.02.001
4. Lv D, Zhu M, Jiang Z et al (2018) Green electrospun nanofibers and their application in air filtration. Macromol Mater Eng 303:1800336. https://doi.org/10.1002/mame.201800336
5. Mirjalili M, Zohoori S (2016) Review for application of electrospinning and electrospun nanofibers technology in textile industry. J Nanostructure Chem 6:207–213. https://doi.org/10.1007/s40097-016-0189-y
6. Zhao K, Zhang X, Wang M et al (2019) Electrospun carbon nanofibers decorated with Pt-Ni2P nanoparticles as high efficiency counter electrode for dye-sensitized solar cells. J Alloys Compd 786:50–55. https://doi.org/10.1016/j.jallcom.2019.01.295
7. Fang J, Niu H, Lin T, Wang X (2008) Applications of electrospun nanofibers. Sci Bull 53:2265–2286. https://doi.org/10.1007/s11434-008-0319-0
8. Ramakrishna S, Fujihara K, Teo W-E et al (2006) Electrospun nanofibers: solving global issues. Mater Today 9:40–50. https://doi.org/10.1016/S1369-7021(06)71389-X
9. Shi X, Zhou W, Ma D et al (2015) Electrospinning of Nanofibers and their applications for energy devices. J Nanomater 2015:140716. https://doi.org/10.1155/2015/140716
10. Wang X, Yu J, Sun G, Ding B (2016) Electrospun nanofibrous materials: a versatile medium for effective oil/water separation. Mater Today 19:403–414. https://doi.org/10.1016/j.mattod.2015.11.010
11. Haider A, Haider S, Kang I-K (2018) A comprehensive review summarizing the effect of electrospinning parameters and potential applications of nanofibers in biomedical and biotechnology. Arab J Chem 11:1165–1188. https://doi.org/10.1016/j.arabjc.2015.11.015
12. Schreuder-Gibson H, Gibson P, Senecal K et al (2002) Protective textile materials based on electrospun nanofibers. J Adv Mater 34:44–55
13. Gao X, Han S, Zhang R et al (2019) Progress in electrospun composite nanofibers: composition, performance and applications for tissue engineering. J Mater Chem B 7:7075–7089. https://doi.org/10.1039/c9tb01730e
14. Gopal R, Kaur S, Ma Z et al (2006) Electrospun nanofibrous filtration membrane. J Memb Sci 281:581–586. https://doi.org/10.1016/j.memsci.2006.04.026
15. Thavasi V, Singh G, Ramakrishna S (2008) Electrospun nanofibers in energy and environmental applications. Energy Environ Sci 1:205–221. https://doi.org/10.1039/b809074m
16. Sofi HS, Abdal-hay A, Ivanovski S et al (2020) Electrospun nanofibers for the delivery of active drugs through nasal, oral and vaginal mucosa: current status and future perspectives. Mater Sci Eng C 111:110756. https://doi.org/10.1016/j.msec.2020.110756
17. Khan N, Dilshad S, Khalid R et al (2019) Review of energy storage and transportation of energy. Energy Storage 1. https://doi.org/10.1002/est2.49
18. Lin SC, Lu YT, Chien YA et al (2017) Asymmetric supercapacitors based on functional electrospun carbon nanofiber/manganese oxide electrodes with high power density and energy density. J Power Sources 362:258–269. https://doi.org/10.1016/j.jpowsour.2017.07.052
19. Mukhiya T, Dahal B, Ojha GP et al (2019) Engineering nanohaired 3D cobalt hydroxide wheels in electrospun carbon nanofibers for high-performance supercapacitors. Chem Eng J 361:1225–1234. https://doi.org/10.1016/j.cej.2019.01.006
20. Raulo A, Bandyopadhyay S, Ahamad S et al (2019) Bio-inspired poly(3,4-ethylenedioxythiophene): poly(styrene sulfonate)-sulfur@polyacrylonitrile electrospun

nanofibers for lithium-sulfur batteries. J Power Sources 431:250–258. https://doi.org/10.1016/j.jpowsour.2019.05.055

21. Xia J, Yuan Y, Yan H et al (2020) Electrospun SnSe/C nanofibers as binder-free anode for lithium–ion and sodium-ion batteries. J Power Sources 449:227559. https://doi.org/10.1016/j.jpowsour.2019.227559

22. Dissanayake SS, Dissanayake MAKL, Seneviratne VA et al (2016) Performance of dye sensitized solar cells fabricated with electrospun polymer nanofiber based electrolyte. Mater Today Proc 3:S104–S111. https://doi.org/10.1016/j.matpr.2016.01.014

23. Li Y, Li H, Cao K et al (2018) Electrospun three dimensional Co/CoP@nitrogen-doped carbon nanofibers network for efficient hydrogen evolution. Energy Storage Mater 12:44–53. https://doi.org/10.1016/j.ensm.2017.11.006

24. López-Covarrubias JG, Soto-Muñoz L, Iglesias AL, Villarreal-Gómez LJ (2019) Electrospun nanofibers applied to dye solar sensitive cells: a review. Materials (Basel, Switzerland) 12:3190. https://doi.org/10.3390/ma12193190

25. Gong J, Liang J, Sumathy K (2012) Review on dye-sensitized solar cells (DSSCs): fundamental concepts and novel materials. Renew Sustain Energy Rev 16:5848–5860. https://doi.org/10.1016/j.rser.2012.04.044

26. Aboagye A, Elbohy H, Kelkar AD et al (2015) Electrospun carbon nanofibers with surface-attached platinum nanoparticles as cost-effective and efficient counter electrode for dye-sensitized solar cells. Nano Energy 11:550–556. https://doi.org/10.1016/j.nanoen.2014.10.033

27. Gnanasekar S, Kollu P, Jeong SK, Grace AN (2019) Pt-free, low-cost and efficient counter electrode with carbon wrapped VO2(M) nanofiber for dye-sensitized solar cells. Sci Rep 9:5177. https://doi.org/10.1038/s41598-019-41693-1

28. Mohan K, Dolui S, Nath BC et al (2017) A highly stable and efficient quasi solid state dye sensitized solar cell based on polymethyl methacrylate (PMMA)/carbon black (CB) polymer gel electrolyte with improved open circuit voltage. Electrochim Acta 247:216–228. https://doi.org/10.1016/j.electacta.2017.06.062

29. Yousef A, Brooks RM, El-Newehy MH et al (2017) Electrospun Co-TiC nanoparticles embedded on carbon nanofibers: active and chemically stable counter electrode for methanol fuel cells and dye-sensitized solar cells. Int J Hydrogen Energy 42:10407–10415. https://doi.org/10.1016/j.ijhydene.2017.01.171

30. Zhang S, Jin J, Li D et al (2019) Increased power conversion efficiency of dye-sensitized solar cells with counter electrodes based on carbon materials. RSC Adv 9:22092–22100. https://doi.org/10.1039/C9RA03344K

31. Dinesh VP, Sriram kumar R, Sukhananazerin A, et al (2019) Novel stainless steel based, eco-friendly dye-sensitized solar cells using electrospun porous ZnO nanofibers. Nano-Struct Nano-Objects 19:100311. https://doi.org/10.1016/j.nanoso.2019.100311

32. Salam Z, Vijayakumar E, Subramania A et al (2015) Graphene quantum dots decorated electrospun TiO_2 nanofibers as an effective photoanode for dye sensitized solar cells. Sol Energy Mater Sol Cells 143:250–259. https://doi.org/10.1016/j.solmat.2015.07.001

33. Bandara TMWJ, Weerasinghe AMJS, Dissanayake MAKL et al (2018) Characterization of poly (vinylidene fluoride-co-hexafluoropropylene) (PVdF-HFP) nanofiber membrane based quasi solid electrolytes and their application in a dye sensitized solar cell. Electrochim Acta 266:276–283. https://doi.org/10.1016/j.electacta.2018.02.025

34. Kaschuk JJ, Miettunen K, Borghei M et al (2019) Electrolyte membranes based on ultrafine fibers of acetylated cellulose for improved and long-lasting dye-sensitized solar cells. Cellulose 26:6151–6163. https://doi.org/10.1007/s10570-019-02520-y

35. Kanimozhi G, Vinoth S, Kumar H et al (2019) Electrospun nanocomposite Ag–ZnO nanofibrous photoanode for better performance of dye-sensitized solar cells. J Electron Mater 48:4389–4399. https://doi.org/10.1007/s11664-019-07199-2

36. Patil JV, Mali SS, Patil AP et al (2019) Electrospun TiO2 nanofibers for metal free indoline dye sensitized solar cells. J Mater Sci Mater Electron 30:12555–12565. https://doi.org/10.1007/s10854-019-01616-2

37. Vikraman D, Arbab AA, Hussain S et al (2019) Design of WSe2/MoS2 heterostructures as the counter electrode to replace Pt for dye-sensitized solar cell. ACS Sustain Chem Eng 7:13195–13205. https://doi.org/10.1021/acssuschemeng.9b02430

38. Zhai Y, Liu H, Li L et al (2019) Electrospun nanofibers for lithium-ion batteries, pp 671–694

39. Duan J, Tang X, Dai H et al (2019) Building safe lithium-ion batteries for electric vehicles: a review. Electrochem Energy Rev. https://doi.org/10.1007/s41918-019-00060-4

40. Li X, Wei B (2013) Supercapacitors based on nanostructured carbon. Nano Energy 2:159–173. https://doi.org/10.1016/j.nanoen.2012.09.008

41. Li Y, Zhang W, Dou Q et al (2019) Correction: Li7La3Zr2O12 ceramic nanofiber-incorporated composite polymer electrolytes for lithium metal batteries. J Mater Chem A 7:4190. https://doi.org/10.1039/C9TA90038A

42. Tian X, Xin B, Lu Z et al (2019) Electrospun sandwich polysulfonamide/polyacrylonitrile/polysulfonamide composite nanofibrous membranes for lithium-ion batteries. RSC Adv 9:11220–11229. https://doi.org/10.1039/C8RA10229E

43. Samad YA, Asghar A, Hashaikeh R (2013) Electrospun cellulose/PEO fiber mats as a solid polymer electrolytes for Li ion batteries. Renew Energy 56:90–95. https://doi.org/10.1016/j.renene.2012.09.015

44. Aravindan V, Suresh Kumar P, Sundaramurthy J et al (2013) Electrospun NiO nanofibers as high performance anode material for Li-ion batteries. J Power Sources 227:284–290. https://doi.org/10.1016/j.jpowsour.2012.11.050

45. Jin J, Shi Z, Wang C (2014) Electrochemical Performance of Electrospun carbon nanofibers as free-standing and binder-free anodes for sodium-ion and lithium-ion batteries. Electrochim Acta 141:302–310. https://doi.org/10.1016/j.electacta.2014.07.079

46. Xiang J, Wu Z, Zhang X, Yao S (2018) Enhanced electrochemical performance of an electrospun carbon/MoO2 composite nanofiber membrane as self-standing anodes for lithium-ion batteries. Mater Res Bull 100:254–258. https://doi.org/10.1016/j.materresbull.2017.12.045

47. Qiu L, Shao Z, Yang M et al (2013) Electrospun carboxymethyl cellulose acetate butyrate (CMCAB) nanofiber for high rate lithium-ion battery. Carbohydr Polym 96:240–245. https://doi.org/10.1016/j.carbpol.2013.03.062

48. Wu Q, Tran T, Lu W, Wu J (2014) Electrospun silicon/carbon/titanium oxide composite nanofibers for lithium ion batteries. J Power Sources 258:39–45. https://doi.org/10.1016/j.jpowsour.2014.02.047

49. Yanilmaz M, Dirican M, Zhang X (2014) Evaluation of electrospun SiO2/nylon 6,6 nanofiber membranes as a thermally-stable separator for lithium-ion batteries. Electrochim Acta 133:501–508. https://doi.org/10.1016/j.electacta.2014.04.109

50. Zhao Y, Li X, Dong L et al (2015) Electrospun SnO 2–ZnO nanofibers with improved electrochemical performance as anode materials for lithium-ion batteries. Int J Hydrogen Energy 40:14338–14344. https://doi.org/10.1016/j.ijhydene.2015.06.054

51. Thirugunanam L, Kaveri S, Etacheri V et al (2017) Electrospun nanoporous TiO2 nanofibers wrapped with reduced graphene oxide for enhanced and rapid lithium-ion storage. Mater Charact 131:64–71. https://doi.org/10.1016/j.matchar.2017.06.012

52. Wu Y, Reddy MV, Chowdari BVR, Ramakrishna S (2013) Long-term cycling studies on electrospun carbon nanofibers as anode material for lithium ion batteries. ACS Appl Mater Interf 5:12175–12184. https://doi.org/10.1021/am404216j

53. Zhou S, Chen J, Gan L et al (2016) Scalable production of self-supported WS2/CNFs by electrospinning as the anode for high-performance lithium-ion batteries. Sci Bull 61:227–235. https://doi.org/10.1007/s11434-015-0992-8

54. Yin H, Li Q, Cao M et al (2017) Nanosized-bismuth-embedded 1D carbon nanofibers as high-performance anodes for lithium-ion and sodium-ion batteries. Nano Res 10:2156–2167. https://doi.org/10.1007/s12274-016-1408-z

55. Gupta A, Dhakate SR, Gurunathan P, Ramesha K (2017) High rate capability and cyclic stability of hierarchically porous Tin oxide (IV)–carbon nanofibers as anode in lithium ion batteries. Appl Nanosci 7:449–462. https://doi.org/10.1007/s13204-017-0577-8

56. Park JS, Oh YJ, Kim JH, Kang YC (2020) Porous nanofibers comprised of hollow SnO_2 nanoplate building blocks for high-performance lithium ion battery anode. Mater Charact 161:110099. https://doi.org/10.1016/j.matchar.2019.110099

57. Hausbrand R, Cherkashinin G, Ehrenberg H et al (2015) Fundamental degradation mechanisms of layered oxide Li-ion battery cathode materials: methodology, insights and novel approaches. Mater Sci Eng B 192:3–25. https://doi.org/10.1016/j.mseb.2014.11.014

58. Miao Y-E, Liu T (2019) Electrospun nanofiber electrodes: a promising platform for supercapacitor applications. In: Ding B, Wang X, Yu JBT-EN and A (eds) Micro and nano technologies. William Andrew Publishing, pp 641–669

59. Kumuthini R, Ramachandran R, Therese HA, Wang F (2017) Electrochemical properties of electrospun MoS_2@C nanofiber as electrode material for high-performance supercapacitor application. J Alloys Compd 705:624–630. https://doi.org/10.1016/j.jallcom.2017.02.163

60. Agyemang FO, Tomboc GM, Kwofie S, Kim H (2018) Electrospun carbon nanofiber-carbon nanotubes coated polyaniline composites with improved electrochemical properties for supercapacitors. Electrochim Acta 259:1110–1119. https://doi.org/10.1016/j.electacta.2017.12.079

61. Tian D, Lu X, Nie G et al (2018) Growth of polyaniline thorns on hybrid electrospun CNFs with nickel nanoparticles and graphene nanosheets as binder-free electrodes for high-performance supercapacitors. Appl Surf Sci 458:389–396. https://doi.org/10.1016/j.apsusc.2018.07.103

62. Yu B, Gele A, Wang L (2018) Iron oxide/lignin-based hollow carbon nanofibers nanocomposite as an application electrode materials for supercapacitors. Int J Biol Macromol 118:478–484. https://doi.org/10.1016/j.ijbiomac.2018.06.088

63. Pech O, Maensiri S (2019) Electrochemical performances of electrospun carbon nanofibers, interconnected carbon nanofibers, and carbon-manganese oxide composite nanofibers. J Alloys Compd 781:541–552. https://doi.org/10.1016/j.jallcom.2018.12.088

64. Yang C-M, Kim B-H (2019) Incorporation of MnO_2 into boron-enriched electrospun carbon nanofiber for electrochemical supercapacitors. J Alloys Compd 780:428–434. https://doi.org/10.1016/j.jallcom.2018.11.347

65. Zhao X, Nie G, Luan Y et al (2019) Nitrogen-doped carbon networks derived from the electrospun polyacrylonitrile@branched polyethylenimine nanofibers as flexible supercapacitor electrodes. J Alloys Compd 808:151737. https://doi.org/10.1016/j.jallcom.2019.151737

66. Tyagi A, Singh N, Sharma Y, Gupta RK (2019) Improved supercapacitive performance in electrospun TiO_2 nanofibers through Ta-doping for electrochemical capacitor applications. Catal Today 325:33–40. https://doi.org/10.1016/j.cattod.2018.06.026

67. Dai Z, Ren P-G, Jin Y-L et al (2019) Nitrogen-sulphur Co-doped graphenes modified electrospun lignin/polyacrylonitrile-based carbon nanofiber as high performance supercapacitor. J Power Sources 437:226937. https://doi.org/10.1016/j.jpowsour.2019.226937

68. Samir M, Ahmed N, Ramadan M, Allam NK (2019) Electrospun mesoporous Mn-V-O@C nanofibers for high-performance asymmetric supercapacitor devices with high stability. ACS Sustain Chem Eng 7:13471–13480. https://doi.org/10.1021/acssuschemeng.9b03026

69. Jung KH, Kim SJ, Son YJ, Ferraris JP (2019) Fabrication of carbon nanofiber electrodes using poly(Acrylonitrile-co-vinylimidazole) and their energy storage performance. Carbon Lett 29:177–182. https://doi.org/10.1007/s42823-019-00035-x

70. Smirnov MA, Tarasova EV, Vorobiov VK et al (2019) Electroconductive fibrous mat prepared by electrospinning of polyacrylamide-g-polyaniline copolymers as electrode material for supercapacitors. J Mater Sci 54:4859–4873. https://doi.org/10.1007/s10853-018-03186-w

71. Hu Q, Yue B, Shao H et al (2020) Facile syntheses of cerium-based $CeMO_3$ (M = Co, Ni, Cu) perovskite nanomaterials for high-performance supercapacitor electrodes. J Mater Sci. https://doi.org/10.1007/s10853-020-04362-7

72. Khuspe GD, Navale YH, Chougule MA et al (2020) Electrodeposition synthesised PAni-MnO_2 hybrid electrode for energy storage applications. In: Pawar PM, Ronge BP, Balasubramaniam R et al (eds) Techno-Societal 2018. Springer, Cham, pp 645–652

73. Tamura T, Kawakami H (2010) Aligned electrospun nanofiber composite membranes for fuel cell electrolytes. Nano Lett 10:1324–1328. https://doi.org/10.1021/nl1007079

74. Haichao L, Li H, Bubakir MM et al (2019) Engineering nanofibers as electrode and membrane materials for batteries, supercapacitors, and fuel cells BT—Handbook of nanofibers. In: Bechelany M, Makhlouf ASH (eds) Barhoum A. Springer, Cham, pp 1105–1130

75. Woo Park J, Wycisk R, Lin G et al (2017) Electrospun Nafion/PVDF single-fiber blended membranes for regenerative H2/Br 2 fuel cells. J Memb Sci 541:85–92. https://doi.org/10.1016/j.memsci.2017.06.086

76. Jindal A, Basu S, Chauhan N et al (2017) Application of electrospun CNx nanofibers as cathode in microfluidic fuel cell. J Power Sources 342:165–174. https://doi.org/10.1016/j.jpowsour.2016.12.047

77. Salahuddin M, Uddin MN, Hwang G, Asmatulu R (2018) Superhydrophobic PAN nanofibers for gas diffusion layers of proton exchange membrane fuel cells for cathodic water management. Int J Hydrogen Energy 43:11530–11538. https://doi.org/10.1016/j.ijhydene.2017.07.229

78. Ahn M, Han S, Lee J, Lee W (2019) Electrospun composite nanofibers for intermediate-temperature solid oxide fuel cell electrodes. Ceram Int. https://doi.org/10.1016/j.ceramint.2019.11.057

79. Powers D, Wycisk R, Pintauro PN (2019) Electrospun tri-layer membranes for H2/Air fuel cells. J Memb Sci 573:107–116. https://doi.org/10.1016/j.memsci.2018.11.046

80. Barakat NAM, Amen MT, Al-Mubaddel FS et al (2019) NiSn nanoparticle-incorporated carbon nanofibers as efficient electrocatalysts for urea oxidation and working anodes in direct urea fuel cells. J Adv Res 16:43–53. https://doi.org/10.1016/j.jare.2018.12.003

81. Ekrami-Kakhki M-S, Naeimi A, Donyagard F (2019) Pt nanoparticles supported on a novel electrospun polyvinyl alcohol-CuO Co3O4/chitosan based on Sesbania sesban plant as an electrocatalyst for direct methanol fuel cells. Int J Hydrogen Energy 44:1671–1685. https://doi.org/10.1016/j.ijhydene.2018.11.102

82. Chan S, Jankovic J, Susac D et al (2018) Electrospun carbon nanofiber catalyst layers for polymer electrolyte membrane fuel cells: fabrication and optimization. J Mater Sci 53:11633–11647. https://doi.org/10.1007/s10853-018-2411-4

83. Liu G, Bonakdarpour A, Wang X et al (2019) Antimony-doped tin oxide nanofibers as catalyst support structures for the methanol oxidation reaction in direct methanol fuel cells. Electrocatalysis 10:262–271. https://doi.org/10.1007/s12678-019-00524-7

84. Abdullah N, Kamarudin SK, Shyuan LK (2018) Novel anodic catalyst support for direct methanol fuel cell: characterizations and single-cell performances. Nanoscale Res Lett 13:90. https://doi.org/10.1186/s11671-018-2498-1

85. Lee HF, Wang PC, Chen-Yang YW (2019) An electrospun hygroscopic and electron-conductive core-shell silica@carbon nanofiber for microporous layer in proton-exchange membrane fuel cell. J Solid State Electrochem 23:971–984. https://doi.org/10.1007/s10008-019-04198-5

86. Parbey J, Wang Q, Lei J et al (2020) High-performance solid oxide fuel cells with fiber-based cathodes for low-temperature operation. Int J Hydrogen Energy. https://doi.org/10.1016/j.ijhydene.2019.12.125

87. Kenry LCT (2017) Nanofiber technology: current status and emerging developments. Prog Polym Sci 70:1–17. https://doi.org/10.1016/j.progpolymsci.2017.03.002

88. Akduman C, Akumbasar P, Elemen Morsümbül S (2017) Electrospun nanofiber membranes for adsorption of dye molecules from textile wastewater. IOP Conf Ser Mater Sci Eng 254:102001. https://doi.org/10.1088/1757-899X/254/10/102001

89. Xu T, Zheng F, Chen Z et al (2019) Halloysite nanotubes sponges with skeletons made of electrospun nanofibers as innovative dye adsorbent and catalyst support. Chem Eng J 360:280–288. https://doi.org/10.1016/j.cej.2018.11.233

90. Bai L, Jia L, Yan Z et al (2018) Plasma-etched electrospun nanofiber membrane as adsorbent for dye removal. Chem Eng Res Des 132:445–451. https://doi.org/10.1016/j.cherd.2018.01.046

91. Qureshi UA, Khatri Z, Ahmed F et al (2017) Electrospun Zein nanofiber as a green and recyclable adsorbent for the removal of reactive black 5 from the aqueous phase. ACS Sustain Chem Eng 5:4340–4351. https://doi.org/10.1021/acssuschemeng.7b00402

92. Kaerkitcha N, Sagawa T (2018) Amplified polarization properties of electrospun nanofibers containing fluorescent dyes and helical polymer. Photochem Photobiol Sci 17:342–351. https://doi.org/10.1039/C7PP00413C

93. Hamad AA, Hassouna MS, Shalaby TI et al (2019) Electrospun cellulose acetate nanofiber incorporated with hydroxyapatite for removal of heavy metals. Int J Biol Macromol. https://doi.org/10.1016/j.ijbiomac.2019.10.176

94. Zhang S, Shi Q, Christodoulatos C, Meng X (2019) Lead and cadmium adsorption by electrospun PVA/PAA nanofibers: batch, spectroscopic, and modeling study. Chemosphere 233:405–413. https://doi.org/10.1016/j.chemosphere.2019.05.190

95. Guan X, Lv X, Huang K et al (2019) Adsorption of Cu(II) ion by a novel hordein electrospun nanofiber modified by β-cyclodextrin. Int J Biol Macromol 135:691–697. https://doi.org/10.1016/j.ijbiomac.2019.05.107

96. Supraja P, Tripathy S, Krishna Vanjari SR et al (2019) Label free, electrochemical detection of atrazine using electrospun Mn2O3 nanofibers: towards ultrasensitive small molecule detection. Sensors Actuators B Chem 285:317–325. https://doi.org/10.1016/j.snb.2019.01.060

97. Mehrani Z, Ebrahimzadeh H, Asgharinezhad AA, Moradi E (2019) Determination of copper in food and water sources using poly m-phenylenediamine/CNT electrospun nanofiber. Microchem J 149:103975. https://doi.org/10.1016/j.microc.2019.103975

98. Du L, Quan X, Fan X et al (2019) Conductive CNT/nanofiber composite hollow fiber membranes with electrospun support layer for water purification. J Memb Sci 117613. https://doi.org/10.1016/j.memsci.2019.117613

99. Zhang W, Yang P, Li X et al (2019) Electrospun lignin-based composite nanofiber membrane as high-performance absorbent for water purification. Int J Biol Macromol 141:747–755. https://doi.org/10.1016/j.ijbiomac.2019.08.221

100. Wu T, Li H, Xie M et al (2019) Incorporation of gold nanocages into electrospun nanofibers for efficient water evaporation through photothermal heating. Mater Today Energy 12:129–135. https://doi.org/10.1016/j.mtener.2018.12.008

101. de Medeiros KM, Araújo EM, de Lira HL et al (2017) Hybrid membranes of polyamide applied in treatment of waste water. Mater Res 20:308–316

102. Bortolassi ACC, Nagarajan S, de Araújo LB et al (2019) Efficient nanoparticles removal and bactericidal action of electrospun nanofibers membranes for air filtration. Mater Sci Eng C 102:718–729. https://doi.org/10.1016/j.msec.2019.04.094

103. Zhu M, Han J, Wang F et al (2017) Electrospun nanofibers membranes for effective air filtration. Macromol Mater Eng 302:1600353. https://doi.org/10.1002/mame.201600353

104. Liang W, Xu Y, Li X et al (2019) Transparent polyurethane nanofiber air filter for high-efficiency PM2.5 capture. Nanoscale Res Lett 14:361. https://doi.org/10.1186/s11671-019-3199-0

105. Huang X, Jiao T, Liu Q et al (2019) Hierarchical electrospun nanofibers treated by solvent vapor annealing as air filtration mat for high-efficiency PM2.5 capture. Sci China Mater 62:423–436. https://doi.org/10.1007/s40843-018-9320-4

106. Zhou W, He J, Cui S, Gao W (2011) Preparation of electrospun silk fibroin/cellulose acetate blend nanofibers and their applications to heavy metal ions adsorption. Fibers Polym 12:431–437. https://doi.org/10.1007/s12221-011-0431-7

107. Li Z, Li T, An L et al (2016) Highly efficient chromium(VI) adsorption with nanofibrous filter paper prepared through electrospinning chitosan/polymethylmethacrylate composite. Carbohydr Polym 137:119–126. https://doi.org/10.1016/j.carbpol.2015.10.059

108. Naseri A, Samadi M, Mahmoodi NM et al (2017) Tuning composition of electrospun ZnO/CuO nanofibers: toward controllable and efficient solar photocatalytic degradation of organic pollutants. J Phys Chem C 121:3327–3338. https://doi.org/10.1021/acs.jpcc.6b10414

109. Mohamed A, Osman TA, Toprak MS et al (2016) Visible light photocatalytic reduction of Cr(VI) by surface modified CNT/titanium dioxide composites nanofibers. J Mol Catal A Chem 424:45–53. https://doi.org/10.1016/j.molcata.2016.08.010
110. Zhang S, Shi Q, Christodoulatos C et al (2019) Adsorptive filtration of lead by electrospun PVA/PAA nanofiber membranes in a fixed-bed column. Chem Eng J 370:1262–1273. https://doi.org/10.1016/j.cej.2019.03.294

Polymer Nanofibers via Electrospinning for Flexible Devices

Subhash B. Kondawar, Chaitali N. Pangul, and Mahelaqua A. Haque

Abstract Flexible devices promise numerous applications in modernization of human life. Accompanying flexibility into planar devices has greatly improved the applicability of the devices. Electrospun polymer nanofibers have shown the major breakthrough due to their high flexibility overcoming the rigidness of the conventional planar devices. Due to innovative methods and manufacturing processes of novel materials, the flexible devices enabled the design of new architectures that are not possible with conventional planar devices. Novel synthesis techniques for flexible nanofibers have bright future prospects toward academic studies and research in one-dimensional nanomaterials. Electrospun polymer nanofibers have emerged as exciting one-dimensional nanomaterials and empowered as a building material into flexible devices. Tremendous efforts have been focused on exploring the electrospun nanofibers for potential functional applications. Electrospun polymer nanofibers with embedded nanoparticles can be easily fabricated to be used for flexible devices by electrospinning technique. The purpose of this chapter is to explore the capability of electrospinning to fabricate various polymer nanofibers for flexible devices. In this chapter, the fabrication of electrospun polymer nanofibers and their potential applications in flexible devices including light-emitting diodes (LEDs), sensors, UV photodetectors, transparent electrodes and nanogenerators are systematically reviewed along with challenges in synthesizing and designing nanofibers for flexible devices.

Keywords Flexible devices · Polymers · Nanofibers · Electrospinning

1 Introduction

The word "nano" has become familiar with almost each and everyone in today's world due to large-scale miniaturation of our daily life utilities. This has not only helped in incorporation of multifaceted components but also increased their functionability, durability, and efficiency. The nanostructure materials are better than

S. B. Kondawar (✉) · C. N. Pangul · M. A. Haque
Department of Physics, Rashtrasant Tukadoji Maharaj Nagpur University, Nagpur, India

© The Author(s), under exclusive license to Springer Nature Switzerland AG 2021   53
S. K. Tiwari et al. (eds.), *Electrospun Nanofibers*, Springer Series on Polymer
and Composite Materials, https://doi.org/10.1007/978-3-030-79979-3_3

their bulk counterparts in terms of their physical, chemical, physiochemical, electrical, magnetic, thermal, optical, biological, catalytic, or mechanical properties. The past decades have witnessed tremendous research being carried out in the field of nanoscience and technology. This has offered nanomaterials an extensive application in medicine, biology, textiles, and electronics [1, 2]. Nanomaterials can be broadly classified on account of quantum confinement as 0D (quantum dots), 1D (quantum wires), and 2D (quantum wells). Among these broad spectrum of nanomaterials, one-dimensional (1D) materials became attractive candidates for many advanced applications. One-dimensional (1D) nanostructures especially nanofibers are continuous with sufficient length, high aspect ratio (length to diameter) and extreme flexibility as compared to that of other 1D materials including nanotubes, nanowires, nanorods, etc. Apart from these advantageous, the multiple extra functions incorporated into nanofibers can enhance their structural and mechanical properties beneficial for wide variety of applications [3, 4]. Owing to this specialty, nanofibers have gained popularity as smart materials in designing flexible electronics including cells, displays, telecom, sensors, medical, and memories (Fig. 1) [5]. This quest has led researchers in search of flexible devices that are based on polymer nanofibers. Flexible electronics currently became the topic of interest to modify devices as per requirement

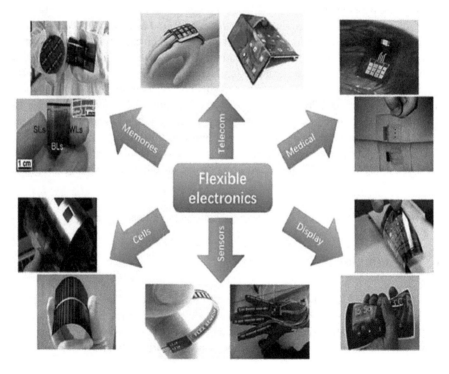

Fig. 1 Applications of flexible electronics (Reproduced with permission from Ref. [5], Copyright (2020) Elsevier)

owing to the stretchability and foldability of the nanofibers. Polymer nanofibers have facilitated the union of metals, organics, inorganics, ceramics, composites, and metal oxides in polymer matrices. This has led to new generation of technological advancement.

Due to extreme flexibility of nanofibers, they are capable of forming networks with high surface area and highly porous structure beneficial for the designing of advanced flexible devices. The significant impact of nanofiber technology has emerged from the variety of fundamental materials such as natural polymers, synthetic polymers, carbon-based materials, semiconducting materials, and composite materials that can be used for the synthesis of nanofibers. The emergence of fabrication of nanofibers techniques has been rapidly reported with the currently available preparation strategies. The fabrication of nanofibers has attracted a lot of researchers due to unique properties such as surface morphology, porosity, and geometry can be tailored or functionalized for flexible devices. Polymer nanofibers can be prepared by a variety of techniques out of which electrospinning is a quite suitable technique with controllable diameter of the nanofibers useful to enhance the properties along with extreme flexibility due to which, electrospun nanofibers are considered an excellent candidate for a variety of flexible purposes. Additionally, electrospinning can be used for fabricating polymer composite fibers by blending additives to get the desired properties. Considering these benefits, electrospinning has gained a remarkable popularity for sharp rise in scientific publications in recent years. Despite the advantages of electrospinning technique for fabrication of polymer nanofibers such as inexpensive setup, ability to control fiber diameter, orientation, and composition, etc., the applications of electrospun nanofibers especially for flexible devices also face some challenges which include the use of organic solvents, the limited control of pore structures and low mechanical strength. More advanced nanofiber configurations, such as core–shell, multilayer, and multicomponent nanofibers, may be prepared through methods like co-axial electrospinning to overcome those problems. Therefore, the variations of this method have been developed as co-axial electrospinning to synthesize core–shell and multilayer composite nanofibrous structures. Looking to inherent applicability especially the flexibility characteristics of electrospun nanofibers, this chapter particularly focuses on the fabrication of a variety of electrospun polymer nanofibers and their application for flexible devices including light-emitting displays, sensors, UV photodetectors, transparent electrodes, and nanogenerators.

2   Electrospun Polymer Nanofibers

Nanofibers are favored in recent times as one of the best 1D nanomaterials on account of their ease of production and miniaturized applications. The versatility of nanofibers, being lightweight and strong structures, has proved its worth as the most successful nanomaterial. High-value sectors are making the best use of nanofibers in their technological innovations. Nanofibers have discovered altogether new properties of the materials already known for decades. The decrease in the fiber diameter results into a much larger surface area. High aspect ratio makes the nanofibers equipped with particular functions like enhanced solubility, compatibility and recognition that portray these nanofibers as the best candidates in the current technological uprising. Electrospinning provides many controllable parameters that can produce nanofibers of desired diameter with utmost uniformity. The obtained electropsun polymer nanofibers can then be subjected to other treatments to obtain the required end product. Electrospinning has eased the fabrication of various functional materials having different functions and technical aspects. Polymer nanofibers being stretchable and bendable due to flexibility are finding emerging applications. Polymer nanofibers are produced by electrospinning, the simplest method of obtaining nanofibers as a result of uniaxial stretching of a viscoelastic solution by electrical interaction [6–8]. Nanofibers have been synthesized via electrospinning by incorporation of the precursor in polymer solution or by the conventional sol–gel process. The so obtained nanofibers have homogeneous dispersion of the metal salts and appropriate thermal treatment yields the nanofibers of the desired metals or metal oxides (Fig. 2). Another class of electrospun nanofibers exhibiting photonic and magnetic functions include luminescent nanofibers, conjugated polymer-based fibers, and nanofibers with specific magnetic properties. The basic approach in preparation of these polymers is the incorporation of components of polymer blends into conjugated polymers and doping organic and inorganic materials with rare earth complexes which control the emissive property. The resultant nanofibers obtained thereafter are flexible, light-weight, hard, thermally stable, chemical resistant, and optically active [9].

Electrospun nanofibers are found to be the best materials for sensors including polymer composite sensors and inorganic composite sensors. Sensing mechanism requires a large surface for quick sensing and detection activity. Nanofibers fit in the

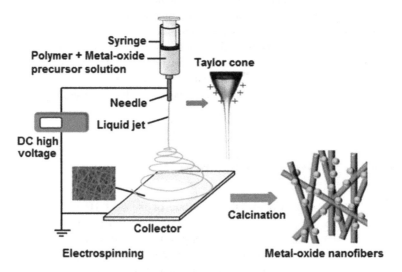

Fig. 2 Preparation of metal oxide nanofiber using electrospinning and calcination

best way and enhance the sensors compared to thin films. Besides these, nanofibers are also used in photocatalytic applications. Thus electrospinning has proved a very significant technique in the preparation of noble polymer nanofibers [10]. Although electrospinning has shown the advancement for preparation of nanofibers, the dimension, quality, design, control and pattern of electrospun nanofibers is equally dependent on the choice of polymer used for electrospinning and hence the name "polymer nanofibers". Polymers form the backbone of nanofibers which gives these nanofibers their characteristic uni-dimensionality. Polymers are basically composed of large number of macromolecules, known as monomers. These macromolecules have chain-like molecular geometry having random orientation and strong Van der Waals forces held them together and cross-linking gives them mechanical strength. It may happen that the monomers may be either all alike or mixed, called as copolymers. However, the electrospinning process requires a solution form of the precursor and the polymer to be fed into the syringe for drawing fibers and hence comes to light the concept of polymer solutions [11]. A solution is obtained by the molecular-level interactions between different chemical species, resulting into a homogeneous phase. Polymer solution basically consists of a polymer material and an appropriate solvent. Anymore addition of other molecular species, polymer, dopants or other materials to this mixture results in the formation of polymer "blends". Dispersion, suspension, emulsion, aerosols, and colloids are some of the other ways of preparing the preliminary polymer solution depending upon the physio-chemical property of the targeted application, but most of the polymer nanofibers are synthesized via solution method [12]. When the polymer is mixed with a suitable solvent, the monomer units interact with the solvent molecules and dissolve in it to produce a homogeneous solution. The dissolution of polymer in the solvent is based on certain factors such as the selection of appropriate solvent, temperature, and the average molecular weight and polydispersity of polymer.

The polymers that are commonly used in electrospinning are mainly of three types- natural polymers, synthetic polymers, and polymer blends. Natural polymers such as silk, chitosan, collagen, gelatine, cellulose find extensive applications in healthcare, biomedical, and pharmaceutical research. Polymers such as polyamides- Nylon-6,12 [13], Polyamide-6 [14], Polyacrylamide [15]; polyacids-Poly(acrylic acid) (PAA) [16], Poly(2-acrylamido- 2-methyl-1-propane sulfonic acid) (PAMPS) [17]; Polyacrylonitrile (PAN) [18], Polyaniline (PANI) [19]; Poly-caprolactone (PCL) [20]; Polyoxides-Polyethylene oxide (PEO) [21]; Polyesters-Poly(ethyleneterephthalate) (PET) [22], Poly(butylenes succinate) (PBS) [23]; polyacids; Acrylic polymers Poly(methyl methacrylate) (PMMA) [24]; Styrenics-Polystyrene (PS) [25]; Polyurethane [26]; Polyalcohols-Poly(vinyl alcohol) [27]; Vinyl polymer-Poly(vinyl chloride) (PVC) [28], Poly(vinylidene Fluoride) (PVDF) [29, 30]; Polyamine-Poly(vinyl pyrrolidone) (PVP) [31]; and other polymers such as Polysulfone (PSU) [32], Poly(etherimide) (PEI) [33], Poly(ferrocenyl dimethyl-silane) [34], Poly(meta-phenylene isophthalamide) [35], Poly(vinylphenol) [36], Polypyrrole [37], Polybenzimidazol (PBI) [38], Aramid (Kevlar49) [39], etc., are widely used in electrospinning. Conducting polymers are extensively used in smart

clothing as supercapacitor, battery and sensors owing to their extraordinary electrical conductivity. Redox polymers, ionic conducting polymers and conjugated
double-bond structured polymers are used for this purpose [40]. The choice of
solvent is extremely important so as to yield a homogeneous polymer solution and its
spinnability. Large numbers of solvents are utilized for this purpose in accordance
with the polymer possessing characteristics like surface tension, vapor pressure,
boiling point, dielectric constant and good rate of evaporation (volatility). Some
of the commonly used solvents are water [41–43], chloroform [37], dimethylformamide (DMF) [44], dimethyl sulfoxide (DMSO) [45], acetone [46], trifluoro acetic
acid (TFA) [47], N,N'-dimethyl acetamide (DMAc) [48], tetrahydrofuran (THF) [49,
50], formic acid [51], etc. Along with the electrospinning processing parameters, the
morphology of electrospun nanofibers is largely affected by some other parameters
of polymer solution such as molecular weight of polymer, polymer concentration
(viscosity), solution conductivity and solvent volatility as shown in the Fig. 3.

Researchers have studied the effect of these parameters on the electrospun
nanofibers morphology and depending upon the desired application the appropriate
parameters need to be fixed. It has been observed that an increase in electrical conductivity results in reduction of fiber diameter, while increase in concentration, that is
viscosity of the polymer solution leads to an increase in fiber diameter.

History of fibers can be dated far back in time but significant developments have
been witnessed in the past 100 years. Cotton was known as most useful fiber and led
to the silver fiber revolution in the eighteenth and nineteenth centuries. With the emergence of three synthetic fibers-polyamide fibers (Nylon 66, Nylon 6), poly(ethylene
terephthalate) (PET) fibers and acrylic fibers, along with spandex fiber materializing
in the first half of the twentieth century has now become the basis of our everyday
clothing. The manipulation of design and control at the nano-level is an important
feature of any useful fiber. Nanofiber has the essence of possessing its diameter in the
range of nanometer in addition to a large surface area. Robust characteristics can be

Fig. 3 Parameters of
polymer solution for
electrospinning

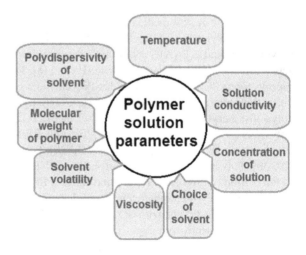

Fig. 4 Relation between fiber diameter sizes in terms of nanometer

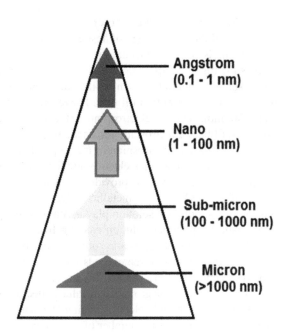

Angstrom (0.1 - 1 nm)

Nano (1 - 100 nm)

Sub-micron (100 - 1000 nm)

Micron (>1000 nm)

identified by controlling the surface area and diameter of these fibers which will add on to multifunctionality of the materials. Nanofibers can be defined as a nano-sized fiber having nanometer order in its dimension. A fiber may have diameter of various sizes ranging from angstrom, nano, sun-micron, and micron.

Figure 4 shows the relation between these sizes in terms of nanometer. Fibers having diameter size in angstrom are basically molecular chains. Fibers having diameters in nano and sub-micron range are classified as "nanofibers", while those in micron range are micron fibers. Conventional fibers have diameter more than few micrometers [52]. Nanosized materials enhance the properties of the materials as a result of size effects caused due to increase in the surface area per volume, along with the reactivity and selectivity ascribed due to decrease in volume. The arrangement of supermolecules is regular, coherent, and self-organized. Nanofibers provide hierarchal structure effect and recognition to cells and biomaterials, and facilitate them to interact and combine. Easy processing, controllability, and design of these nanofibers render enormous potential applications in flexible devices. The scope of nanofibers has increased recently in recent technological advancements. Many physical, chemical, and biological techniques have been reported in literature such as grinding, milling, cryo-crushing, and high-pressure homogenization, laser ablation, physical vapor deposition (PVD), and spinning techniques-flash spinning, centrifugal spinning, electrospinning, electrochemical deposition, interfacial polymerization, polyol, phase separation, microemulsion, hydrothermal, chemical vapour deposition (CVD), sol–gel, hydrothermal, template-assisted synthesis, sonochemical, microwave synthesis, bacterial cellulose synthesis, enzymatic hydrolysis of wood pulp. This has led to increase in the fabrication of nanofibers. The appropriate fabrication technique can be chosen on the basis of the characteristics of precursor involved and the morphology of desired nanofibers.

Spinning is an age-old technology that uses thin filaments or fibers to form yarns by twisting together. Spinning is basically a physical method but it can be combined with chemical methods to synthesize nanofibers. Various types of spinning techniques have been reported till date. However, they have been broadly classified as electrospinning and non-electrospinning techniques on the basis of forces involved in drawing of fibers. Electrostatic forces are responsible for fiber formation in electrospinning, while non-electrospinning techniques make use of forces (centrifugal, gravitational, air, etc.) other than electrostatic force. Electrospinning set-up comprises of a pipette or plastic syringe provided with a metallic needle, high voltage power supply (20–50 kV), and a metallic collector. The electrostatic force and surface tension of the polymer solution play a vital role in spinning of nanofibers. A drop of viscoelastic polymer solution ejecting from the tip of the needle is uniaxially stretched under the electrostatic force exerted by the high tension power supply. When the electrostatic force exceeds the surface tension of the polymer drop, the drop is stretched to form a charged Taylor cone thereby forming a polymer jet. This jet gets attracted toward the grounded metallic collector forming continuous, non-woven nanofiber mat. Although, electrospinning mechanism is simple and easy, the quality of nanofibers is ruled by a number of parameters concerned with the polymer solution, experimental set-up, and environmental conditions [12, 53]. Figure 5 represents a schematic diagram of electrospinning set-up and the physical (distance, voltage, flow rate, and collector), chemical (concentration, conductivity, molecular weight, viscosity, solvent volatility, and molecular structure), and environmental (humidity and temperature) parameters on which the diameter of the fibers depends.

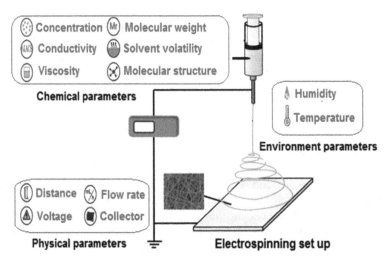

Fig. 5 Schematic diagram of a conventional electrospinning setup with chemical, physical, and environment parameters

3 Application of Electrospun Polymer Nanofibers for Flexible Devices

The concept of wearable and portable electronics has brought about a new aspect to the current material research by investigation of additional functionality of stretchability, foldability, and flexibility of these materials. The devices available today are heavy, rigid, and bulky. When these devices are incorporated in textiles, they alter the fundamental property of these textiles, that is, flexibility. A flexible device is an electronic device that remains stable and retains its electrical behavior, even when it undergoes mechanical deformation in the presence of any kind of stress and strain such as twisting, bending, compressing, folding, and stretching. This gives rise to the concept of electronic devices mounted on flexible substrates to be combined with textiles. However, a major drawback faced with such devices is that laundering and wear and tear of the textile may affect their functionality and affect is reusability. This led to the idea of incorporating the devices within the textile. Since the fibers that are known to us and spun into fabric are mainly insulators; researchers considered it worthwhile to integrate electronic materials into the fiber itself. Fibers are excellent flexible materials and the addition of electrical behavior into these fibers started the notion of fiber-assisted flexible devices which are extremely lightweight and can be easily woven into the fabric to produce flexible textile. This has made the textiles "smart" by providing them with additional functionalities. The amalgamation of textile and electronic technology has brought together the positive qualities of each of these technologies as one. Although fibers act as a substrate for these electronic materials, but it does not pose any danger to the functionality of these fibers and goes hand-in-hand with each other. Nanotechnology has assisted in bringing the concept of flexible textiles on a global technological platform by the introduction of one-dimensional nanomaterials. Nano-engineered functional textiles have gathered perpetual applause due to their ease of application in devices. Large numbers of materials have been investigated for manufacturing fibers. Among these materials, polymers possess the intrinsic property of being mechanically flexible, and are most commonly used to yield different types of fibers. Electrospinning technique has helped in producing flexible and multifunctional nanofibers with the help of polymers and hence called polymeric electrospun nanofibers.

Flexible devices based on polymeric electrospun nanofibers are biocompatible, durable, soft, flexible, stretchable, hydrophobic, antibacterial, conductive, antiwrinkle, and antistatic. The applications of these flexible devices are innumerable [54–56]. In this chapter, application of electrospun polymer nanofibers for flexible devices (Fig. 6) such as light-emitting diodes (LEDs), sensors, UV photodetectors, transparent electrodes, and nanogenerators are explored.

Fig. 6 Application of
electrospun nanofibers for
various flexible devices

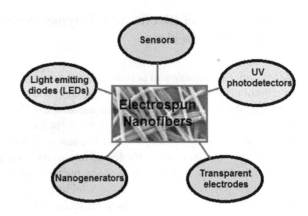

3.1   Light-Emitting Diodes (LEDs)

The light-emitting properties can be obtained in polymer nanofibers by electrospinning and accordingly such electrospun nanofibers can be used as light-emitting diodes
(LEDs) [57]. Phosphors emit light due to occurrence of radiative transitions. Non-
radiative transitions also occur at the same time which leads to a loss in efficiency of
luminescence. Activators and sensitizers act as wheels to keep the process of luminescence on track. Activators are responsible for emission and are doped into the
host material. A sensitizer or co-activator is co-doped in the similar fashion, which
gathers energy from the source of excitation and passes on to the activator. Innumerable phosphors have been reported with rare earth (RE) elements acting as activators
[58]. Inorganic materials have gained enormous popularity in lighting applications.
Of these inorganic materials, oxides have attracted remarkable interest. RE activated
oxide phosphors have been recorded in many research articles. Several RE-doped
metal oxides have been detailed in extensive literature such as phosphates, tungstates,
molybdates, aluminates, borates, niobates, titanate, vanadate, titania, fluorides, etc.,
for innumerable color display applications [59–65].

Electrospun luminescent nanofibers have spurred application for flexible light-
emitting display devices by incorporation of optically active molecules, metallic
nanoparticles, conjugated polymer nanoparticles, colloidal nanoparticles, fluorescent dyes, bio-chromophores, quantum dots, hybrid (organic–inorganic), composite,
conducting, and semiconducting polymers. Carbon quantum dots (CDQs) are excellent luminescent materials emitting red, blue, and green color at appropriate excitation energies. Alam et al. [66] examined transparency and photoluminescent properties of composite electrospun nanofibers of CDQs in polyacrylonitrile (PAN) blend
with polyacrylic acid (PAA). The process of co-electrospinning for the fabrication of
PAN/PAA/CQDs composite nanofibers and its optical behavior are shown in Fig. 7.
They have observed that CDQs/PAN/PAA composite nanofibers exhibited luminescence in agreement with the CDQs, while PAN/PAA fibers showed no luminescence
in absence of CDQs. PAN/PAA nanofibers absorb UV light and transmit in the visible

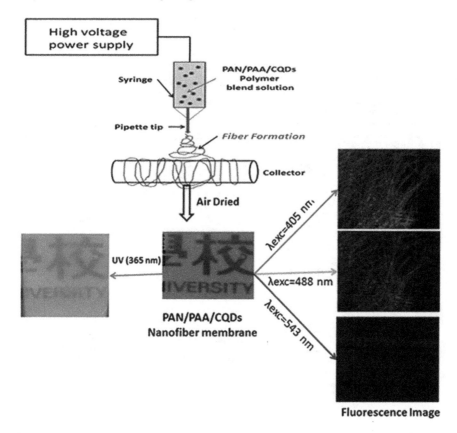

Fig. 7 The process of co-electrospinning for the fabrication of PAN/PAA/CQDs composite nanofibers and its optical behavior (Reproduced with permission from Ref. [66], Copyright (2015) Elsevier)

range with transparency of 65%. However, the addition of CDQs in the polymer array reduced the transparency to 60% for PAN/PAA/CQD composite nanofibers. Electrospun PAN/PAA/CQD composite nanofibers can be the potential candidate for flexible light-emitting display devices.

The fabrication of Keratin/polyvinylalcohol (PVA) blended nanofibers by electrospinning for flexible ZnO@graphene quantum dots (ZGQDs) LEDs demonstrated by Lee et al. [67]. Keratin/PVA fibers have random, bead-free orientation with an average diameter of 180 nm before dipping in water. As the keratin/PVA nanofibers coated with glyoxal were dipped in water, the average diameter increased from 180 to 210 nm along with smoothening of the fiber. This increase in average diameter contributed to increased transparency. Spin coating method was used for the fabrication of ZGQDs LEDs. The main emission peak of ZGQDs is observed at 393 nm (3.15 eV). Figures 8 and 9 depict the fabrication of transparent textile using electrospinning process and fabrication process of ZGQDs LEDs embedded in textile.

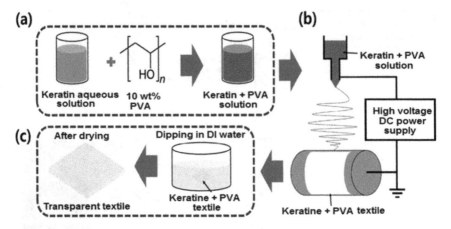

Fig. 8 Schematic illustration of **a** Keratin + PVA solution preparation, **b** loading solution into syringe for electrospinning, and **c** transparent textile formation (Reproduced with permission from Ref. [67], Copyright (2017) Elsevier)

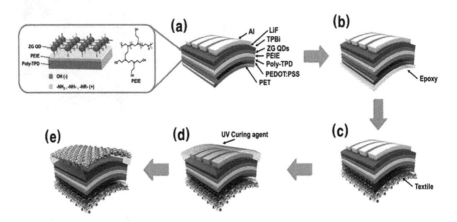

Fig. 9 Fabrication process of ZGQDs LEDs embedded in textile (Reproduced with permission from Ref. [67], Copyright (2017) Elsevier)

For the development of photoluminescence properties in electrospun nanofibers, rare earth ions RE-doped semiconductor are the best suitable as promising hosts for incorporation in nanofibers to improve upon textile industries for designing flexible display devices. Fang et al. [68] reported the preparation of Eu^{3+} doped cerium oxide (CeO_2) nanofibers using electrospinning and studied photoluminescence under excitation wavelength 290 nm. The emission bands ranging from 310–530 nm were reported. The highest peak observed at 367 nm accounted for the band-gap of CeO_2 (3.4 eV). The emission intensities of the Eu^{3+}:CeO_2 nanofibers were found to be increased with increase in Eu^{3+} content as well as heating temperature. The effect of annealing showed improved results.

Ge et al. [69] demonstrated the fabrication of electrospun $Yb_2Ti_2O_7$:Er nanofibers and its red upconversion luminescence properties. The fabrication of electrospun $Yb_2Ti_2O_7$:Er nanofibers by electrospinning is shown in Fig. 10a. For the preparation of electrospun $Yb_2Ti_2O_7$:Er nanofibers, the solution of titanium butoxide (TBOT), N, N-dimethylformamide (DMF), acetic acid (HAC), polyvinylpyrrolidone (PVP), ytterbium trinitrate pentahydrate ($Yb(NO)_3.5H_2O$) and erbium trinitrate pentahydrate ($Er(NO)_3.5H_2O$) prepared and transferred to syringe for electrospinning. As seen from the emission spectrum of $Yb_2Ti_2O_7$:Er nanofibers (Fig. 10b, UCL spectrograph exhibits two peaks corresponding to red and green color 658 nm and 546 nm respectively. Remarkable improvement in red emission occurred due to an increment in the content of Yb host. This dominated the energy transfer from $^2F_{5/2}$ of Yb^{3+} ions to $^4I_{11/2}$ of Er^{3+} followed by relaxation by the transition from $^4F_{7/2}$ to $^4F_{9/2}$, thereby suppressing the transitions of green-emitting level as shown in Fig. 10c.

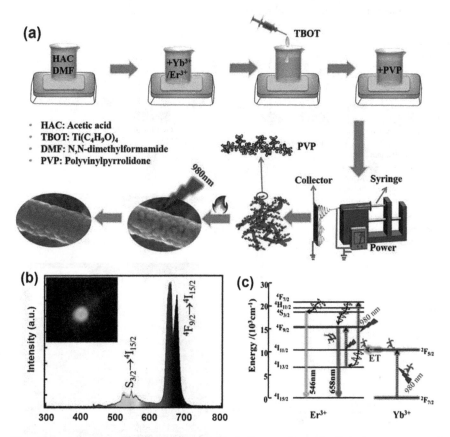

Fig. 10 **a** Preparation of electrospun $Yb_2Ti_2O_7$:Er nanofibers, **b** Emission spectrum of $Yb_2Ti_2O_7$:Er nanofibers, and **c** Schematic energy level diagram for Er^{3+} and Yb^{3+} (Reproduced with permission from Ref. [69], Copyright (2019) Elsevier)

Wei et al. [70] demonstrated the UCL characteristics of electrospun rare earth oxyfluoride GdOF:Er^{3+} nanofibers fabricated by simple electrospinning combined with fluoro-oxidation technique for obtaining nanofibers with variable molar ratios of Er^{3+} as represented in Fig. 11a. GdF_3:Er^{3+} and GdOF:Er^{3+} nanofibers exhibit orthorhombic and rhombohedral phases, respectively, and this corresponding change is attributed to oxidation of GdF_3:Er^{3+} nanofibers.

UC emission spectra of GdOF:9%Er^{3+} nanofibers was characterized with various pumping powers from 330 to 1393 mW as shown in Fig. 11b, the stronger emission intensity of UC luminescence was found at higher pump power. UC emission spectra with 980 nm excitation were observed intense peaks at 523 and 543 nm corresponding to $^{2}H_{11/2} \rightarrow \, ^{4}I_{15/2}$, $^{4}S_{3/2} \rightarrow \, ^{4}I_{15/2}$ transitions of Er^{3+} for respectively. The effect of dopant on UCL is clearly visible from the spectra. The UC intensity increased with a relative increase in dopant concentration. The maximum intensity is observed at a concentration of 9% and further increase caused a substantial decrease in intensity. CIE chromaticity analysis suggests potential application

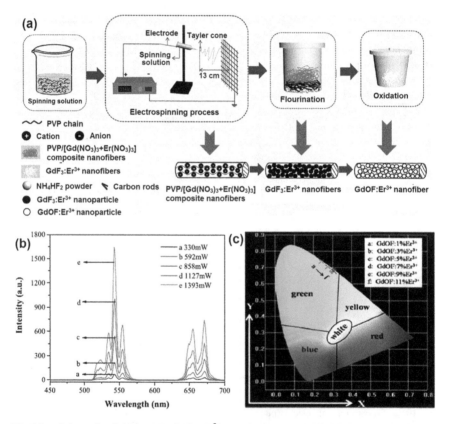

Fig. 11 **a** Schematic of electrospun GdOF:Er^{3+} nanofibers preparation, **b** UC emission spectra of GdOF:Er^{3+} nanofibers, and **c** CIE chromaticity coordinates of GdOF:Er^{3+} nanofibers (Reproduced with permission from Ref. [70], Copyright (2019) Elsevier)

of electrospun GdOF:Er^{3+}nanofibers in light-emitting display devices. Figure 11c represents GdOF:Er^{3+} nanofibers with varied concentration of Er^{3+} marking CIE coordinates (x,y) at (0.242, 0.738) (1%), (0.275, 0.706) (3%), (0.295, 0.688) (5%), (0.301, 0.682) (7%), (0.305, 0.679) (9%), and (0.307,0.676) (11%). Electrospun GdOF:Er^{3+} nanofibers are the potential luminescent materials useful for flexible light-emitting display devices.

With the help of electrospinning, Pangul et al. [71] prepared electrospun Dy^{3+}-doped ZnO nanofibers and studied their photoluminescence characteristics. The intensity of emissions peaks of dopant Dy^{3+} was found to be increased when the concentration of Dy^{3+} increased due to a reduction in fiber diameter. They have also reported the fabrication of electrospun Sm^{3+}-doped ZnO nanofibers and showed orange-red luminescence when excited at 394 nm [72]. They had concluded that the color tunability in electrospun nanofibers can be done by the type of dopant. Different dopant with a specific concentration in electrospun doped ZnO nanofibers can be the potential candidate for designing visible light-emitting diodes as flexible devices.

One approach for the preparation of electrospun polymeric materials for light-emitting displays is to electrospin the polymer blends such as polyfluorene (PF), polyphenylene vinylene (PPV), polystyrene (PS), poly(methyl methacrylate) (PMMA), and poly(vinylidene fluoride) (PVDF) using a single solution spinneret for the purpose of reducing aggregation-induced quenching to enhance lumines-cence efficiency. In another approach, polymeric materials have been synthetically modified with aggregation-induced emission active pendants and subsequently elec-trospun into flexible light-emitting displays. Itankar et al. [73] demonstrated the fabrication of electrospun Eu^{3+}-doped polystyrene (PS) and polyvinylidene fluo-ride (PVDF) nanofibers and showed more polarized chemical environment of these polymers for the Eu^{3+} ions due to which bright red emission was observed in photo-luminescence spectra. Such electrospun polymer nanofibers can be the potential candidates for flexible visible light-emitting diodes.

Dandekar et al. [74] demonstrated the fabrication of electrospun polymer blend nanofibers using electrospinning technique for photoluminescent fabric designing. They used europium complex Eu(TTA)$_3$phen (TTA = 2-thenoyltrifluoroacetone, phen = 1,10-phenanthroline) and blending of various polymers such as polystyrene (PS), poly(methyl methacrylate) (PMMA) and poly(vinylidene fluoride) (PVDF) and prepared Eu(TTA)3phen/PVDF-PS, Eu(TTA)3phen/PVDF-PMMA and Eu(TTA)3phen/PS-PMMA nanofibers. They showed high-intensity peaks in emis-sion spectra obtained due to polymer blends with PVDF and concluded that the increased in fluorescent intensity was due to smaller fiber diameter for PVDF-PMMA, and PVDF-PS blends in composites nanofibers (Fig. 12). CIE chromaticity coordinates for Eu(TTA)3phen/polymer blends nanofibers showed the emission of red color which can be tuned with the various polymer blend combination with PVDF.

Qin et al. demonstrated the fabrication of electrospun Janus nanofibers as flexible white-light-emitting nanofibers prepared by electrospinning. White-light-emitting Janus nanofiber film was consisting of PAN on one side and PVP on other side of electrospun nanofibers [75]. Two different polymer matrix solutions containing certain mass of fluorescent dyes anthracene and rhodamine-B mixed with PAN

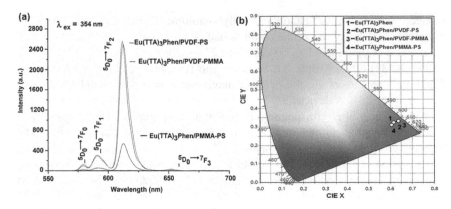

Fig. 12 **a** Emission spectra and **b** Chromaticity of Eu(TTA)3phen with polymer blends nanofibers (Reproduced with permission from Ref. [74], Copyright (2018) Elsevier)

and coumarin-6 mixed with PVP were prepared for the fabrication of fluorescent nanofibers. Such electrospun nanofibers are the potential candidates for flexible light-emitting devices.

3.2 Sensors

The improvement in the sensors with respect to their sensitivity, selectivity, response and recovery rate, and detection limit has been obtained with the advent of nanotechnology particularly one-dimensional nanomaterials such as nanofibers. High flexibility of electrospun nanofibers made them right candidates for the development of flexible sensors. The performance of electrospun nanofibers-based sensors was found to be highly sensitive than that of film-based sensors [55]. Electrospun nanofibers-based sensors have been widely used in flexible devices. Flexible sensors designed by electrospun nanofibers have shown great improvement in their sensor performance due to high absorptive power of nanofibers for analyte molecules to detect [76]. The functional nanofibers for sensors have been well prepared by electrospinning as compared to methods currently available for the preparation of nanofibers [77].

In recent years, different strategies have been applied for doping of noble metals to form heterojunction structures beneficial for tuning low detection limit and various sensing parameters obtained from electrospun composite/hybrid nanofibers [78]. There had been considerable interest in the fabrication of excellent gas sensing materials with the help of metal oxide semiconductors (ZnO, In_2O_3, SnO_2, TiO_2, and Fe_2O_3). Many approaches have been proposed to overcome the disadvantages of metal oxide semiconductors such as its poor selectivity, need of high operating temperature, and low response at low gas concentration. With this background, the construction of core@shell nanofibers became the smart electrospun nanostructure materials due to their individual components for the development of flexible sensors.

Huang et al. [79] demonstrated the preparation of core–shell nanofibers made of two metal oxide semiconductors ZnO@In$_2$O$_3$ core–shell nanofibers by coaxial electrospinning for the study of ethanol gas sensing and found that such nanofibers are highly sensitive for ethanol as compared to pure ZnO or pure In$_2$O$_3$ nanofibers. Schematic diagram of formation of ZnO@In$_2$O$_3$ core–shell nanofibers is shown in Fig. 13a. Two syringes with different nozzle diameters were used for polymer solutions made of precursors of Zn and In. The collected fibers were annealed at 600 °C to obtain ZnO@In$_2$O$_3$ core–shell nanofibers.

Schematic diagram and response of ZnO@In$_2$O$_3$ core shell nanofibers when exposed to air and ethanol are, respectively, shown in Fig. 13b, c. The sensor demonstrated excellent response and recovery characteristics for ethanol sensing. The high-performance ZnO@In$_2$O$_3$ core–shell nanofibers gas sensor for ethanol was due to shell depletion of In$_2$O$_3$ layer for increased in electron conduction channel leads to decrease in sensor resistance.

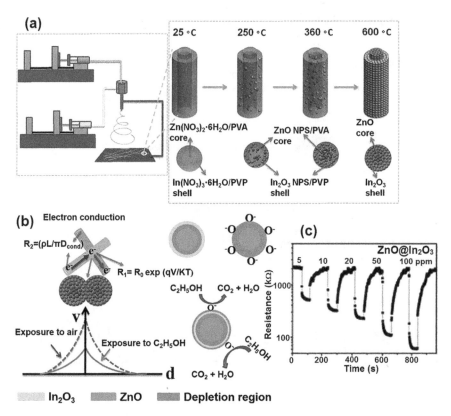

Fig. 13 **a** Formation of core–shell ZnO@In$_2$O$_3$ nanofibers, **b** exposure of ZnO@In$_2$O$_3$ nanofibers to air and ethanol and **c** sensing performance of ZnO@In$_2$O$_3$ nanofibers (Reproduced with permission from Ref. [79], Copyright (2018) Elsevier)

Kim et al. prepared SnO_2–NiO composite nanowebs by electrospinning and studied their gas sensing properties for the detection of hazardous gases [80]. For the electrospinning, they had prepared PVA solution mixed with precursor materials with different concentrations. Fabrication of SnO_2–NiO composite nanowebs with various compositions is shown in Fig. 14a. To investigate the composition dependency of the responses, all compositions were exposed to NO_2, and C_6H_6 gases and their dynamic normalized response was tested with different ppm concentration are shown in Fig. 14b, c, respectively. This shows that such electrospun composites nanowebs-based flexible sensor forms the p-n heterojunction to easily detect oxidizing or reducing gases.

Pang et al. [81] reported the preparation of electrospun cellulose/TiO_2/polyaniline (PANI) nanofibers by electrospinning and polymerization as shown in Fig. 15A. They have studied the dynamic response of the cellulose/TiO_2/PANI composite nanofibers for ammonia sensing (Fig. 15B). The sensitivity of the cellulose/TiO_2/PANI composite nanofibers was found to be significantly improved by TiO_2 nanoparticles as compared to that with out TiO_2. Wang et al. [82] reported the preparation of Pd-SnO_2 composite nanofibers-based nanosensor using electrospinning. They

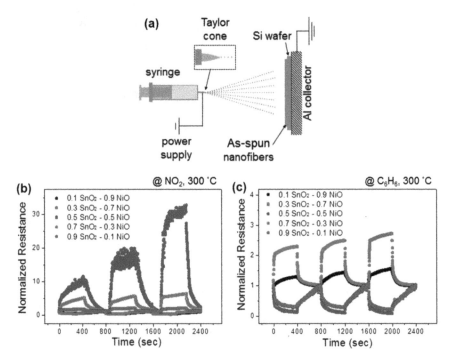

Fig. 14 **a** Electrospinning set-up for preparation of SnO_2-NiO nanofibers, **b** and **c** Dynamic normalized response of SnO_2–NiO composite nanoweb sensors with various composition for NO_2 gas and C_6H_6 (Reproduced with permission from Ref. [80], Copyright (2018) Elsevier)

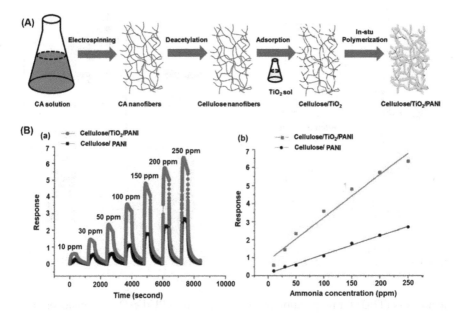

Fig. 15 **A** Fabrication of cellulose/TiO$_2$/PANI composite nanofibers, **B** Dynamic response for ammonia sensing Reproduced with permission from Ref. [81], Copyright (2016) Elsevier Ltd)

have shown the change in resistance when the material is exposed to hydrogen gas achieving faster response and shorter recovery.

Xu et al. demonstrated the fabrication of NiO/ZnO core–shell composite nanofibers by co-axial electrospinning [83] and studied this nanofibers-based sensor for the detection of H$_2$S gas. They have shown excellent response and recovery of sensor based on NiO/ZnO core–shell composite nanofibers as compared to individual ZnO nanofibers-based sensor. Yang et al. [84] developed a sensor device based on 9-chloromethylanthracene (9-CMA) fluorophore doped cellulose acetate (CA) electrospun nanofibers. From its sensing performance toward methyl violet, a novel fluorescent sensor based on electrospun nanofibers could be designed.

Al-doped SnO$_2$ composite nanofibers fabricated by electrospinning showed improved hydrogen sensing as compared to pure SnO$_2$ nanofibers. It was concluded that the doping of Al into SnO$_2$ could strongly affect the sensing performances [85]. Sharma et al. [86] reported the fabrication of SnO$_2$/PANI composite nanofibers by electrospinning technique for hydrogen gas sensing at low temperature. More et al. [87] prepared Ag-SnO$_2$/PANI nanofibers by electrospinning and found improved sensing performance toward hydrogen gas at low temperature. In all these electrospun nanofibers, the polymer polyvinyl pyrrolidone (PVP) was used during electrospinning process.

Recently, Beniwal and Sunny [88] investigated the analyte sensing properties of Fe$_2$O$_3$ and polypyrrole (PPy) infused in electrospun thermoplastic polyurethane

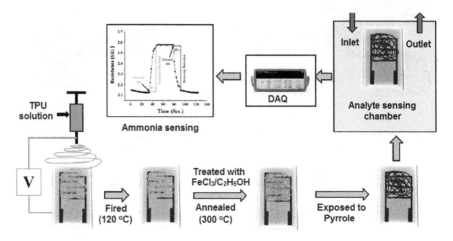

Fig. 16 Fabrication of TPU/Fe$_2$O$_3$/PPy nanofibers-based sensor (Reproduced with permission from Ref. [88], Copyright (2020) Elsevier)

nanofibers (TPU) for preparation of TPU/Fe$_2$O$_3$ and TPU/Fe$_2$O$_3$/PPy nanocomposites. Figure 16 shows the schematic of fabrication of TPU/Fe$_2$O$_3$/PPy sensor for ammonia sensing. Using electrospinning, they have systematically collected TPU nanofibers on interdigitated gold electrodes and then treated with FeCle/C$_2$H$_5$OH solution so as to form nanofibers of TPU/Fe$_2$O$_3$ and exposed to pyrrole for the formation of TPU/Fe$_2$O$_3$/PPy sensor. Utility of highly stable ferric oxide and conducting polypyrrole with the thermoplastic polyurethane electrospun nanofibers had been tested for the enhancement of the analyte sensors at room temperature. Inclusion of fillers ferric oxide and polypyrrole had turned out to be advantageous for improving the sensing performance of the thermoplastic polyurethane at room temperature. Busacca et al. [89] investigated the performance of CO gas sensor based on electrospun Co$_3$O$_4$ nanofibers. They have reported different morphology of electrospun Co$_3$O$_4$ obtained by using ethanol and dimethylformamide (DMF) as solvent mixed with polyvinylpyrrolidone (PVP) during electrospinning for obtaining Co$_3$O$_4$ nanofiber by ethanol and Co$_3$O$_4$ nanosheets by DMF. While comparing the sensing properties of these Co$_3$O$_4$ nanostructures for CO gas sensing, Co$_3$O$_4$ nanofibers showed fast response and recovery compared to that of Co$_3$O$_4$ nanosheets.

3.3 UV Photodetectors

An emerging need of highly responsive ultraviolet photodetector at high temperature and harsh conditions has been consistently there in the fields of military, optical communication, and space research [90]. Hence, fabrication of UV photodetectors with such operating features enduring visible region of wavelength is an unceasing

course. Superfast imaging, monitoring ultrasonic vibrations, remote optical communication network, homodyne detection of polarized state of weakest signal field are
some commercially available photodetection system [91]. The ultraviolet photodetectors are classically identified as vacuum UV photodetectors and solid-state UV
photodetectors. Vacuum UV photodetectors based on photomultiplier tubes are
heavyweight, low efficient engaging high-power consumption but are relatively
mature and high gain devices. Whereas, solid-state UV photodetectors primarily
semiconductor UV photodetector devices are classified as photoconductive and
photovoltaic UV photodetectors [92, 93]. Researchers have been successively able
to achieve these properties in conventional UV photodetectors but with the progression of age and technology the prerequisite for the wearable UV photodetectors are at
surge and so is the research progressing in that phase [94]. Polat et al. [95] were positively able to fabricate a variety of flexible prototype devices which were efficaciously
integrated with the user-friendly devices and are able to monitor real-time health-
related readings even at low power consumption for long range of periods. Another
successful attempt was attained by Núñez et al., wherein semiconductor-based flexible UV photodetector devices were fruitfully fabricated and confirmed under various
conditions [96]. Utilization of low-cost and simple manufacturing processes for
different innovative wide band-gap semiconductor nanomaterials having 1–100 nm
dimensions have been discussed and studied for fabricating UV photodetectors. High
surface area to volume ratio, a major asset held by electrospun nanomaterials holds
the likelihood and valor to show higher responsivity and photoconductivity gain [97].

Reddy et al. [98] prepared electrospun NiO–p/Si–n heterojunction nanofibers and
studied its characteristics as UV photodetector (Fig. 17A). Their study had proved the
utilization of electrospinning method to fabricate p-NiO nanowire-based photodetector device to be an adequate choice exhibiting optimum properties. In a typical
procedure, polyvinyl alcohol and $Ni(NO_3)_2$ composite nanofiber were allowed to
deposit over heavily doped n-type Si substrates previously cleaned with hydroflu-
oric acid and acetone and water. The as-deposited electrospun composite nanofibers
were allowed to undergo calcination at 450 °C for 1 h. The calcination process
removes the organic residual of polyvinyl alcohol and remains the high crystalline
p-NiO nanowires. A shadow mask utilizing electron beam technique was used to
deposit aluminum top contacts on p-NiO nanowire—n-Si heterojunction device. The
resultant device was annealed for another 1 h which improved the electrical performance. Electrical performance of the p-NiO nanowire—n-Si heterojunction device
was studied and then compared with the assortment of the heterojunction devices.
Upon calculating the critical parameter to estimate the act of the photodetectors, i.e.,
responsivity and external quantum efficiency, the highest value of responsivity was
achieved for weaker ultraviolet light intensity at a wavelength of 350 nm. The detec-
tivity of the device was also found to pursue the higher value. Both the values are
comparable to that of other similar heterojunction photodetectors. The fast photodetection of this heterojunction photodetector device at zero bias has been devoted due
to the p-NiO nanowires carrier diffusion length as studied from Fig. 17B. Overall, the
electrospun one-dimensional p-NiO nanowire—n-Si heterojunction for self-powered

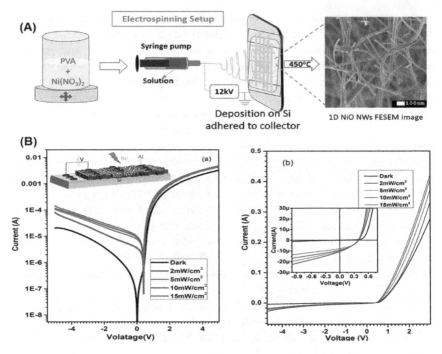

Fig. 17 **A** Fabrication of electrospun NiO nanowires and its SEM image, **B** I–V characteristics of NiO/Si under dark and different illumination of light (Reproduced with permission from Ref. [98], Copyright (2020) Elsevier)

UV photodetector has proved itself as the demanding device owing to their astounding properties.

Kim et al. [99] had innovatively utilized the *Kirigami* pattern for flexible and transparent electrospun nanofiber network-based UV photodetectors. Conceptualizing *Kirigami* into metalized nanofiber network has actually enhanced the mechanical performance of the as-fabricated UV photodetectors. A very precise method was adapted to fabricate the stretchable photodetectors where a zinc oxide seed layer was deposited over the electrospun PVP nanofibrous mat by RF sputtering method. Over these zinc oxide layered PVP nanofibers gold coating was deposited which further carefully transferred onto the Kirigami polydimethylsiloxane proprietary mold. Figure 18 displays the detailed process of the fabrication of *Kirigami* UV photodetector along with the SEM images of the hybrid zinc oxide layered nanofibers. This device was further examined for UV detection against strain parameter and it is noteworthy to mention that photoresponsivity for the device was dynamic even under the 80% of the strained conditions. Integration of flexibility into the UV photodetectors has encompassed countless ways for the utilization in real-time monitoring and assortment of applications.

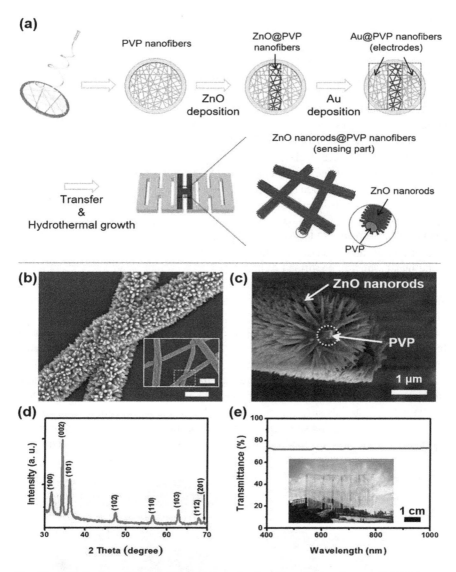

Fig. 18 **a** Fabrication of ZnO@PVP nanofibers for the kirigami photodetector, **b** SEM image of a junction of two ZnO nanorods@PVP nanofibers, **c** Cross-sectional SEM image for the ZnO nanorods@PVP nanofibers, **d** XRD image for ZnO nanorods grown on PVP nanofiber surfaces, and **e** Optical transmittance spectrum of the photodetector (Reproduced with permission from Ref. [99], Copyright (2020) Elsevier)

3.4 Transparent Electrodes

Serving as cathode or anode for the optoelectronic devices and to allow the passage of light with minimal losses is the basic utility of the transparent electrodes. Transparent electrode needs to be compatible with the harsh environments. Electrodes that are electrically conductive and optically transparent had manifest its place in the grounds of the light-emitting diodes, photodetectors, integrated modulators, electrochromic devices, solar cells, etc. [100]. Indium-tin-oxide (ITO) thin films have been commercially used in electronic devices as transparent electrode owing to the virtuous performance and chemical stability. However, the rigidness and brittleness had hindered their utilization in the application point of view with the mounting technology. Continuous efforts have been carried out to overcome this problem. Lee et al. [101] investigated the formation of multilayered transparent electrodes composed of ZnO:Ga (GZO)/Ag/GZO layers as shown in Fig. 19.

Jeong et al. have successfully fabricated the hybrid transparent electrode with metal mesh which is confirmed to have good transmittance and lower sheet resistance [102]. Essaidi et al. have studied indium-free transparent electrode made of WO_3/Ag/WO_3 multilayer structures which when used as transparent anode in organic photovoltaic cells performs as good as ITO transparent electrode [103]. Scopes of transparent electrodes have been further explored in a variety of fields. A novel biosensor based on nanostructured ITO transparent electrode has been fabricated and tested for monitoring cardiovascular diseases by Pruna and the team [104]. Thus, the introduction of nanostructure into transparent electrodes provides more surface area to react and thus improving the outcome of the device. Ultra-thin layers of alternative materials such as conducting polymers, carbon nanotubes, graphene, and metal nanowire grids show unique electronic transport and optical properties. Furthermore, the intrinsic conductivity of metals categorized high conductivity and

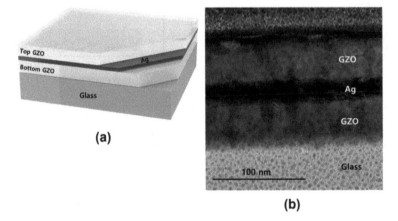

Fig. 19 a Schematic and **b** TEM image of a fabricated GZO/Ag/GZO transparent electrode (Reproduced with permission from Ref. [101], Copyright (2018) Elsevier)

ultra-fine diameter of each electrospun nanofiber leads to high transparency which allows metallic nanofibrous used for transparent electrodes. Fuh et al. [105] reported the fabrication of a new combination of additive manufacturing and electrospinning called near-field electrospinning (NFES) for random metal nano/microwire networks used as flexible transparent electrodes. This has proved to be a powerful technique which recently developed to print out uniaxially aligned fibers with precise control of fiber size in fabrication of polymeric flexible transparent electrodes.

Yousefi et al. [106] premeditated electrospun flexible electrodes which are found to exhibited notable transparency over the entire visible light range and relatively trivial sheet resistance. A combination of additive manufacturing method along with electrospinning termed as near-field electrospinning has been utilized by the team to obtain uniaxially aligned fibers with controlled size of fiber diameter and its placements. In a multi-process of manufacturing, polymeric fibers of polystyrene were allowed to deposit over the gold-coated polyethylene terephthalate in dimension-specific pattern. To enhance the connectivity within the gold and the specific pattern of the polystyrene layer, the as-obtained electrospun fibrous sheet was annealed and thereafter the etching of the gold-coated polyethylene terephthalate layer was carried out while the polystyrene acts as the mask for the gold particles to remain beneath the specific pattern. Figure 20 represents the (a) schematic of NFES, (b) Wet chemical etching, (c) removal of PS fibers, (d) prepared 2 × 1.5 cm electrodes, and (e) representation of optical clarity of electro-printed sample. The so obtained gold-coated polyethylene terephthalate electrodes having a specific pattern were then examined via conductivity measurements, transparency measurements, and bending tests. The transparency tests of the substrates with the lateral distance of 500 μm exhibits constant transmittance over the visible wavelength range. With the increase in the number of the printed fibers, the lateral distance is condensed and this roots the final electrode to parties lower resolution which ultimately hampers the transparency of the electrodes. However, when the impact of bending tests was studied as a function of the sheet resistance on the substrate with lateral distance of 500 μm, it was observed that the looked-for flexibility and the sheet resistance remains steady up to 600 cycles. The learning of the electrode size varying from 5 to 25 mm was supported on the conductivity of the sample and the sheet resistance was found to increase, respectively. Thus, a transparent and conductive electrospun fibrous network was fabricated by the team and concluded it to be the finest aspirant as a transparent electrode which could be used as an alternate for conventional rigid electrodes.

Jiang et al. [107] introduced a novel transparent electrode based on copper silver core–shell metal electrospun nanofibers prepared by combination of redox heating process and electrospinning. A very simple procedure was utilized to fabricate the copper-silver core–shell electrospun nanofibers. Firstly, the copper nanofibers were electrospun which after oxidation and reduction were allowed to coat by silver ink. The copper=silver nanofibers were thoroughly studied and characterized and core–shell nanofibers are found to have good transmittance and low resistance. Transparent copper silver nanofiber electrodes were simply transferred over polydimethyl-siloxane substrates. This transparent electrode was used as anode, polydimethyl-siloxane substrates as hole transport layer while eutectic indium–gallium material

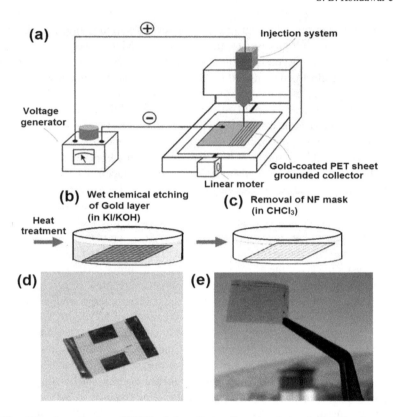

Fig. 20 **a** The schematic view of NFES printing of microfibers, **b** wet chemical etching for removal of Au coating, **c** removal of PS fibers, **d** prepared 2 × 1.5 cm electrodes and **e** representation of optical clarity of electro-printed sample (Reproduced with permission from Ref. [106], Copyright (2019) Elsevier)

was used as cathode. Such a formation was used to produce flexible optoelectronic device which shows improved luminous efficacy and high degree of transparency and flexible bending while electrospinning improves the transparency and conductivity of the transparent electrodes.

Wu et al. [108] fabricated and investigated continuous nanotrough networks from a wide variety of functional materials, including silicon, indium-tin-oxide, and metals such as gold, silver, copper, platinum, aluminum, chromium, nickel, and their alloys. Polymeric ultra-long and continuous nanofibers were drawn through electrospinning which was then coated with thin layer of the active materials. This nanofibrous network was then carefully transferred onto the substrates. The polymer templates were then submerged into organic solvents thereby remaining the continuous nanotrough networks of the active materials. The metal nanotrough networks show good transmittance owing highest to the copper nanotrough networks. The metal nanotrough networks demonstrate flat transmittance spectra from 300 to 2000 nm which is a desirable quality in transparent electrodes. The bending and stretching ability of the metal nanotrough were also studied. A real-time application device has also been setup using gold nanotrough network on 178 mm thick polyethylene terephthalate substrate as a replacement of ITO/PET transparent film in four-wire analogue-resistive touch-screen display device which ensures the real-time application of the metal nanotrough network.

3.5 Nanogenerators

Sustainable and stable energy source has been in request to cope with the evolving technology of portable electronics. Owing to the limitations like limited battery lifetime, probable health, and environmental hazards, batteries are becoming hostile as power source. Nanogenerators are broadly classified according to the source of energy harvested into three types namely triboelectric nanogenerators, piezoelectric nanogenerators, and pyroelectric nanogenerators. Triboelectric nanogenerators are devices that harvest friction into static electricity using electrostatic induction. These kinds of nanogenerators are usually robust and environmentally friendly [109, 110]. External energy of piezoelectric materials is harvested by piezoelectric nanogenerators [111, 112]. Piezoelectric nanomaterials are generally used in nanogenerators because such materials can harvest small mechanical vibrations from ambient environment and easily convert the vibrational energy into electric energy. Piezoelectric-based nanogenerators generally use thin films, nanowires, and nanofibers. Piezoelectric polymer PVDF nanofiber prepared by electrospinning has been extensively studied due to its flexibility, ultralong length, and long-term stability compared with normal inorganic piezoelectric materials for further improvement in nanogenerators. The zinc oxide piezoelectric nanogenerator on solid and flexible substrate formation as shown in Fig. 21, is demonstrated by Heever et al. [113].

As the name suggests, nanostructures have always been the part of nanogenerators but the variation in the nanomaterials, variety of nanostructures have been a part of constant research. Ding et al. [114] studied piezoelectric nanogenerator based on formamidinium lead halide perovskite nanoparticles combined with poly (vinylidene fluoride) polymer. These piezoelectric nanogenerators show the uppermost

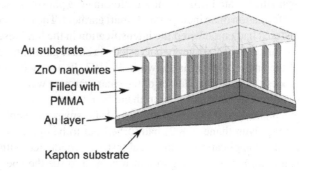

Fig. 21 ZnO piezoelectric nanogenerator formation on Kapton film (Reproduced with permission from Ref. [113], Copyright (2012) Elsevier)

Au substrate

ZnO nanowires

Filled with PMMA

Au layer

Kapton substrate

outstanding outputs with voltage and good current density. Bakar et al. [115] studied the probability of indium-tin-oxide free, barium titanate, and graphene quantum dots doped polyvinylidene fluoride polymer nanocomposite-based flexible nanogenerator. The presence of graphene dots is supposed to enhance the output of the device. The output voltage, current, and power density of nanogenerators could be enhanced by directing the surface charge density. Usage of high-performance dielectric materials like polarized nanoparticles, polymers or polymer nanocomposites has been effectively proven to improve charge density [116].

Shi et al. [117] reported the fabrication of electrospun polyvinylidene fluoride nanofiber mat composed of graphene and barium titanate nanoparticles for high-performance flexible piezoelectric nanogenerators. Barium titanate and graphene embedded polyvinylidene fluoride nanocomposite nanofibers were fabricated through electrospinning whereas, barium titanate/polyvinylidene fluoride nanofiber, and graphene/polyvinylidene fluoride nanofiber were also fabricated in order to study the effect of nanoparticles on the device performance. Interfacial interactions within the barium titanate and graphene with the polyvinylidene fluoride tend to the crystallization of polyvinylidene fluoride chains. The fabricated nanogenerator was also tested to examine the stain and frequency parameter and showed that output remains constant for almost over 1800 cycles. The device is further investigated for real-time application for harvesting biomechanical from the ambient environment of daily life. Figure 22 displays the real-time use of the nanogenerator.

Veeramuthu et al. [118] fabricated smart garment energy generator using electrospun nanofibers. A very facile method was utilized for the study of the energy-generating smart garment. Inter weaving of two kinds of fibers was optimized to form a smart garment where one kind is three-dimensional elastic nanoconductive fibers (NCF) and the latter is three-dimensional elastic coated nanoconductive fibers (CNCF). Polystyrene butadiene styrene elastic fibers were electrospun to form nanoconductive fibers and pristine polytetrafluroethylene, pristine polyurethane and different blends of polytetrafluroethylene and polyurethane were dip-coated with as prepared nano conductive fibers. Also, silver nanoparticles were dip coated over the nano conductive fibers A suitable combination of blends of polytetrafluroethylene and polyurethane was examined and the compatible composition was used for the fabrication of coated nano conductive fibers. Moreover, this smart garment fabricated by interweaving of the nanoconductive fibers and coated nano conductive fibers was again then coated with strontium-aluminate, a phosphorescent material that results in the formation of an incorporated smart garment. The incorporation of phosphorescent material had empowered the intensification in the luminescence of the light which is the outcome of the energy harvested via smart garment. The manufacturing process of the smart garment is shown in Fig. 23. In a very comprehensive manufacturing process, five various classes of the smart garment were fabricated reliant to the polytetrafluroethylene and polyurethane blends ratio. After the detailed morphological, mechanical, and electrical studies of the smart garment, SG2 (polytetrafluroethylene: polyurethane = 8:2) had turned out to be optimum and most reliable for the practical application. A practical test was conducted with SG2 where the mechanical deformation of the garment was used to see the energy harvesting in the form

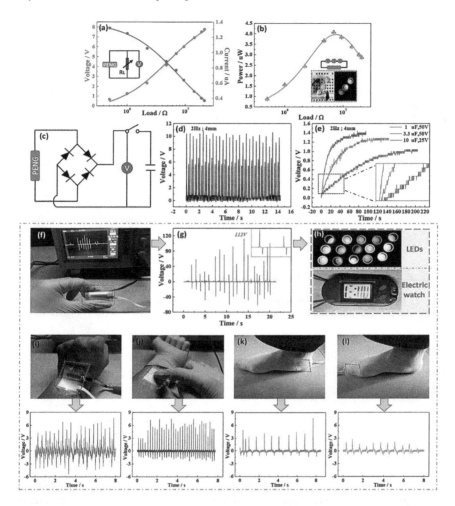

Fig. 22 Real-time use of nanogenerator based on electrospun barium titanate and graphene embedded polyvinylidene fluoride nanocomposite nanofibers (Reproduced with permission from Ref. [117], Copyright (2018) Elsevier)

of glowing red LED. This SG2 further was coated with phosphorescent material to achieve incorporated smart garment which is again allowed to undergo another experiment.

They have also demonstrated the application of strontium-aluminate-incorporated smart garment (SG) in the footwear and showed the exceptional use of the electrospun nanofibers in the field of energy harvesting which had proven its real-time application along with the durability [118]. They have prepared strontium-aluminate-incorporated smart garment (Fig. 24) by depositing strontium-aluminate on the surface of nonconductive fibers as-fabricated by electrospinning. It was observed that the light was emitted from strontium-aluminate-incorporated smart

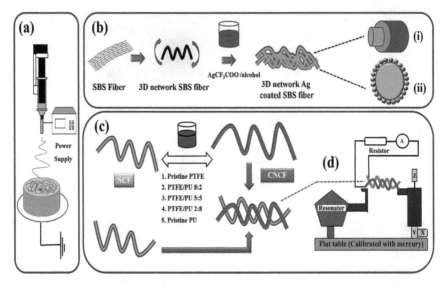

Fig. 23 Smart garment energy generator based on electrospun nonconductive and coated nonconductive nanofibers (Reproduced with permission from Ref. [118], Copyright (2019) Elsevier)

garment when exposed to sunlight and hence they have concluded that the strontium-aluminate-incorporated smart garment can be well useful in footwear designing.

4   Conclusions

Functional polymeric nanofibers have been easily fabricated by using electrospinning technique which can be further modified for the fabrication of polymer composites or core–shell nanofibers. Electrospun polymer nanofibers owing to their amazing properties like high mechanical strength with porous fibrous structure are the potential candidates for wide range applications. The exceptional performance and multifunctionality of electrospun nanofibers with high flexibility and stretchability useful for designing flexible devices which have brought about a cutting edge in wearable and portable electronics. Due to innovative methods and manufacturing processes of smart materials, the flexible devices enabled design of new architectures that are not possible with conventional planar devices. Accompanying flexibility into planar devices has greatly improved the applicability of the devices. The current chapter has projected the capability of electrospinning to fabricate varied polymeric nanofibers by modifications in design, instrumentation, precursor form, polymer matrix, environmental conditions, etc. Various controllable parameters stated have given researchers the power to examine every possible aspect of nanofibers. Polymer nanofibers and

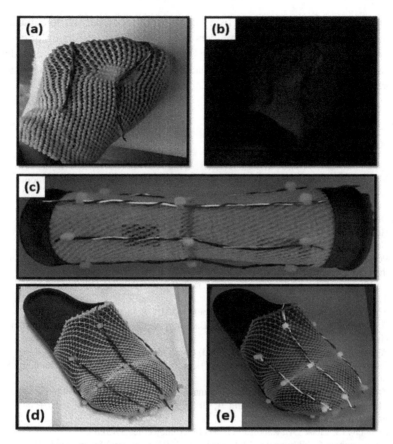

Fig. 24 Strontium-aluminate-based smart garment (Reproduced with permission from Ref. [118], Copyright (2019) Elsevier)

polymer hybrid composite structures are still fabricating by well cost-effective electrospinning. Despite the advantages of electrospinning technique for fabrication of polymer nanofibers such as inexpensive setup, ability to control fiber diameter, orientation, and composition, etc., the applications of electrospun nanofibers especially for flexible devices also face some challenges which include the use of organic solvents, the limited control of pore structures and low mechanical strength. More advanced nanofiber configurations, such as core–shell, multilayer, and multicomponent nanofibers, may be prepared through methods like co-axial electrospinning, electro-hydrodynamic direct writing or mechano-electrospinning, to overcome those problems. Mechano-electrospinning technique uses electrical and mechanical forces to drive the polymer solution in a viscous form for direct writing of polymer ink link on the substrate. As-fabricated nanofibers can be directly or indirectly modified with the help of coating, decorating or embedding nanoparticles over/into the fibers, which can be enormously useful in flexible devices due to their extreme flexibility with high mechanical strength apart from very large surface area. The potential applications of electrospun polymer nanofibers prepared by electrospinning or co-axial electrospinning methods in light-emitting diodes (LEDs), sensors, UV photodetectors, transparent electrodes, and nanogenerators are enormous and systematically explored in this chapter.

References

1. Mishra R, Militky J (ed) (2019) Nanotechnology in textiles: theory and application, 1st edn. Elsevier
2. Capek I (ed) (2019) Nanocomposite structures and dispersions. Elsevier
3. Ding B, Wang M, Wang X, Yu J, Sun G (2010) Mater Today 13:16
4. Huang Y, Song J, Yang C, Long Y, Wu H (2019) Mater Today 28:98
5. Gao W, Zhu Y, Wang Y, Yuan G, Liu JM (2020) J Materiomics 6:1
6. Peijs T (2018) In: Beaumont PWR, Zweben CH (eds) Electrospun polymer nanofibers and their composites. Comprehensive composite materials II, vol 6, p 162
7. Barhoum A, Pal K, Rahier H, Uludag H, Kim IS, Bechelany M (2019) Appl Mater Today 17:1
8. Baji A, Mai YW, Wong SC, Abtahi M, Chen P (2010) Compos Sci Technol 70:703
9. Arumugam V, Moodley KG (2019) In: Barhoum A, Bechelany M, Makhlouf ASH (eds) Handbook of nanofibers, vol 346
10. Agarwal S, Greiner A, Wendorff JH (2013) Prog Polym Sci 38:963
11. Pisignano D (ed) (2013) Polymer nanofibers: building blocks for nanotechnology. Royal Society of Chemistry
12. Zahmatkeshan M, Adel M, Bahrami S, Esmaeili F, Rezayat SM, Saeedi Y, Mehravi B, Jameie SB, Ashtari K (2019) In: Barhoum A, Bechelany M, Makhlouf ASH (eds) Handbook of nanofibers, vol 261
13. Stephens JS, Chase DB, Rabolt JF (2004) Macromolecules 37(3):877
14. Mit-uppatham C, Nithitanakul M, Supaphol P (2004) Macromol Symp 216(1):293
15. Zhao Y, Yang Q B, Lu X F, C Wang, Wei Y (2005) J Polym Sci Part B: Polym Phys 43(16):2190
16. Ding B, Yamazaki M, Shiratori S (2005) Sens Actuators, B Chem 106(1):477
17. Kim B, Park H, Lee SH, Sigmund WM (2005) Mater Lett 59(7):829
18. Mack J, Viculis LM, Ali A, Luoh R, Yang GL, Hahn HT, Ko FK, Kaner RB (2005) Adv Mater 17(1):77
19. Kahol PK, Pinto NJ (2002) Solid State Commun 124(5–6):195
20. Kahol PK, Pinto NJ (2004) Synth Met 140(2–3):269
21. Zeng J, Chen XS, Liang QZ, Xu XL, Jing XB (2004) Macromol Biosci 4(12):1118
22. Fennessey SF, Farris RJ (2004) Polymer 45(12):4217
23. Chronakis IS, Milosevic B, Frenot A, Ye L (2006) Macromolecules 39(1):357
24. Jeong EH, Im SS, Youk JH (2005) Polymer 46(23):9538
25. Czaplewski D, Kameoka J, Craighead HG (2003) Nonlithographic approach to nanostructure fabrication using a scanned electrospinning source. J Vac Sci Technol, B 21(6):2994
26. Ji Y, Li BQ, Ge SR, Sokolov JC, Rafailovich MH (2006) Structure and nanomechanical characterization of electrospun PS/clay nanocomposite fibers. Langmuir 22(3):1321
27. Cha DI, Kim HY, Lee KH, Jung YC, Cho JW, Chun BC (2005) J Appl Polym Sci 96(2):460
28. Shenoy SL, Bates WD, Wnek G (2005) Polymer 46(21):8990
29. Lee KH, Kim HY, La YM, Lee DR, Sung NH (2002) J Polym Sci Part B: Polym Phys 40(19):2259
30. Kim JR, Choi SW, Jo SM, Lee WS, Kim BC (2004) Electrochim Acta 50(1):69
31. Kim JR, Choi SW, Jo SM, Lee WS, Kim BC (2005) J Electrochem Soc 152(2):A295

32. Shenoy SL, Bates WD, Frisch HL, Wnek GE (2005) Polymer 46(10):3372
33. Ma ZW, Kotaki M, Ramakrishna S (2006) J Membr Sci 272(1–2):179
34. Lee SG, Choi SS, Joo CW (2002) J Korean Fiber Soc 39(1):1
35. Chen ZH, Foster MD, Zhou WS, Fong H, Reneker DH, Resendes R, Manners I (2001) Macromolecules 34(18):6156
36. Liu WX, Graham M, Evans EA, Reneker DH (2002) J Mater Res 17(12):3206
37. Kenawy ER, Abdel-Fattah YR (2002) Macromol Biosci 2(6):261
38. Kang TS, Lee SW, Joo J, Lee JY (2005) Synth Met 153(1–3):61
39. Kim C, Kim JS, Kim SJ, Lee WJ, Yang KS (2004) J Electrochem Soc 151(5):A769
40. Srinivasan G, Reneker DH (1995) Polym Int 36(2):195
41. Son WK, Youk JH, Lee TS, Park WH (2004) Polymer 45(9):2959
42. Theron SA, Zussman E, Yarin AL (2004) Polymer 45(6):2017
43. Theron SA, Yarin AL, Zussman E, Kroll E (2005) Polymer 46(9):2889
44. Shawon J, Sung CM (2004) J Mater Sci 39(14):4605
45. Chua KN, Chou CP, Lee C, Tang YN, Ramakrishna S, Leong KW, Mao HQ (2006) Biomaterials 27(36):6043
46. Ma ZW, Kotaki M, Ramakrishna S (2005) J Membr Sci 265(1–2):115
47. Ma ZW, Kotaki M, Yong T, He W, Ramakrishna S (2005) Biomaterials 26(15):2527
48. Smit E, Buttner U, Sanderson RD (2005) Polymer 46(8):2419
49. Zhao SL, Wu XH, Wang L, Huang Y (2003) Cellulose 10(4):405
50. Zhao SL, Wu XH, Wang L, Huang Y (2004) J Appl Polym Sci 91(1):242
51. Ki CS, Baek DH, Gang KD, Lee KH, Um IC, Park YH (2005) Polymer 46(14):5094
52. Hongu T, Phillips GO, Takigami M (ed) (2005) New millenn, fibers. Elsevier
53. Kenry LCT (2017) Prog Polym Sci 70:1
54. Peng SX, Weng W, Fang X (ed) (2017) Polymer materials for energy and electronic applications. Academic Press
55. Chinnappan A, Baskar C, Baskar S, Ratheesh G, Ramakrishna S (2017) J Mater Chem C-5:12657
56. Sun B, Long YZ, Chen ZJ, Liu SL, Zhang HD, Zhang JC, Han WP (2014) J Mater Chem C-2:1209
57. George G, Luo Z (20202) Curr Nanosci 16:321
58. Erdem R, İlhan M, Ekmekçi MK, Erdem Ö (2017) Appl Surf Sci 421:240
59. Hou Z, Wang L, Lian H, Chai R, Zhang C, Cheng Z, Lin J (2009) J Solid State Chem 182:698
60. Du P, Song L, Xiong J, Cao H, Xi Z, Guo S, Wang N, Chen J (2012) J Alloys Compd 540:179
61. Cheng Y, Zhao Y, Zhang Y, Cao X (2010) J Colloid Interface Sci 344:321
62. Shen H, Feng S, Wang Y, Gu Y, Zhou J, Yang H, Feng G, Li L, Wang W, Liu X, Xu D (2013) J Alloys Compd 550:531
63. Yu H, Song H, Pan G, Qin R, Fan L, Zhang H, Bai X, Li S, Zhao H, Lu S (2008) J Nanosci Nanotechnol 8:1432
64. Cacciotti I, Bianco A, Pezzotti G, Gusmano G (2011) Mater Chem Phys 126:532
65. Xu X, Zhao S, Liang K, Zeng J (2014) J Mater Sci Mater Electron 25:3324
66. Alam AM, Liu Y, Park M, Park SJ, Kim HY (2015) Polymer 59:35
67. Lee KS, Shim J, Park M, Kim HY, Son DI (2017) Compos Part B Eng 130:70
68. Fang D, Zhang M, Luo Z, Cao T, Wang Q, Zhou Z, Jiang M, Xiong C (2014) Opt Mater 38:1
69. Ge W, Shi J, Xu M, Chen X, Zhu J (2019) J Alloys Compd 788:993
70. Wei Y, Zhong L, Li D, Ma Q, Dong X (2019) Opt. Mater 95:109261
71. Pangul CN, Anwane SW, Kondawar SB (2018) Luminescence 33:1087
72. Pangul CN, Anwane SW, Kondawar SB (2019) Mater Today Proc 15:464
73. Itankar SG, Dandekar MP, Kondawar SB, Bahirwar BM (2017) Luminescence 32:1535
74. Dandekar MP, Itankar SG, Kondawar SB, Nandanwar DV, Koinkar P (2018) Opt Mater 85:483
75. Qin Z, Wang Q, Wang C, Xu D, Ma G, Pan K (2019) J Mater Chem C-7:1065
76. Liu Q, Ramakrishna S, Long YZ (2019) J Semicond 40:111603
77. Zhang S, Jia Z, Liu T, Wei G, Su Z (219) Sensors 19:3977
78. Mercante LA, Andre RS, Mattoso LHC, Correa DS (2019) ACS Appl. Nano Mater 2:4026

79. Huang B, Zhang Z, Zhao C, Cairang L, Bai J, Zhang Y, Mu X, Du J, Wang H, Pan X, Zhou J, Xie E (2018) Sens Actuators B 255:2248
80. Kim JH, Lee JH, Mirzaei A, Kim HW, Kim SS (2018) Sens Actuators B 258:204
81. Pang Z, Yang Z, Chen Y, Zhang J, Wang Q, Huang F, Wei Q (2016) Colloids Surf A: Physicochem Eng Aspects 494:248
82. Wang Z, Li Z, Jiang T, Xu X, Wang C (2013) ACS Appl Mater Interfaces 5:2013
83. Xu L, Zheng R, Liu S, Song J, Chen J, Dong B, Song H (2012) Inorg Chem 51:7733
84. Yang Y, Fan X, Long Y, Su K, Zou, Li N, Zhou J, Li K, Liu F (2009) J Mater Chem 19:7290
85. Xu X, Sun J, Zhang H, Wang Z, Dong B, Jiang T, Wang W, Li Z, Wang C (2011) Sens Actuators B 160:858
86. Sharma H, Sonwane N, Kondawar S (2015) Fibers Polym 16:1527
87. More AM, Sharma HJ, Kondawar SB, Dongre SP (2017) J Mat NanoSci 4:13
88. Beniwal A, Sunny. (2020) Sens Actuators B Chem 304:127384
89. Busacca C, Donato A, Faro M Lo, Malara A, Neri G, Trocino S. (2020) Sens Actuators B Chem 303:127193
90. Shen Y, Yan X, Bai Z, Zheng X, Sun Y, Liu Y, Lin P, Chen X, Zhang Y (2015) RSC Adv 5:5976
91. Alaie Z, Nejad SM, Yousefi MH (2015) Mater Sci Semicond Proc 29:16
92. Shi L, Nihtianov S (2012) IEEE Sens J 12:2453
93. Zou Y, Zhang Y, Hu Y, Gu H (2018) Sensors 18:2072
94. Cai S, Xu X, Yang W, Chen J, Fang X (2019) Adv Mater 31:1808138
95. Polat EO, Mercier G, Nikitskiy I, Puma E, Galan T, Gupta S, Montagut M, Piqueras JJ, Bouwens M, Durduran T, Konstantatos G, Goossens S, Koppens F (2019) Sci Adv 5:7846
96. Núñez CG, Vilouras A, Navaraj WT, Liu F, Dahiya R (2018) IEEE Sens J 18:7881
97. Chen H, Liu K, Hu L, Ghamdi A, Fang X (2015) Mater Today 18:493
98. Reddy KCS, Sahatiya P, Sauceda IS, Cortázar O, Bon RR (2020) Appl Surf Sci 145804
99. Kim J, Park H, Jeong SH (202) J Indust Eng Chem 82:144
100. Martínez L, Ghosh DS, Giurgola S, Vergani P, Pruneri V (2009) Optical Mater 31:1115
101. Lee SH, Kim G, Lim JW, Lee KS, Kang MG (2018) Solar energy mater. Solar Cell 186:378
102. Jeong WL, Min JH, Kwak HM, Jeon YJ, Lee HJ, Kim KP, Lee JS, Kang SJ, Kim DY, Lee DS (2019) J Alloy Comp 794:114
103. Essaidi H, Cattin L, Jouad ZE, Touihri S, Blais M, Ortega E, Louarn G, Morsli M, Abachi T, Manoubi T, Addou M, del Valle MA, Diaz F, Bernède JC (2018) Vacuum 153:225
104. Pruna R, Baraket A, Bonhomm A, Zine N, Errachid A, Lopez M (2018) Electrochim Acta 283:1632
105. Fuh YK, Lien LC (2013) Nanotechnology 24:55301
106. Yousefi AA, Mohebbi AR, Moghadam SF, Poursamar SA, Hao L (2019) Solar Energy 188:1111
107. Jiang DH, Tsai PC, Kuo CC, Jhuang FC, Guo HC (2019) ACS Appl Mater Interfaces 11:10118
108. Wu H, Kong D, Ruan Z, Hsu PC, Wang S, Yu Z, Carney TJ (2013) Nat Nanotechnol 8:421
109. Niu S, Wang ZL (2015) Nanoenergy 14:61
110. Wang S, Lin L, Wang ZL (2015) Nanoenergy 11:436
111. Wang ZL (2006) J Song, Science 312:242
112. Yang Y, Guo W, Pradel KC, Zhu G, Zhou Y, Zhang Y, Hu Y, Lin L, Wang ZL (2012) Nano Lett 12:2833
113. Heever TS, Perold WJ (2012) Microelect Eng 112:41
114. Ding R, Zhang X, Chen G, Wang H, Kishor R, Xiao J, Gao F (2017) Nanoenergy 37:126
115. Bakar EA, Mohamed MA, Ooi PC, Wee MFMR, Dee CE, Majlis BY (2018) Org Electron 61:289
116. Liu L, Yang X, Zhao L, Xu W, Wang J, Yang Q, Tang Q. (202) Nanoenergy 73:104797
117. Shi K, Sun B, Huang X, Jiang P (2018) Nanoenergy 52:153
118. Veeramuthu L, Li WL, Liang FC, Cho CJ, Kuo CC, Chen WC, Lin JH, Lee WY, Wang CT, Lin WY, Rwei SP (2019) React Funct Polym 142:96

Electrospun Nanofibers for Wastewater Treatment

Jyotendra Nath, Kashma Sharma, Shashikant Kumar, Vishal Sharma, Vijay Kumar, and Rakesh Sehgal

Abstract Electrospinning (ES) is a flexible and straightforward strategy that permits the creation of ultrathin fibers of various compositions. The need for helpful advancements in filtration ability has led to little thought of cutting-edge materials, for example, electrospun nanofibers (NFs) for wastewater treatment. Electrospun NFs play a significant part in numerous fields because of their high surface area, high porosity, and good functional capabilities. These properties make them encouraging materials for a range of applications, most explicitly water treatment. Furthermore, electrospun nanofibers can be simply functionalized by joining multifunctional materials to meet extraordinary water treatment effects. This chapter focuses on the most recent progress of electrospun NFs with special attention on wastewater treatment applications. In particular, we discuss the synthesis and functionalization of electrospun NFs. The various process parameters involved during ES have also been discussed. Finally, the points of view are introduced in regards to difficulties, openings, and new prospects for the use of electrospun NFs.

Keywords Nanofiber · Electrospinning · Nanocatalyst · Wastewater treatment

J. Nath · S. Kumar (✉)
Department of Chemical Engineering, National Institute of Technology Srinagar, Srinagar, Jammu and Kashmir 190006, India

K. Sharma
Department of Chemistry, DAV College, Sector-10, Chandigarh, India
e-mail: Kashma@davchd.ac.in

V. Sharma
Institute of Forensic Science & Criminology, Panjab University, Chandigarh 160014, India

V. Kumar
Department of Physics, National Institute of Technology Srinagar, Hazratbal, Jammu and Kashmir 190006, India

Department of Physics, University of the Free State, P.O. Box 339, Bloemfontein ZA9300, South Africa

R. Sehgal (✉)
Department of Mechanical Engineering, National Institute of Technology Srinagar, Srinagar, Jammu and Kashmir 190006, India

© The Author(s), under exclusive license to Springer Nature Switzerland AG 2021
S. K. Tiwari et al. (eds.), *Electrospun Nanofibers*, Springer Series on Polymer and Composite Materials, https://doi.org/10.1007/978-3-030-79979-3_4

1   Introduction

Direct disposal of industrial wastewater in natural water bodies is one of the main sources of water pollution. There are many contaminants such as pesticides, surfactants, heavy metals, dyes and other contaminants present in wastewater that are harmful to the living organisms so these need to be removed before disposal to the direct water body. In recent years, there has been increased interest in the use of nanofibers (NFs) with unusual structures for water and wastewater treatment. NFs are mainly produced by electrospinning (ES) due to their basic arrangement, cost-effectiveness, and adaptability. ES is being characterized by the application of a high electric field voltage in a cycle that produces fibers from their polymer arrangement. Electrospun NFs are important in many fields due to their high specific surface area, high as well as fine porosity, high permeability, and special composite structures. These properties make it possible to use electro-sensitive NFs to discharge paint particles from wastewater, and it is preferably used for wastewater treatment. In this chapter, we are mainly focusing on the factors influencing NFs and the advances in research on the placement of different forms of NFs. Electro manipulation of NFs with different polymers is used to adsorb dyes and heavy metals from water and wastewater. As a result, electrospun functional NFs have water treatment capabilities such as separation, adsorption, photocatalysis, and antimicrobial properties. Summarize the production of electrospun NFs and their specific applications in the field of wastewater treatment. Electrospun NFs are the cutting-edge materials for nanocatalyst that has distinctive properties to lower down the water contamination calamity as compared with other conventional catalysts. ES is the most efficient technique for the production of NFs. Metal oxide NFs such as ZnO, TiO_2, ZrO_2, etc. are possible materials for wastewater treatment owing to their ability to complete mineralization of organic pollutants under atmospheric conditions with maximum elimination efficacy.

In the twenty-first century, water pollution is the foremost challenge because of the expanding population and better way of life. All advanced and emerging countries have grieved with water contamination. Direct disposal of industrial wastewater in the natural water body is additionally one of the primary hotspots for water pollution [1]. The most serious issue in the world is water pollution and no nation is left immaculate by this issue. The earth is covered with two-third of water and it is available in different structures and sources like the ocean, river, underground water, and Ica Cap [2]. Expansion in human population and environmental misuse has brought about an increase in the burden on various sources of water and this causes a wastewater problem. As a result, a huge part of the global population had confronted severe water cataclysm together in the term of amount and nature of drinkable water [3]. In wastewater, toxins may contain heavy metals, nutrients, microbes, and biological matter, and many more [4]. The debasement of the nature of water and development in contaminations has expanded the danger to human prosperity and to the climate. Different sorts of wastewater poisons have existed like colloidal particles (1 nm–1 μm in diameter) and the other kind of contaminations are made as basically a consequence of modern activities, household activities, etc. [5]. The fundamental sources of wastewater are industries, for example, acid and alkali manufacturing plants, pesticide unit, paper production factories and boiler in thermal power plants for coolant use, and different ventures. The extra wastewater sources are civil wastewater like dark water, natural squanders, and wastewater from a toilet flush, cooking, washing, and other residential wastewater [6]. Apart from these, other sources of water pollution are highways, agricultural, building construction, and clinic wastewater [7]. The other toxin of wastewater contains metallic particles, rock-solid wastes, mineral poisons, organic contaminations, natural waste, microorganisms, soil particles, and different solvents [8]. The wastewater arrival of different current frameworks, for example, magnificence care items, plastics, leather, creation of paper, textile, printing, drug, and food ventures has become a basic normal worry because of the presence of natural dyes [9]. During the gathering method, different factories discharge wastewater containing a huge load of utilized dyes, which may cause different health and biological issues [10, 11].

A couple of strategies have been made for squander water de-colorization, for example, oxidation, reverse osmosis, adsorption, ozonation, photocatalytic, coagulation, electrochemical treatment, membrane separation, ultrafiltration, etc. [12]. Membrane separation plays a significant role in the cleaning of wastewater and it has high separation efficiency, the least handling cost, and treating of various types of wastewater. The most common process in membrane separation is reverse osmosis, microfiltration, ultrafiltration, and nanofiltration. Polymeric and ceramic-based membranes are available in the market yet face various difficulties regarding fouling, degradation, and their complete stability [12]. Subsequently, there is a flood of examination done on refining membrane efficiency by tuning their design, structure, and physicochemical properties. Lately, the utilization of nanotechnology has extensively enhanced the advancement of membranes and particularly on the utilization of electrospun NFs. The electrospun NFs have demonstrated immense potential for adsorption application because of their high porosity and the enormous surface-to-volume ratio [13, 14]. The ES technique is one of the methods that has helped in the advancement of successful adsorbents specifically polymer nanofibers and carbon nanofibers, which have been utilized to expel different contaminants [15–18]. NFs can be made from almost any polymer. NFs can be supplied from a wide range of polymers and in all cases, they are unique in their properties and applications. The diameter of NFs depends on the type of polymer used and the manufacturing process. Electrospun carbon nanofibers (ECNFs) were prepared from their precursor polymeric materials, for example, polyacrylonitrile (PAN), polyvinyl chloride (PVC), polysulfone (PSN), polyimide (PI), cellulose, and poly(vinyl alcohol) (PVA), by means of ES procedure followed by adjustment and carbonization [19]. NFs have surprisingly interesting properties of high surface area to volume ratio due to their small diameter, extremely high porosity, high frictional permeability, enough mechanical strength, and good pore size interconnectivity. In addition to the functionality of the polymer itself, these unique properties give many NF attractive properties for innovative applications (e.g., biomedical, apparel, smart materials, energy generation, sensors, water, and wastewater treatment, etc.) [20, 21]. In any case, the versatility of NFs as water filtration agents, including the removal of metal particles and dyes from water sources, is highly appreciated [22]. Much exploratory work has also been carried out on the preparation, modification, and potential of NFs in wastewater treatment. This is mainly due to the filtering properties of NF mats, similar to those of wastewater from the water and wastewater treatment sector. NFs as a membrane play an important role in this report. In this section, we report on the continued progress in the improvement of NFs and their applications in water and wastewater treatment. Several strategies for the production of NFs are available, including design, melting and blowing, model synthesis, phase separation, and ES [23, 24].

2 Electrospinning

ES is the most common method to produce fibers with widths ranging from submicron to nanometer by using electrical energy to induce a set of polymeric fibers loaded with fibers. This method does not require high temperatures or scientific coagulation to obtain NFs from a polymer network, thus allowing the production of superior fibers with solid or composite particles. ES is a very old process for spinning fibers. However, it is emerging as the most limiting dynamic strategy for obtaining huge NFs. "The term "electrospinning" dates back to 1897 because of the use of electrostatic fields to produce fibers [23, 25]. Rayleigh developed the process in the early nineteenth century in 1897, and later in 1914 by Zeleny with the intention of developing it in stages [26]. In the early twentieth century, Cooley and Morton had the opportunity to produce electrospun yarns. Anton Formhals continued with nonwoven ES and brought it to market, receiving a patent for the process and equipment for producing woven yarn in 1930 [27–29]. Subsequently, Formhals patented a different technique for the production of electrostatic polymerized fibers to produce fibers made from different polymers. The company established an experimental process for the production of yarn materials using counterfeit fibers that generate an electrical charge. Factors affecting fiber stability such as electric field, flow rate, and experimental conditions were also considered [30–32]. In 1988, Simon explained that electrical testing could be used to produce carpets of polystyrene (PS) and polycarbonate (PC) fibers in the nano- and submicron range using an in vitro cellular matrix. He made a clear suggestion that the chemical adjustment of the textile surface depended on the ES process end of the medium electric field [33]. Therefore, prior to 1990, few trades were fascinated in the ES process for textile production. Fusion spinning was the ideal strategy for textile production using both common and synthetic polymers. Fusion spinning does not allow fiber measurements in the nano-domains [28, 34, 35]. Dalton et al. [36] maintained networks of electron-spun NFs in tissue cells to kill cytotoxic solvents and for tissue engineering applications [37, 38]. Today, there is a growing enthusiasm for electron-spun NFs in a wide range of potential applications [43, 49]. There are several preferred locations for electron spin NFs. Many flame retardant polymers such as polyether ketone ether (PEEK), PVC, PAN, and PS can be electrospun. The surface of nanofibers is 100 to several times larger than that of conventional fibers.

2.1 Electrospinning Process

ES is a means of producing large diameter submicron and nanofibers with excellent physical properties (e.g., mechanical, aesthetic, electrical, and thermal). The application of high voltages to the liquefaction and assembly of the polymer generates a large electrostatic field that leads to the formation of NFs. At the end of the capillary, the substrate or polymer liquefies under the influence of its surface tension. The high charge of the liquid is due to the electric field, and a perceived increase in force can be generated due to the reduced surface tension and the aversion to ground charge. Advantageously, in the semi-circular region of the assembly, the highest point of the capillary is reached by applying an electric field, thus causing the aging of another structure recognized as the Taylor cone [38]. ES is believed to generate micro-nanofibers in polymer devices under a strong electric field (kV) at atmospheric pressure and room temperature. There are two main arrangements of ES devices: vertical and planar [28, 34, 39]. High voltages generate electrical charges on the polymer components, which accumulate on the polymer surface. These charges exceed the surface tension of the assembly in the basic electric field and repel each other, eventually reaching the structure Taylor cone. The tip of the Taylor cone emits a charged beam of light, which expands further in the electric field. Due to the loss of its solubility, the package eventually grows into stronger fibers [40, 41]. In particular, as the electric field increases, the electrostatic repulsion force expands beyond the surface tension. The placement/formation of the charged beam polymer is displaced from the Taylor cone discharge. Finally, due to the proportional repulsive force of the charge on the jet, polymerization process jets undergo a circular or pyramidal development, commonly called electrojet winding inconsistency. Due to the inconsistency of the winding, the package can be long, narrow, or small. Charged polymer fibers form and accumulate due to loss of resolving force and bending instability. The distance between the loaded polymer fibers and the capillary is very short [42, 43]. When an electrostatic field is applied to the polymer solution, the outside of the set becomes charged and the particles move through the solution, forming a jet that carries an extra charge. Once out of the capillary, the beam first follows a straight path and then finds a series of unsteady forward and backward bends, which unfold in three stages of amplitude. As the beam loop becomes longer and narrower, the width and limits of the circle increase. After a while, the loop produces another bending change. Each mode of bending change can be described in three steps [44].

i.   A smooth beam is created across the circle, accumulating multiple curves.
ii.   The cross-section of each curved beam widens into a series of helical loops, increase in width in curvature diversity [44].

iii. As the loop edges increase, the distance between the beam cross-sections decreases, and in a less meaningful way, the state of step 1 is established, and the following bending destabilization diagram begins.

As the number of cycles increases, the diameter of the beam becomes smaller and smaller, resulting in the formation of NFs. After the next cycle, the axis of a given fragment can point to any path. The light beam is scattered as it flies from the capillary to the authoritative display and the electrically sensitive NFs are collected [44]. The application of NF thin films in water and wastewater treatment has generated a great deal of enthusiasm in the last two decades. This is a direct result of the flexible arrangement of NFs, which allows controlling porosity, pore size, structure, surface properties, mechanical and thermal stability [45].

2.2 Effecting Parameter

The parameters of ES are essential because they affect the width and morphology of the electrospun fibers. The mode of ES is that the polymer spheres reduced the surface tension to form a jet, and then the solubility disappears or solidifies to construct the microfibers. The surface morphology of electrospun NFs can be severely limited by numerous constraints, for example, processing parameters, solution parameters, and environmental parameters [46]. The production of smooth, drip-free filaments depends on these variables. Therefore, the production of drip-free electrostatic NFs requires a comprehensive understanding of these factors. Rotational voltage is the most fundamental process parameter in electrodeposition. Applied stress, transport/flow rate, and collector and tip separation are other important parameters that affect the morphology of electrostatic NFs. The process parameters also have a strong influence on the morphology and diameter of the NFs. The "Taylor cone" is set when the baseline voltage is reached. The applied voltage is an important factor in electrical testing because it controls the quality of the electric field between the needle tip and the collector as well as the tensile strength [47]. Increasing the voltage inevitably creates a frame packing of the NFs. In most cases, high tension favors the stretching of the fibers, increases the thickness of the surface loading of the beam, decreases the size of the fibers, and the solvent dissolves faster without forming droplets [39, 48–50]. Zhang et al. [51] found that at higher stresses, the polymer ejects more and the measurement of large diameter fibers is improved. In the electrofiltration method, the flow of the polymer network is another dynamic parameter that is likely to affect the jet speed and material exchange. It has been proposed to slow down the feed rate to give enough opportunity for the removal of soluble materials, but at high flow rates, pearl fiber structures can be obtained. As the feed rate of the assembly increases, the thickness of the load decreases. Increasing the thickness of the filler material may cause instability in the assisted bending of the electrofiltration bundle, which helps to form filaments in a smaller filament space [52–55]. The injection speed of the centrifuge assembly is another important parameter of the electrospinning process, which can be limited by varying the speed of the needle pump. If the speed is too slow, the duration of the ES process will be longer. However, if the speed is too fast, the diameter of the fibers will increase, forming huge beads that will fall directly onto the board. The separation of the capillary and the cluster also affects the solidification and expansion of the beam. As the rotational separation expands, the chance of dice jet separation increases, radically shortening the distance between them [56]. The distance between the collector and the tip is another parameter that affects the diameter and morphological control of the NFs. It is important to choose complete separation before the fibers reach the collector to avoid the disappearance of polymer alignment. Since the drying of electrospun NFs is important for the formation of drip-free filaments, it is important to have an optimal separation between the needle and the collector to promote adequate soluble evaporation [57]. Solution handling is another handling parameter in electrical testing, and polymer concentration is one of the most important parameters in electrical testing. At low concentrations, the beam becomes unstable due to transcontinental capture and surface tension reduces the width of the beam, leading to the aggregation of beads and strands. However, if the overall concentration is too high to create a liquid jet, an electrical connection may not be possible. Increasing the concentration of polymer generally results in a uniform diameter of the filament. Eventually, an additional increase in polymer concentration leads to the formation of pure fibers [58–60]. Viscosity is another parameter that affects the development of fibers in electrical tests. If the consistency of the solution is very low, the fibers will not be structured, where a solution with a particularly high thickness will cause problems in the release of polymer bundles from the matrix. It has also been stated that viscosity is extremely important for the morphology of electrostatic filaments. A low concentration polymer solution will consist of droplet-like fibers, while a high concentration matrix will consist of thread-like filaments. The measurement of the fibers corresponds to the concentration of the polymer. The thickness of the polymer solution is relative to the concentration of the polymer [27, 53, 60–64]. The surface tension of the solvent has a significant impact on the electrical testing process. The electrical power filament released by the beads is obtained by reducing the overall surface tension. When the surface tension of the set is expanded, the electrical filtration is prevented due to the instability of the nozzle and resulted in the creation of droplets. Similarly, the low surface tension of the set facilitates the framing of electro-optic fibers under low magnetic fields [57, 65]. The conductivity of the solution is another vital parameter that can influence the developmental morphology of electro-optical fibers. The polymer matrix is considered to be electrically conductive and its charged particles have an incredible influence on the production of light. The conductivity of the substrate is controlled by the type of polymer and the soluble and ionic salts used. As the conductivity of the substrate increases, the diameter of the fibers decreases, and vice versa. It is hypothesized that highly conductive polymer devices are unstable in the solid-state region, leading to bending variations [27, 52, 55, 66, 67]. Molecular weight has a tremendous influence on the morphology of electrophoretic NFs. It essentially affects the surface tension, viscosity, and conductivity of the polymer arrangement. Many scientists believe that the structure of low atomic weight polymers forms droplets rather than filaments, and smooth fibers are obtained by expanding the molecular weight. In addition, at higher molecular weights, larger diameter electromagnetic filaments can be obtained [53, 54, 68]. Ecological conditions are also vital variables that can disturb the morphology of NFs [69]. In most cases, low relative humidity accelerates the rate of dissipation of soluble fibers in the beam and facilitates the placement of finer fibers. The temperature has two opposite effects on the normal diameter of the fibers. Temperature expansion accelerates the rate of evaporation of soluble fibers, limiting further expansion of the beam. Low temperature reduces the thickness of the beam and promotes a thinner filament. Therefore, to get the desired electrofiltration conditions, the natural temperature and humidity must be significantly altered [53, 54].

2.3 Electrospun Nanofibers with Structure

One of the advantages of electroforming innovations is that by changing the boundaries of electroforming or changing the design solution, you can get a structure that solves a clear problem. It is always incredibly interesting to see how NFs have more necessary properties such as higher performance, better functionality, larger surfaces, higher porosity, and higher absorption capacity. As a result, several tweaks were made to the basic layout to create the embedded fiber structure. The refinements focused on the direction of the fiber economy and the morphology of the fibers. Four types of specially oriented NFs are described in detail below, including adaptive NFs, center-coated NFs, porous NFs, and hollow NFs [69–71].

3 Nanofibers for Wastewater Treatment

3.1 Polyacrylonitrile

Mokhtari-Shourijeh et al. [83] reported a nanofiber permeable compound consisting of amine PAN/polyvinylidene fluoride (PVDF). This hybrid nanomaterial was used as an adsorbent to expel direct red 23 (DR23) from the liquid set. The effects of working parameters such as sorbent content, pH, contact time, and initial color concentration were evaluated. The materials used here were PAN, PVDF, sodium carbonate ($NaHCO_3$), hydrochloric acid, and dimethylformamide (DMF). The resistive technique involves several steps, including the combination of PAN/PVDF/$NaHCO_3$ nanofibers using an ES strategy, salt extraction to increase porosity, and finally the functionality of the nanofibers. The results showed that amine-based PAN/PVDF NF permeable compounds were used as adsorbents to remove DR23 [83]. Almasian and coworkers [84] synthesized NFs consisting of PAN and PA by electrospinning and demonstrated their structures by FTIR, SEM, and BET strategies and further investigated their decolorizing ability. It was investigated. Materials such as PAN, PA, N, N-dimethylformamide, DR80, and DR20 were used for these NFs. The adsorption of composite NFs on color was followed by Langmuir isotherms. The results showed that direct red 80 (DR80) and direct red 20 (DR20) were adsorbed/deleted from the liquid shading matrix [84]. Nei et al. [85] reported that NFs composed of PAN/CuS were prepared by a simple aqueous technique, and monodisperse CuS nanoparticles (NPs) were continuously flowing on the outer surface of PAN electrospun PAN fibers. The materials used included PAN, acrylamide, N, N-DMF, copper(III) acetate monohydrate (CuAc$_2$-H$_2$O), thioacetamide (C$_2$H$_5$NS), and hydrogen peroxide (H$_2$O$_2$). The reaction has been theorized and must be treated according to the following equation.

$$CH_3CSNH_2 + H_2O \rightarrow CH_3CONH_2 + H_2S \qquad (1)$$

$$H_2S + Cu_2+ \rightarrow CuS + 2H^+ \qquad (2)$$

$$CH_3CSNH_2 + Cu^{2+} + H_2O \rightarrow CuS + CH_3CONH_2 + 2H^+ \qquad (3)$$

The whole liquefaction process of CuS-NP can be registered roughly in the domestic (3). The results show that the generated PAN/CuS composite nanofibers have high catalytic activity and can be repeatedly used for methylene blue (MB) reduction to H$_2$O$_2$ [85]. Kampalanonwat et al. [75] studied polyacrylonitrile amine (PANAM) NF mats as surface modification of electrospun PAN with diethylenetriamine (DETA). This was done in order to drain Ag(I), Cu(II), Pb(II), and Fe(II) from the liquid aggregates. The materials used for these electron spin NFs were PAN, methyl acrylate monomer, DMF, DETA, and aluminum chloride hexahydrate. The adsorption behavior of Ag(I), Cu(II), Pb(II), and Fe(II) was determined at different pH values. It was found that the adsorption isotherms of these metal particles could be better represented using the Langmuir model: the estimated adsorption limits of Ag(I) and Cu(II) particles were larger in a mixture of four metal particles wound on APAN [75]. PAN-based materials for wastewater treatment are reviewed and summarized in Table 1.

3.2 Polyvinyl Alcohol

In 2018, Mahmoodi and co-workers [95] prepared PVA/chitosan/silica NFs with mesoporous structures by sol–gel/electron detection and used them for the removal of direct red 80 (DR80). The results show that the ready-to-use composite NFs have better adsorption limits than direct red 80 (DR80). The adsorption of DR80 is suitable for Langmuir isotherms and pseudo-quadratic kinetic models and the adsorption rate of DR80 reaches 98% within 90 min. Therefore, PVA/chitosan/silica is considered a

Table 1 Polyacrylonitrile-based nanofibers for wastewater treatment

S. no.	Materials	Nature	Applications	References
1	Cibacron blue-attached polyacrylonitrile using with N,N-dimethylacetamide	Nanofiber	High bromelain adsorption capacity	[72]
2	Titanium dioxide functionalized with polyacrylonitrile	Nanofiber	Adsorbent lead (II) (PB^{2+}) ions in aqueous solutions	[73]
3	Polyacrylonitrile containing with TiO_2	Nanofiber	Degradation of dye rhodamine B	[74]
4	Aminated polyacrylonitrile with diethylenetriamine	Nanofiber	Adsorption of Ag(I), Cu(II), Pb(II), and Fe(II). Extreme adsorption Ag(I) and Cu(II) ions	[75]
5	Polyacrylonitrile with titania	Nanofiber	Degradation of DR23 and DR 80, DR-Direct red	[76]
6	Polyacrylonitrile with carbon nanofibers and titanium dioxide	Nanofiber	Maxmium rejection of heavy metals Pb2 +, Cu2 +, and Cd2 + and dye methylene blue (MB)	[77]
7	Polyacrylonitrile with zinc oxide	Nanofiber	In this aqueous solution removes Pb(II) ions	[78]
8	Polyacrylonitrile functionalized with polyethylenediaminetetraacetic acid using ethylenediamine	Nanofiber	Effectively removal of Cd(II) and Cr(VI)	[79]
9	Polyacrylonitrile functionalized with hexamethylenediamine	Nanofiber	Absorbent of heavy metal Cu (II), Pb (II), and Ni (II) ions	[80]
10	Polyacrylonitrile functionalized with polyethylenediaminetetraacetic acid using ethylene diamine	Nanofiber	Maximum adsorption of methyl orange (MO) and reactive red (RR)	[81]
12	Polyacylonitile loaded with zinc oxide	Nanofiber	Maximum adsorption of Pb(II) and Cd(II)	[82]
13	Polyacrylonitrile functionalized with polyvinylidene fluoride	Nanofiber	Adsorbent for Direct Red 23 (DR23) dye	[83]
14	Polyacrylonitrile functionalized with polyamidoamine	Nanofiber	Dyes (DR80, DR23) removal of wastewater	[84]
15	Polyacrylonitrile (PAN)/CuS	Nanofiber	Degradation of methylene blue dye	[85]
16	Polyacrylonitrile with polyvinylidene fluoride	Nanofiber	Maximum dyes adsorption	[86]
17	Polyacrylonitrile functionalized with diethylenediamine	Nanofiber	Maximum adsorption of anionic metal ions Cr(VI)	[87]

(continued)

Table 1 (continued)

S. no.	Materials	Nature	Applications	References
18	Polyethyleneimine functionalized with magnetic iron oxide and polyacrylonitrile	Nanofiber	High removal efficiencies towards Cr(VI)	[88]
19	Polyacrylonitrile functionalized with tannins	Nanofiber	Absorbed chromium ions	[89]
20	Polyacrylonitrile functionalized with amine compound dimethylamine, diethylenetriamine, and triethylenetetramine	Nanofiber	Maximum Absorption of DR80 and DR23 DR—Direct red	[90]
21	PAN-CNT/TiO$_2$-NH$_2$	Nanofiber	Remove chromium	[91]

promising adsorbent for the removal of DR80 from wastewater [95]. Various PVA-based NFs have been summarized in Table 2. Yan et al. [104] reported electrospun unbound PVA/polyacrylic acid (PAA) layers, in cooperation with PDA as an organic dye adsorbent (Fig. 1). The adsorption properties of the film were radically improved compared with MB. The results showed that, due to the permeation structure of the PVA/PAA@PDA layer and the binding properties obtained by the PDA layer, within 30 min, the PVA/PAA@PDA layer is able to absorb more than 93% of the methylene blue in the matrix. It was concluded that PVA/PAA@PDA membrane is found to be a very capable adsorbent for dye wastewater treatment.

3.3 Polyamide

Recently, increasing attention has been given to hydrophilic membranes such as polyamide 6 (PA6) in applications of microfiltration and reverse osmosis. Yang et al. [112] reported the atrazine exclusion with nylon6/polypyrrole (PPY) core–shell NFs mat. They have also studied the possible mechanism and characteristics of the prepared NFs. Yalcinkaya et al. [113] prepared a novel PA6 NF hybrid membrane for liquid filtration. They have increased the mechanical strength of the prepared NF layers by including various supporting layers with different adhesion methods. ES is a technique to produce nanofibrous nonwovens. Such nonwovens can be applied to various applications such as water filtration. Vrieze et al. [114] prepared PI nanofibrous membranes and further evaluated them for their pore size, which is an important aspect in water filtration, and for their elimination of microorganisms. To boost the elimination efficiency to values above the state of the art, innovative functionalization of the NFs is studied. Various authors prepared PA, PAA, and nylon-based NFs for various wastewater treatments, and the work is summarized in Tables 3, 4 and 5.

Table 2 Polyvinyl alcohol-based nanofibers for wastewater treatment

S. no.	Materials	Nature	Applications	References
1	Polyvinyl alcohol functionalized with SiO_2	Nanofiber	Highly effective at absorbing Cu(II) ions	[92]
2	Polyethyleneimine functionalized with polyvinyl alcohol	Nanofiber	Immobilization of AuNPs	[93]
3	Chitosan/polyvinyl alcohol polyacrylonitrile	Nanofiber	Maximum removal of COD, TDS, Turbidity	[94]
4	Polyvinyl alcohol/chitosan/silica	Nanofiber	Excellent adsorption of Direct Red 80 (DR80)	[95]
5	Poly(vinyl alcohol)/poly(acrylic acid)/ titanium(IV) oxide	Nanofiber	High efficiency of dye removal	[96]
6	Chitosan/polyvinyl alcohol	Nanofiber	Methyl orange, Congo red, and chromium (VI) are adsorbed	[97]
7	Poly(vinyl alcohol) and triethylenetetramine functionalized with glutaraldehyde	Nanofiber	High adsorption of anionic dye	[98]
8	Polyvinyl alcohol (PVA)/titanium oxide (TiO_2) functionalized with f 3-mercaptopropyltrimethoxysilane	Nanofiber	Absorption of uranium and thorium	[99]
9	Polyvinyl alcohol (PVA)/(SiO_2) functionalized with cyclodextrin	Nanofiber	Maximum adsorption of indigo carmine dye	[100]
10	Polyvinyl alcohol/chitosan NH_2-functionalized with ZnO	Nanofiber	Removal of Cd(II) and Ni(II)	[101]
11	Polyvinyl alcohol and hydroxyapatite cross-linked with glutaraldehyde	Nanofiber	Absorption of contaminant Zn^{2+} ions	[102]
12	Polyvinyl alcohol/bovine serum albumin cross-linked with glutaraldehyde	Nanofiber	Immobilized horseradish peroxidase	[103]
13	Poly(vinyl alcohol)/ poly(acrylic acid) (PVA/PAA) functionlized with polydopamine	Nanofiber	Maximum adsorption of methyl blue dye	[104]
14	Polyvinyl alcohol and chitosan	Nanofiber	Adsorption of Pb(II) and Cd(II)	[105]
15	Polyvinyl alcohol functionalized with cobalt metal-organic framework	Nanofiber	Adsorption of Pb^{2+} ions	[106]

(continued)

Table 2 (continued)

S. no.	Materials	Nature	Applications	References
16	Mesoporous polyvinyl alcohol/silica functionalized with sulfhydryl	Nanofiber	Removal of bisphenol A and Cu(II)	[107]
17	Chitosan/polyvinyl alcohol	Nanofiber	High adsorption of Cr (VI), Fe (III), and Ni (II) ions	[108]
18	Poly(vinyl) alcohol/ poly(acrylic) acid	Nanofiber	Removal of Pb(II) from water	[109]
19	Polyethyleneimine/polyvinyl alcohol	Nanofiber	Absorbed dye methylene blue	[110]
20	Chitosan/poly(vinyl alcohol)	Nanofiber	Dyes removal of colored wastewater	[111]

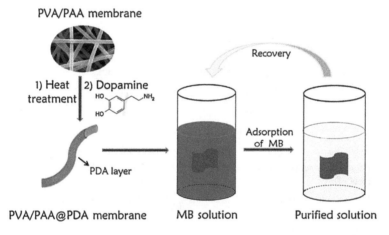

Fig. 1 Schematic illustration of the preparation of PVA/PAA@PDA membranes and their applications for dye adsorption. Reprinted with permission from Ref. [104]

3.4 Some Other Polymer-Based Nanofibers

Zhao et al. [147] reported that cyclodextrin-based fiber β highly water-insoluble material has been combined with several different materials such as N, N-DMF, and citric acid to help PAA polymers. The composite NFs were efficiently produced by an electrical measurement process. After in situ thermal cross-linking of the composite NFs, an independent water-soluble NF membrane based on cyclodextrin β was investigated to check its adsorption properties in relation to the cationic MB dye. The results showed that the cyclodextrin β-based filaments were stable, impermeable to

Table 3 Polyamide 6 (PA6)-based nanofibers for wastewater treatment

S. no.	Materials	Nature	Applications	References
1	Polyamide 6 (PA6)	Nanofiber	Filtration of wastewater pitch and tar oils, engine oil, and kitchen oil/water mixture	[113]
2	Polyamide (PA)	Nanofiber	Removal of microorganism	[114]
3	Polyamide 6 (PA6)	Nanofiber	Filtration performance (clean water permeability (CWP), bacterial removal)	[115]
4	Polyamide 6 (PA6)	Nanofiber	Acid dye removal (acid blue 41, Acid yellow 42 and C.I acid blue 78)	[116]
5	Polyamide 6/chitosan (PA6/CHIT)	Nanofiber	Immobilization of laccase (T. versicolor)	[117]
6	Polyamide/polyethylenimine (PA/PEI)	Nanofiber	Immobilization of laccase (T. versicolor)	[118]
7	PA6/chitosan functionalized with Fe_XO_Y	Nanofiber	Excellent performance of Cr(II) removal	[119]
8	Polyamide-6/chitosan	Nanofiber	Anionic dye removal solophenyl red 3BL and polar yellow GN	[120]
9	Polyamide 6 and polypyrrole (PA6/PPy)	Nanofiber	Adsorption of atrazine	[112]

Table 4 Polyacrylic acid-based nanofibers for wastewater treatment

S. no.	Materials	Nature	Applications	References
1	Poly(acrylic acid)/poly(vinyl alcohol)	Nanofiber	Adsorbent of Cu(II) from wastewater	[121]
2	Poly (acrylic acid)/SiO_2	Nanofiber	Absorption of indigo carmine dye	[122]
3	Poly(acrylic acid)/SiO_2	Nanofiber	Removal of malachite green	[123]
4	Poly(acrylic acid)/silica	Nanofiber	Removal of lanthanide ions	[124]
5	Poly(acrylic acid)–sodium alginate	Nanofiber	Removal of Cu^{2+} ions from wastewater	[125]
6	Poly(vinyl alcohol)/poly(acrylic acid) functionalized polydopamine	Nanofiber	Methyl blue (MB) adsorption percentage is 93%	[104]

liquid, and had increased elastic modulus after thermal reticulation. Compared with the highly recoverable cationic MB dye, the filaments exhibited superior adsorption performance. It has been shown that β-cyclodextrin-based electrospun silk has a broad application in the adsorption and stripping of colored wastewater treatment [147]. Bai et al. [148] studied the removal of harmful coatings from wastewater.

Table 5 Nylon acid-based nanofibers for wastewater treatment

S. no.	Materials	Nature	Applications	References
1	Nylon-6	Nanofiber	Removal of dye and salt of wastewater	[126]
2	Nylon-6	Membrane	Remove 95% of suspended from brewery wastewater	[127]
3	Nylon-6	Nanofiber	Successfully remove reactive dyes remazol red RR	[128]
4	Nylon-6	Nanofiber	Bisphenol A (BPA) in water sample	[129]
5	Nylon 6,6	Nanofiber	Adsorbent for Bisphenol A (BPA) removal	[130]
6	Nylon-6/poly(propylene imine) (N6-PPI)	Nanofiber	Remove anionic dye acid red 252(AR252)	[131]
7	Nylon-6 with polyaniline	Nanofiber	Maximum efficient dye adsorbent	[132]
8	Nylon 6,6/ZIF-8	Nanofiber	Filtered produced water treatment	[133]
9	Polyvinyl acetate (PVAc) coated with nylon 6/silica (N6/SiO_2)	Nanofiber	Separation of oil–water emulsion	[134]
10	Chitosan/nylon 6	Nanofiber	Removal of heavy metals and antibacterial activity	[135]
11	Nylon 6,6/graphene oxide	Nanofiber	Adsorption of Cr(VI)	[136]

In this study, PAA electrospun NF layers were treated with air plasma; the appearance of NFs was immobilized and modified using oxygen electrode collection. The large surface area and the generated barrier target played a crucial role in the rapid release of MB from the wastewater. It was shown that the MB cationic dye could be electrostatically adsorbed onto the plasma-treated NF membranes. The adsorbent was also shown to be highly reusable. The resulting adsorbent can also be used to remove various pollutants from wastewater [148]. San et al. [155] studied the nanofiber network against electrically induced cellulose acetate and found it to be highly convincing in immobilizing bacterial cells. In this study, three microscopic organisms (*Aeromonas eucrenofila*, *Clavibacter michiganensis*, and *Pseudomonas aruginosa*) immobilized on cellulose acetate NFs were utilized to get MB dye color change in aqueous media. Therefore, the above microorganisms were immobilized on permeability electrophoresis cellulose acetate NFs, respectively, for the treatment of liquid media containing MB. During the dyeing process, the process was considered harmless to aquatic organisms. Productive staining was achieved within 24 h, with a discharge rate of about 95%. The reuse of immobilized microorganisms in untreated water was clear after four cycles, reaching about 45% of the staining limit at the end of the fourth cycle, indicating that these immobilized microorganisms in untreated water can be reused for repeated staining of industrial wastewater [155]. Sethuraman et al. [156] produced high prospective content, Zn-CuO nanofibers in the form of an aqueous solution of copper acetate, NaOH, and polyethylene glycol (PEG) using a newly planned ES wrapping unit. The Zn-CuO nanofibers were generated by electrophoresis using a laboratory-scale structured NF generation unit. Therefore, discarded Zn-CuO NFs were tried as photo catalysts for photo degradation of harmful toxic dyes in wastewater. The results showed that the anomalously requested Zn-CuO NFs were used for photocatalytic studies. The Zn-CuO NFs were found to be stable for four consecutive cycles, but gradually lose their properties [156]. Kanjwal et al. [159] reported electrospun NiO, ZnO, and composite NiO-ZnO-NFs for photocatalytic degradation of dairy effluent (DE). Layers of NFs (ZnO and NiO-NF) and composite nanofibers (NiO-ZnO-NF) were prepared by liquid polyvinyl acetate electrophoresis using sol–gel method. The calcined nanomembranes were used as photocatalysts for the degradation of various dyes. The photocatalytic degradation of DE by the NiO-ZnO-NFs compounds improved; NiO-, ZnO-, and NiO-ZnO-NFs achieved 70, 75, and 80% separation after 3 h. The methylene blue staining by photolysis exposed the most extreme adsorbents [159] (Fig. 2).

The hydrolysis of the intrinsically microporous polymer (PIM-1) was carried out in the visible range of NaOH using different alkali concentrations, washing methods, and reaction times by Satilmis et al. [150]. Chemical structure analysis confirmed that using the complex system, PIM-1 could be hydrolyzed at different degrees of hydrolysis (65, 86, 94, and 99%). The increased solubility of hydrolyzed PIM-1 in the examples facilitated the production of ultra-fine filaments with hydrolyzed PIM-1 by ES. The PIM-1 hydrolyzed fibrous membranes were more sensitive to cationic colors (e.g., MB) and anionic colors (e.g., Congo red) and have a low adsorption limit. The highest adsorption limit or the color scheme when PIM-1 is completely hydrolyzed to MB and adsorbed without the use of diluents. A PIM-1 hydrolysis wire was used as a separation layer material to separate MB from the liquid aggregate; the PIM-1 electrolysis wire was able to absorb a certain amount of MB from the holder without

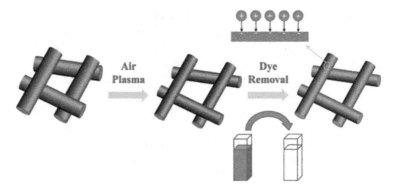

Fig. 2 Schematic representation of MB adsorption on plasma-treated electrospun PLLA membranes. Reused with permission from [148]

using extra power [150]. Liu et al. [157] prepared novel composite thin films based on PE liquid/PAA corrosion/carboxy graphite oxide (GO) tuned silver (Ag) NPs (PVA/PAA/GO-COOH@AgNPs) by heat treatment and electro filtration strategy. The effective design of the process is reported in this paper. The completed composite film can be used as a very effective substrate for organic dye degradation and can be easily separated from the dye solution and reused due to its good mechanical quality. There is an urgent need to check the exposure of the catalyst. Moreover, the labor-saving, easy preparation, and natural availability of directly deployed nanocomposites suggest their potential for important applications in catalysis: ready-to-use nanocomposites of PVA/PAA/GO-COOH help prevent the accumulation and aggregation of AgNPs. On the other hand, the intensive use of composite nanomaterials in wastewater treatment usually requires high recoveries and high resistance. Here, the feasibility of using purchased PVA/PAA/GO-COOH@AgNP nanocomposites for MB removal was investigated. The outcomes show that the reactant sum toward MB remains about 26.32 mg/g (about 99.8%, contrasted with 26.36 mg/g in the primary catalytic process) for PVA/PAA/GO-COOH@AgNPs nanocomposite films after eight back to back cycles at room temperature [157]. Various authors employed different nanofibers for wastewater treatment (Table 6) (Fig. 3).

4 Applications of Electrospinning Nanofibers

Membrane distillation (MD) is a hybrid thermal and membrane separation process that works on the principle of phase change and requires a temperature difference between the feed and cooling sides [161]. It is one of the developing technologies that allow the use of low-grade heat as a fuel. In most cases, the feed side is kept at a high temperature, typically 40–80 C, while the cooling side is kept close to room temperature [162]. When using hydrophobic membranes, only water vapor passes through the membrane and condenses into water on the cold side due to the different vapor partial pressures of the two liquids [163]. These properties can be templated by the membrane structure of NFs by electrospinning. Therefore, research on the use of NF membranes for MD has attracted a lot of attention in the last decade owing to their attractive and eye-catching properties. Studies on NF membranes have shown high water fluxes and high salt retention. On a related note, the membranes typically used for MD testing are PVDF or PTFE flat membrane layers, which run at reasonably low flow rates and have wettability issues [162]. NF membranes generated by electrophoresis have remarkable properties, such as well-defined surface area, high porosity of fine pores, coherent pore structure, and sufficient mechanical mass, which are not difficult to make up and manipulate [164]. The forms of release membranes used for wastewater treatment are limited in different aspects to microfiltration, ultrafiltration, nanofiltration, and feed reverse osmosis, in order of separation limits and increased pore size of specific layers. The various applications of electrospun fibers with multilevel micro-/nanostructures are shown in Fig. 4.

Table 6 Various nanofibers for wastewater treatment

S. no	Material	Nature	Application	References
1	Cellulose functionalized with cobalt tetraaminophthalocyanine	Nanofiber	Adsorption of reactive dyes	[137]
2	Zonal thiol functionalized with silica	Nanofiber	Maximum removal of heavy metals ions of wastewater	[138]
3	Laccase cellulose	Nanofiber	Removal of reactive dyes and simulated dye effluent	[139]
4	Polyindole functionalized with acetonitrile	Nanofiber	Adsorption of Cu(II) metal ions	[140]
5	Polydimethylsiloxane with poly(vinylidene fluoride-co-hexafluoropropene)	Nanofiber	Removal of four dye methylene blue, crystal violet (CV), acid red 18 (AR18), and acid yellow 36 (AY36)	[141]
6	Poly(ethylene terephthalate) functionalized with ethylene diamine	Nanofiber	97% removal of Pb(II) ions in solutions	[142]
7	Diethylenetriaminepentaacetic acid-functionalized with chitosan/ polyethylene oxide	Nanofiber	Adsorption of Cu^{2+}, Pb^{2+}, Ni^{2+} metals ions	[143]
8	Polyacrylonitrile in N,N-dimethylformamide with ZnO	Nanofiber	Adsorption of naphthalene and anthracene dyes	[144]
9	Polyethyleneimine with polyvinyl alcohol	Nanofiber	Maximum absorption of methylene blue dye	[110]
10	MgAl-EDTA-LDH functionalized with polyacrylonitrile	Nanofiber	Adsorption capacity of Cu(II) metal ions	[145]
11	Gelatin/β-cyclodextrin (β-CD)	Nanofiber	Excellent removal of dye methylene blue	[146]
12	β-cyclodextrin functionalized poly (acrylic acid)	Nanofiber	Removal of cationic dye methylene blue	[147]
13	Poly(L-lactic acid) with air plasma	Nanofiber	Adsorption of cationic dye methylene blue	[148]

(continued)

Table 6 (continued)

S. no	Material	Nature	Application	References
14	ZnO/(0.5wt%) CuO	Nanofiber	Photocatalytic MB dye degradation	[149]
15	Hydrolyzed polymer of intrinsic microporosity (PIM-1) functionalized sodium hydroxide	Nanofiber	High adsorption of cationic methylene blue	[150]
16	TiO_2-NP/SiO_2 and TiO_2-NP/Au@ SiO_2	Nanofiber	Photocatalytic MB dye degradation	[151]
17	Poly(vinylidene fluoride), polydopamine, and polypyrrole	Nanofiber	Excellent adsorption of cation (MB) and anion (CR) dyes simultaneously	[152]
18	Electrospun graphene oxide functionalized on polyacrylonitrile fibers	Nanofiber	Filtration (protein rejection)	[153]
19	Polyacylonitile-based Ag/TiO_2 nanofiber	Nanofiber	Excellent photocatalytic degradation of methyl organ, rhodamine B, and methylene blue	[154]
20	Cellulose acetate nanofibers web	Nanofiber	Methylene blue dyes removal successfully	[155]
21	Zn doped copper oxide	Nanofiber	Maximum methylene blue degradation	[156]
22	PVA/PAA/GO-COOH@AgNPs	Nanocomposite membranes	Photocatalytic degradation of methylene blue dyes	[157]
23	TiO_2/CuO composite nanofibers	Nanofiber	Photocatalytic dye wastewater treatment	[158]
24	Zinc oxide nanofiber, nickel oxide nanofiber, composite (ZnO-NiO NFs)	Nanofiber	Maximum degradation of dairy effluent and methylene blue dye	[159]
25	Polysulfone functionalized with organo-montmorillonite (PSF/O-MMT)	Nanofiber	Adsorption of methylene blue dye in wastewater	[160]

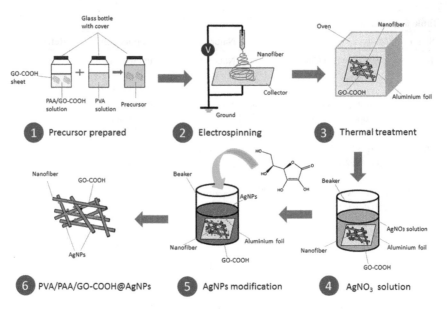

Fig. 3 Schematic illustration of the preparation of PVA/PAA/GO-COOH@AgNPs nanocomposites by electrospinning and thermal treatment [157]

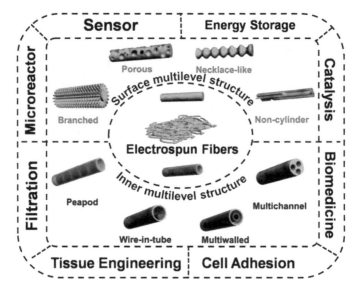

Fig. 4 Illustration of electrospun fibers with multilevel micro-/nanostructures and their applications. These materials are classified into two categories. One is surface multilevel structures, including branched, porous, necklace-like, and non-cylinder structures. The other is inner multilevel structures, such as peapod, wire in tube, multiwalled, and multichannel structures. Reused with permission from [14]

Conventional membranes are limited by a low degree of transition and are susceptible to contamination. In addition, a regular narrow pattern at the preparation stage
can improve the separation efficiency [165]. Finally, electrospun membranes are
efficient adsorbents for heavy metal ions in wastewater due to their unique properties. Solid metal particles are one of the most important pollutants in industrial wastewater. Creative activities such as the production of materials, ignition
of non-renewable energy sources, petroleum refining, production of plastics and
electrical appliances, electroplating, etc. are the main sources of heavy metal ions
[166, 167]. PEO/chitosan and chitosan/polyvinyl liquid nanofiber membranes have
been used to adsorb heavy metals from water. It has been used for the removal of
ions (copper, lead, cadmium, and nickel), especially for nickel, with high discharge
limits in reuse cycles up to five layers [108, 168]. Electrospun coatings of PVC
and chitosan/ZnO/aminopropyltriethoxysilane have been used effectively to remove
cadmium and nickel particles from wastewater [169, 170]. In addition, electrospun
coatings produced with PE bleach and specialty rubber Karaya have been successfully
used to remove metal NPs (Ag, Au, Pt, Fe_3O_4, and CuO) from aqueous solutions
[171–173]. Uranium was discharged from the liquid pool using PE hip/tetraethyl
orthosilicate/aminopropyltriethoxysilane nanofiber membranes, which were found
to have a high recovery limit and showed an uneven decrease in volume after five
treatment cycles [174]. Natural dyes, as one of the toxic pollutants in wastewater,
have received increasing attention in recent years due to rapid industrial development. Most of the natural dyes have deep and complex structures, high atomic weight,
high photothermal stability, and the use of oxidants [175]. Organic dyes are one of
the many new synthetic materials that, depending on the material, can be used in
cosmetics, pulp and paper, medicine, food processing, and horticulture. Due to the
wide range of applications, natural pigments are an important component of water
pollution because more than 10–15% of the pigments are lost as dyes and pigments
during the application process and do not react in wastewater [168, 176]. Therefore,
under characteristic conditions, natural colors are not easily degraded and pose a real
threat to the biological state and well-being of humans. To solve these problems,
various techniques such as biological processes, catalysis, flocculation, ozonation,
electrochemistry, and adsorption are generally used to treat paint pollution [177–
182]. Regarding the structure of the retainer, a high surface area is essential. In
recent years, the use of coatings as substrates for paint absorbents has been further
investigated. ES NF coatings can be used as brilliant carriers for adsorbent materials, with high porosity, large surface area, high porosity, and fantastic formability,
compared to films arranged by basic heat-initiated phase change techniques.

5   Conclusions and Outlooks

Water pollution is one of the biggest environmental problems of the twenty-first
century. Among the advances in the field of wastewater treatment, nanofibers are
attracting attention due to their low energy consumption, ease of treatment, high

permeability, and stability. Innovations in the field of electrical filtration, especially in recent decades, are considered one of the simplest and most versatile ways to produce nanofibers due to their large specific surface area, high pore cross-linking, ease of preparation, generally uniform pore size, and ease of incorporation into functional nanomaterials. It is connected to one of the technologies. Electrospun structures are used to spin high-voltage nanofibers, and the two basic arrangements of electrospinning are vertical and planar. The morphology of electrospun nanofibers is influenced by three basic parameters: solution, process, and environmental parameters. There are many strategies to characterize the morphology of electrospun nanofibers and other frameworks, the conventional ones are SEM as well as AFM, TEM, transmittance, contact angle, and rheology. In fact, significant progress has been made in industrial production to improve electrofiltration measurements of basic polymers and to develop novel devices, leading to the development of new electrospun nanofibers for the treatment of water contaminants. This chapter reports on recent studies using various composite fibers for the extraction of petroleum, heavy metals, and water-based coatings. These electrospun nanofibers exhibit excellent performance in treating oil, heavy metal particles, and dyes in water, and show potential application in wastewater treatment. Numerous studies have shown that electrospun functional nanofibers have become an important component of wastewater treatment. Polymeric materials and electrospinning processes must be continuously improved to enhance the presentation of electrospinning. The properties of polymers have a particularly strong influence on surface roughness, and their hydrophilicity and proximity must be concentrated from top to bottom for the processing of specific layers. In addition, the cost and availability of the polymer must be taken into account. The electrospinning process has so far been difficult to control and cannot be hindered by external variables that limit the scope of electrospinning production. Overall, the electrospinning of nanofibers offers an incredible strategy to successfully improve the structure and design of wastewater treatment materials for further development. Although research activities on electrospinning of nanofibers for wastewater treatment have progressed in recent years, it is still important to discuss the aforementioned research content and directional challenges to advance the rapid improvement of wastewater treatment.

References

1. Pal S, Nasim T, Gosh S, Panda AB (2010) Microwave assisted synthesis of polyacrylamide grafted dextrin (Dxt-g-PAM): Development and application of a novel polymeric flocculant. Int J Biol Macromol 47:623–631
2. Grutzmacher G, Kumar PJ, Rustler M, Hannappel S, Sauer U (2013) Geogenic groundwater contamination—definition occurrence and relevance for drinking water production Zbl.geol.palaont.tl. 11:69–75
3. Srinivasan V, Lambin EF, Gorelick SM, Thompson BH, Rozelle S (2012) The nature and causes of global water crisis: Syndromes from a meta-analysis of coupled human-water studies, water Resour. Res 48:10516

4. Willy JM, Sherwood LM, Woolverton CJ (2011) Prescott's microbiology, 8th edn. McGraw-Hill, New York
5. Spellman FR (2014) Handbook of water and wastewater treatment plant operation, 3rd edn. CRC Press-Taylor Francis
6. Eriksson E, Auffarth K, Henze M, Ledin A (2002) Characteristics of grey wastewater. Urban Water 4(1):85–104
7. Escher BI, Baumgartner R, Koller M, Treyer K, Lienert J, McArdell CS (2011) Environmental toxicology and risk assessment of pharmaceuticals from hospital wastewater. Water Res
8. Muttamara S (1996) Wastewater characteristics. Resour Conserve Recycle 16(1–4):145–159
9. Nandi BK, Goswami A, Purkait MK (2009) Removal of cationic dyes from aqueous solutions by Kaolin: Kinet Equilib Stud 42:583–590
10. Kant R (2012) Textile Dyeing Ind Environ Hazard 4:22–26
11. Liu T, Li Y, Du Q, Sun J, Jiao Y, Yang G, Wang Z, Xia Y, Zhang W, Wang K, Zhu H, Wu D (2012) Adsorption of methylene blue from aqueous solution by graphene. Colloids Surf, B 90:197–203
12. Hashem FS, Amin, MS (2016) Adsorption of methylene blue by activated carbon derived from various fruit peels. Desalin Water Treat 1–12
13. Li X, Zhang C, Zhao R, Lu X, Xu X, Jia X, Wang C, Li L (2013) Efficient adsorption of gold ions from aqueous systems with thioamide-group chelating nanofiber membranes. Chem Eng J 229:420–428
14. Wu J, Wang N, Zhao Y, Jiang L (2013) Electrospinning of multilevel structured functional micro-/nanofibers and their applications. J Mater Chem A 1:7290–7305
15. Wang X, Ge J, Si Y, Ding B (2014) Adsorbents based on electrospun nanofibers. Nanostruct Sci Technol 473–495
16. Pereao OK, Aluko CB, Ndayambaje G, Fatoba O, Petrik LF (2017) Electrospinning: Polymer nanofibre adsorbent applications for metal ion removal. J Polym Environ 25:1175–1189
17. Qureshi UA, Khatri Z, Ahmed F, Khatri M, Kim IS (2017) Electrospun zein nanofiber as green and recyclable adsorbent for the removal of reactive black 5 from aqueous phase. ACS Sustain Chem Eng 5:4340–4351
18. Min LL, Zhong LB, Zheng YM, Liu Q, Yuan ZH, Yang LM (2016) Functionalized chitosan electrospun nanofiber for effective removal of trace arsenate from water. Sci Rep 6:32480
19. Inagaki M, Yang Y, Kang F (2012) Carbon nanofi bers prepared via electrospinning. Adv Mater 24:2547–2566
20. Tijing LD, Woo YC, Yao M, Ren J, Shon HK (2017) Electrospinning for membrane fabrication: strategies and applications. Compr Membr Sci Eng II(1):418–444
21. Khan WS (2010) Fabrication and characterization of polyvinylpyrrolidone and polyacrylonitrile electrospun nanocomposite fibers, Dec 2010
22. Deng S, Bai R, Chen JP (2003) Aminated polyacrylonitrile fibers for lead and copper removal. Langmuir 19:5058–5064
23. Huang ZM, Zhang YZ, Kotaki M, Ramakrishana S (2003) A review on polymer nanofibers by electrospinning and their applications in nanocomposites. Compos Sci Technol 63:2223–2253
24. Ahmed FE, Lalia BS, Hashaikeh R (2015) A review on electrospinning for membrane fabrication: Challenges and applications. Desalination 356:15–30
25. Venugopal J, Ramakrishana S (2005) Appl Polym Nanofibers Biomed Biotechnol 125:147–157
26. Zenely J (1914) The electrical discharge from liquid points, and a hydrostatic method of measuring the electric intensity at their surfaces 3:69–91
27. Bhardwaj N, Kundu SC (2010) Electrospinning: A fascinating fiber fabrication techniq. Biotechnol Adv 28:325–347
28. Teo WE, Ramakrishana S (2006) A review on electrospinning design and nanofibre assemblies. Nanotechnology 17:89–106
29. Han J, Xiong L, Jiang X, Yuan X, Zhao Y, Yang D (2019) Bio-functional electrospun nanomaterials: From topology design to biological applications. Prog Polym Sci 91:1–28

30. Ghorani B, Tucker N (2015) Fundamentals of electrospinning as a novel delivery vehicle for bioactive compounds in food nanotechnology. Food Hydrocoll 51:227–240
31. Niu B, Shao P, Luo Y, Sun P (2020) Recent advances of electrosprayed particles as encapsulation systems of bioactives for food application. Food Hydrocoll 99:105376
32. Patil JV, Mali SS, Kamble AS, Hong CK, Kim JH, Patil PS (2017) Electrospinning: A versatile technique for making of 1D growth of nanostructured nanofibers and its applications: an experimental approach. Appl Surf Sci 423:641–674
33. Simon EM (1988) NIH phase I final report: fibrous substrate for cell culture (R3RR03544A)
34. Pillay V, Dott C, Choonara YE, Tyagi C, Tomar L, Kumar P, Toit LC, Ndesendo VMK (2013) A review of the effect of processing variables on the fabrication of electrospun nanofibers for drug delivery applications. J Nanomater
35. Saleem H, Trabzon L, Kilic A, Zaidai SJ (2020) Recent advances in nanofibrous membranes: Production and applications in water treatment and desalination. Desalination 78:114178
36. Dalton PD, Klinkhammer K, Salber J, Klee D, Moller M (2006) Direct in vitro electrospinning with polymer melts. Biomacromol 7:686–690
37. Daristotle JL, Behrens AM, Sandler AD, Kofinas P (2016) A review of the fndamental principles and applications of solution blow spinning. ACS Appl Mater Interfaces 8:34951–34963
38. Taylor GI (1969) Electrically driven jets. Proc R Soc Lond A313453–475
39. Haider A, Haider S, Kang IK (2018) A comprehensive review summarizing the effect of electrospinning parameters and potential applications of nanofibers in biomedical and biotechnology. Arab J Chem 11:1165–1188
40. Zagho MM, Elzatahry A (2016) Recent trends in electrospinning of polymer nanofibers and their applications as templates for metal oxide nanofibers preparation. Electrospinning Mater Tech Biomed Appl 1
41. Bode-Aluko CA, Laatikainen K, Pereao O, Nechaev A, Kochnev I, Rossouw A, Dobretsov S, Branger C, Sarbu A, Petrik L (2019) Fabrication and characterisation of novel nanofiltration polymeric membrane. Mater Today Commun 20:100580
42. Reneker DH, Yarin AL, Zussman E, Xu H (2007) Electrospinning of nanofibers from polymer solutions and melts. Adv Appl Mech 41:43–195
43. Subramanian S, Seeram R (2013) New directions in nanofiltration applications-are nanofibers the right materials as membranes in desalination. Desalination 308:198–208
44. Reneker DH, Yarin AL, Fong H, Koombhongse S (2000) Bending instability of electrically charged liquid jets of polymer solutions in electrospinning. J Appl Phys 87:4531–4546
45. Tijing LD, Woo YC, Shim WG, He T, Choi JS, Kim SH, Shon HK (2016) Superhydrophobic nanofiber membrane containing carbon nanotubes for high-performance direct contact membrane distillation. J Membr Sci 502:1–42
46. Hou L, Wang N, Wu J, Cui Z, Jiang L, Zhao Y (2018) Bioinspired superwettability electrospun micro/nanofibers and their applications. Adv Funct Mater 1–22
47. Mazoochi T, Jabbari V (2011) Chitosan nanofibrous scaffold fabricated via electrospinning: the effect of processing parameters on the nanofiber morphology. Int J Polym Anal Charact 16:277–289
48. Liu S, White KL, Reneker DH (2019) Electrospinning polymer nanofibers with controlled diameters. IEEE Trans Ind Appl 55:5239–5243
49. Haghi AK, Akbari M (2007) Trends in electrospinning of natural nanofibers. Phys Stat Sol 204:1830–1834
50. Yordem OS, Papila M, Menceloglu YZ (2007) Effects of electrospinning parameters on polyacrylonitrile nanofiber diameter: An investigation by response surface methodology. Mater Des 1–11
51. Zhang C, Yuan X, Wu L, Han Y, Sheng J (2005) Study on morphology of electrospun poly(vinyl alcohol) mats. Eur Polym J 41:423–432
52. Fong H, Reneker DH (1999) Elastomeric nanofibers of styrene-butadiene-styrene triblock copolymer. J Polym Sci Part B: Polym Phys 37:3488–3493
53. Liu Y, He JH, Yu JY, Zeng HM (2008) Controlling numbers and sizes of beads in electrospun nanofibers. Polym Int 57:632–636

54. Beachley V, Wen X (2009) Effect of electrospinning parameters on the nanofiber diameter and length. Mater Sci Eng, C 29:663–668
55. Venugopal J, Ramakrishna S (2005) Applications of polymer nanofibers in biomedicine and biotechnology. Appl Biochem Biotechnol 125:147–157
56. Cui J, Li F, Wang Y, Zhang Q, Ma W, Huang C (2020) Electrospun nanofiber membranes for wastewater treatment applications. Sep Purif Technol 250:117116
57. Doshi J, Reneker DH (1995) Electrospinning process and applications of electrospun fibers. J Electrostat 35:151–160
58. Eda G, Shivkumar S (2007) Bead-to-Fiber transition in electrospun polystyrene. Inc J Appl Polym Sci 106:475–487
59. Khalf A, Madihally SV (2017) Recent advances in multiaxial electrospinning for drug delivery. Eur J Pharm Biopharm 112:1–17
60. Stachewicz U, Dijksman JF, Soudani C, Tunnicliffe LB, Busfield JJC, Barber AH (2017) Surface free energy analysis of electrospun fibers based on Rayleigh-Plateau/Weber instabilities. Eur Polym J 91:368–375
61. Sukigara S, Gandhi M, Ayutsede J, Micklus M, Ko F (2003) Regeneration of Bombyx mori silk by electrospinning-part 1: Processing parameters and geometric properties. Polymer 44:5721–5727
62. Lasprilla-Botero J, Alvarez-Lainez M, Lagaron JM (2018) The influence of electrospinning parameters and solvent selection on the morphology and diameter of polyimide nanofibers. Mater Today Commun 14:1–9
63. Qasim SB, Zafar MS, Najeeb S, Khurshid Z, Shah AH, Hussain S, Rehman IU (2018) Rehman, Electrospinning of chitosan-based solutions for tissue engineering and regenerative medicine. Int J Mol Sci 19:407
64. Lim, CT, Tan EPS, Ng SY (2008) Effects of crystalline morphology on the tensile properties of electrospun polymer nanofibers. Appl Phys Lett 92:141908
65. Taylor G (1964) Disintegration of water drops in an electric field. Math Phys Sci 280:383–397
66. Angammana CJ, Jayaram SH (2016) Fundamentals of electrospinning and processing technologies. Part Sci Technol: Int J 34:72–82
67. Zuo W, Zhu M, Yang W, Yu H, Chen Y, Zhang Y (2005) Experimental study on relationship between jet instability and formation of beaded fibers during electrospinning. Poly Eng Sci 705–709
68. Koski A, Yim K (2004) Shivkumar, Effect of molecular weight on fibrous PVA produced by electrospinning. Mater Lett 58:493–497
69. Chen H, Wang N, Di J, Zhao Y, Song Y, Jiang L (2010) Nanowire-in-microtube structured core/shell fibers via multifluidic coaxial electrospinning. Am Chem Soc 26:11291–11296
70. Dror Y, Salalha W, Avrahami R, Zussman E, Yarin AL, Dersch R, Greiner A, Wendorff JH (2007) One-Step production of polymeric microtubes by co-electrospinning. Small 3:1064–1073
71. Liu Y, Zhang X, Xia Y, Yang H (2010) Magnetic-field-assisted electrospinning of aligned straight and wavy polymeric nanofibers. Adv Mater 22:2454–2457
72. Zhang H, Nie H, Yu D, Wu C, Zhang Y, White CJB, Zhu L (2010) Surface modification of electrospun polyacrylonitrile nanofiber towards developing an affinity membrane for bromelain adsorption. Desalination 256:141–147
73. Shojaei M, Sadjadi S, Rajabi-Hamane M, Ahmadi SJ (2014) Synthesis of TiO2/polyacrylonitrile nanofibers composite and its application to lead ions removal from waste waters. Nanotechnology 1–10
74. Im JS, Kim MI, Lee Y-S (2008) Preparation of PAN-based electrospun nanofiber webs containing TiO2 for photocatalytic degradation. Mater Lett 62:36–3655
75. Kampalanonwat P, Supaphol P (2014) The study of competitive adsorption of heavy metal ions from aqueous solution by aminated polyacrylonitrile nanofiber mats. Energy Procedia 56:142–151
76. Mahmoodi NM, Mokhtari-Shourijeh Z (2017) Preparation of polyacrylonitrile–Titania electrospun nanofiber and its photocatalytic dye degradation ability. Dep Environ Res Inst Color Sci Technol 10:23–30

77. Kumar PS, Venkatesh K, Gui EL, Sundaramurthy J, Singh G, Arthanareeswaran G (2018) Electrospun carbon nanofibers/TiO2-PAN hybrid membranes for effective removal of metal ions and cationic dye. Environ Nanotechnol Monit Manag 1–27
78. Mikal NR, Sadjadi S, Rajabi-Hamane M, Ahmadi SJ, Iravani E (2015) Decoration of electrospun polyacrylonitrile nanofibers with ZnO nanoparticles and their application for removal of Pb ions from waste water. J Iranian Chem Soc 1–9
79. Chauque EFC, Dlamini LN, Adelodun AA, Greyling CJ, Ngila JC (2016) Modification of electrospun polyacrylonitrile nanofibers with EDTA for the removal of Cd and Cr ions from water effluents. Appl Surf Sci 369:19–28
80. Mohamed AS, Fikry NM, Shalaby TI, Aloufy AI, Mohamed MM (2014) The adsorption behaviour of Polycrylonitrilenanofibre by hexamethylenediamine for removing Pb (II), Cu (II) and Ni (II) metal ions from water. Int J Chem Appl Biol Sci 1:52–56
81. Chauque EFC, Ngila JC, Adelodun AA, Greyling CJ, Dlamini LN (2016) Electrospun polyacrylonitrile nanofibers functionalized with EDTA for adsorption of ionic dyes. Phys Chem Earth Parts A/B/C 1–38
82. Haddad MY, Alhardi HF (2018) Enhancement of heavy metal ion adsorption using electrospun polyacrylonitrile nanofibers loaded with ZnO nanoparticles, pp 1–11
83. Mokhtari-Shourijeh Z, Langari S, Montazerghaem L, Mahmoodi NM (2020) Synthesis of porous aminated PAN/PVDF composite nanofibers by electrospinning: Characterization and Direct Red 23 removal. Inc J Appl Polym Sci 8:103876
84. Almasian A, Olya ME, Mahmoodi NM (2015) Synthesis of polyacrylonitrile/polyamidoamine composite nanofibers using electrospinning technique and their dye removal capacity, pp 1–10
85. Nie G, Li Z, Lu X, Lei J, Zhang C, Wang C (2013) Fabrication of polyacrylonitrile/CuS composite nanofibers and their recycled application in catalysis for dye degradation. Appl Surf Sci 84:595–600
86. Mokhtari-Shourijeh Z, Montazerghaem L, Olya ME (2018) Preparation of porous nanofibers from electrospun Polyacrylonitrile/Polyvinylidene fluoride composite nanofibers by inexpensive salt using for dye adsorption. J Polym Environ
87. Zhao R, Li X, Sun B, Ji H, Wang C (2017) Diethylenetriamine-assisted synthesis of amino-rich hydrothermal carbon-coated electrospun polyacrylonitrile fiber adsorbents for the removal of Cr(VI) and 2,4-dichlorophenoxyacetic acid. J Colloid Interface Sci 487:297–309
88. Zhao R, Li X, Li Y, Li Y, Sun B, Zhang N, Chao S, Wang C (2017) Functionalized magnetic iron oxide/polyacrylonitrile composite electrospun fibers as effective chromium (VI) adsorbents for water purification. J Colloid Interface Sci 505:1018–1030
89. Zhang J, Xue C-H, Ma H-R, Ding Y-R, Jia S-T (2020) Fabrication of PAN Electrospun nanofibers modified by tannin for effective removal of trace Cr(III) in organic complex from wastewater. Polymers 12:210
90. Almasian A, Fard GC, Gashti MP, Mirjalili M, Shourijeh ZM (2015) Surface modification of electrospun PAN nanofibers by amine compounds for adsorption of anionic dyes. Desalin Water Treat 1–16
91. Mohamed A, Nasser WS, Osman TA, Topark MS, Muhammed M, Uheida A (2017) Removal of chromium (VI) from aqueous solutions using surface modified composite nanofibers. J Colloid Interface Sci 505:682–691
92. Wu S, Li F, Wu Y, Xu R, Li G (2010) Preparation of novel poly(vinyl alcohol)/SiO2 composite nanofiber membranes with mesostructure and their application for removal of Cu2+ from waste water. Chem Commun 46:1694–1696
93. Fang X, Ma H, Xiao S, Shen M, Guo R, Cao X, Shi X (2011) Facile immobilization of gold nanoparticles into electrospun polyethyleneimine/polyvinyl alcohol nanofibers for catalytic applications. J Mater Chem 21:4493–4501
94. Hejazi F, Mousavi SM (2014) Electrospun nanofibrous composite membranes of chitosan/polyvinyl alcohol-polyacrylonitrile: preparation, characterization, and performance. Desalin Water Treat 1–8
95. Mahmoodi NM, Mokhtari-Shourijeh Z, Abdi J (2018) Preparation of mesoporous Polyvinyl Alcohol/ Chitosan/Silica composite nanofiber and dye removal from wastewater. Environ Prog Sust Energy 1–10

96. Jeon S, Yun J, Lee Y-S, Kim H-I (2012) Preparation of poly(vinyl alcohol)/poly(acrylic acid)/TiO2/carbon nanotube composite nanofibers and their photobleaching properties. J Ind Eng Chem 18:487–491
97. Habiba U, Siddique TA, Joo TC, Salleh A, Ang BC, Afifi AM (2016) Synthesis of chitosan/polyvinyl alcohol/zeolite composite for removal of methyl orange, congo red and chromium (VI) by flocculation/adsorption. Carbohydr Polym 1–28
98. Mahmoodi NM, Mokhtari-Shourijeh Z (2015) Modified poly(vinyl alcohol)-triethylenetetramine nanofiber by glutaraldehyde: preparation and dye removal ability from wastewater. Desalin Water Treat 1–8
99. Abbasizadeh S, Keshtkar AR, Mousavian MA (2013) Preparation of a novel electrospun polyvinyl alcohol/titanium oxide nanofiber adsorbent modified with mercapto groups for uranium(VI) and thorium(IV) removal from aqueous solution. Chem Eng J 220:161–171
100. Teng M, Li F, Zhang B, Taha AA (2011) Electrospun cyclodextrin-functionalized mesoporous polyvinyl alcohol/SiO2 nanofiber membranes as a highly efficient adsorbent for indigo carmine dye 385:229–234
101. Bozorgi M, Abbasizadeh S, Samani F, Mousavi SE (2018) Performance of synthesized cast and electrospun PVA/chitosan/ZnO-NH2 nano-adsorbents in single and simultaneous adsorption of cadmium and nickel ions from wastewater
102. El-aziz AMA, El-Maghraby A, Taha NA (2016) Comparing between polyvinyl alcohol (PVA) nanofiber and polyvinyl alcohol (PVA) nanofiber/Hydroxyapatite (HA) for removal of Zn2+ ions from waste water. Arabian J Chem 1–18
103. Fazel R, Torabi S-F, Naseri-Nosar P, Ghasempur S, Ranaei-siadat S-O, Khajeh K (2016) Electrospun polyvinyl alcohol/bovine serum albumin biocomposite membranes for horseradish peroxidase immobilization. Enzyme Microbial Technol 93–94, 10
104. Yan J, Huang Y, Miao Y-E, Tjiu WW, Liu T (2015) Polydopamine-coated electrospun poly(vinyl alcohol)/poly(acrylic acid) membranes as efficient dye adsorbent with good recyclability. J Hazard Mater 283:730–739
105. Karim MR, Aijaz MO, Alharth NH, Alharbi HF, Al-Mubaddel FS, Awual MR (2019) Composite nanofibers membranes of poly(vinyl alcohol)/chitosan for selective lead(II) and cadmium(II) ions removal from wastewater. Ecotoxicol Environ Saf 169:479–486
106. Shooto ND, Wankasi D, Sikhwivhilu LM, Dikio ED (2016) Modified Electro-spun Polyvinyl Alcohol Nanofibers Used as Super Adsorbing Material for Lead Ions in Aqueous Solution. J Residuals Sci Technol 13(3):233–242
107. Wang D, Wang J (2016) Electrospinning polyvinyl alcohol/silica-based nanofiber as highly efficient adsorbent for simultaneous and sequential removal of Bisphenol A and Cu(II) from water. Chem Eng J 1–30.
108. Habiba U, Afifi AM, Salleh A, Ang BC (2017) Chitosan/(polyvinyl alcohol)/zeolite electrospun composite nanofibrous membrane for adsorption of Cr^{6+}, Fe^{3+} and Ni^{2+}. J Hazard Mater 322:182–194
109. Zhang S, Shi Q, Christodoulatos C, Korfiatis G, Meng X (2019) Adsorptive filtration of lead by electrospun PVA/PAA nanofiber membranes in a fixed-bed column. Chem Eng J 370:1262–1273
110. Fang Xu, Xiao S, Shen M, Guo R, Wang S, Shi X (2011) Fabrication and characterization of water-stable electrospun polyethyleneimine/polyvinyl alcohol nanofibers with super dye sorption capability. New J Chem 35:360–368
111. Hosseini SA, Vossoughi M, Mahmoodi NM, Sadrzadeh M (2019) Clay-based electrospun nanofibrous membranes for colored wastewater treatment. Appl Clay Sci 168:77–86
112. Yang BY, Cao Y, Qi FF, Li XQ, Xu Q (2015) Atrazine adsorption removal with nylon6/polypyrrole core-shell nanofibers mat: possible mechanism and characteristics. Nanoscale Res Lett vol 1–13
113. Yalcinkaya F, Yalcinkaya B, Hruza J (2019) Electrospun polyamide-6 nanofiber hybrid membranes for wastewater treatment. Fibers Polym 20(1):93–99
114. Vrieze SD, Daels N, Lambert K, Decostere B, Hens Z, Hulle SV, Clerck KD (2011) Filtration performance of electrospun polyamide nanofibres loaded with bactericides. Text Res J 82(1):37–44

115. Daels N, Harinck L, Goethals A, Clerck D, Vanhulle SWH (2016) Structure changes and water filtration properties of electrospun polyamide nanofibre membranes. Water Sci Technol 73(980):1920–1926
116. Wiener J, Ntaka S, Ngcobo PS, Knizek R (2012) Sorption process using polyamide nanofibers to remove dye from simulated wastewater. Nanocon
117. Maryskova M, Ardao I, Garcia-Gonzalez CA, Martinova L, Rotkova J, Seveu A (2016) Polyamide 6/chitosan nanofibers as support for the immobilization of Trametes versicolor laccase for the elimination of endocrine disrupting chemicals. Enzyme Microbial Technol 1–29
118. Maryskova M, Schaabova M, Tomankova H, Novotny V, Rysova M (2020) Wastewater treatment by novel polyamide/polyethylenimine nanofibers with immobilized laccase. Wastewater Treat Reuse 12:588
119. Li C-J, Zhang S-S, Wang J-N, Liu T-Y (2014) Preparation of polyamides 6 (PA6)/Chitosan@FeXOY composite nanofibers by electrospinning and pyrolysis and their Cr(VI)-removal performance. Catal Today 224:94–103
120. Ghani M, Gharehaghaji AA, Arami M, Takhtkuse N, Rezaei B (2014) Fabrication of electrospun polyamide-6/chitosan nanofibrous membrane toward anionic dyes removal. J Nanotech 1–12
121. Park J-A, Kang J-K, Lee S-C, Kim S-B (2017) Electrospun poly(acrylic acid)/poly(vinyl alcohol) nanofibrous adsorbents for Cu(II) removal from industrial plating wastewater. RSC Adv 7:18075
122. Xu R, Jia M, Li F, Wang H, Zhang B, Qiao J (2012) Preparation of mesoporous poly (acrylic acid)/SiO2 composite nanofiber membranes having adsorption capacity for indigo carmine dye. Appl Phys A 106:747–755
123. Xu R, Jia M, Zhang Y, Li F (2012) Sorption of malachite green on vinyl-modified mesoporous poly(acrylic acid)/SiO2 composite nanofiber membranes. Microporous Mesoporous Mater 149:111–118
124. Wang M, Li X, Hua W, Shen L, Yu X, Wang X (2016) Electrospun Poly(acrylic acid)/Silica Hydrogel Nanofibers scaffold for highly efficient adsorption of lanthanide ions and its photoluminescence performance. ACS Appl Mater Interfaces 8:23995–24007
125. Wang M, Li X, Zhang T, Deng L, Li P, Wang X, Hsiao BS (2018) Eco-friendly poly(acrylic acid)-sodium alginate nanofibrous hydrogel: A multifunctional platform for superior removal of Cu(II) and sustainable catalytic applications. Colloids Surf A 558:228–241
126. Basiri F, Ravandi SAH, Feiz M, Moheb A (2011) Recycling of Direct Dyes Wastewater by Nylon-6 Nanofibrous Membrane. Curr Nanosci 7:633–639
127. Islam MS, Sultana S, Rahaman MS (2016) Electrospun nylon 6 microfiltration membrane for treatment of brewery wastewater. In: AIP conference proceedings, vol 1754, p 060017
128. Park M, Rabbani MM, Shin HK, Park S-J (2016) Dyeing of electrospun nylon 6 nanofibers with reactive dyes using electron beam irradiation. J Ind Eng Chem 39:16–20
129. Ardekani R, Borhani S, Rezaei B (2018) Simple preparation and characterization of molecularly imprinted nylon 6 nanofibers for the extraction of bisphenol A from wastewater, vol 47112, pp 1–10
130. Jasni MJF, Arulkumar M, Sathishkumar P, Yusoff ARM, Buang NA, Gu FL (2017) Electrospun nylon 6,6 membrane as a reusable nano-adsorbent for bisphenol A removal: Adsorption performance and mechanism. J Colloid Interface Sci 508:591–602
131. Soltan FKM, Hajiani M, Haji A (2020) Nylon-6/poly(propylene imine) dendrimer hybrid nanofibers: an effective adsorbent for the removal of anionic dyes. J Textile Inst 1–11
132. Zarrini K, Rahimi AA, Alihosseini F, Fashandi H (2016) Highly efficient dye adsorbent based on polyaniline-coated nylon-6 nanofibers. J Clean Prod 1–35
133. Halim NSA, Wirzal MDH, Bilad MR, Nordin NAHM, Putra ZA, Yusoff ARM, Narkkun T, Faungnawakij K (2019) Electrospun nylon 6,6/ZIF-8 nanofiber membrane for produced water filtration. Adv Water Wastewater Monit Treat Technol 11:2111
134. Islam MS, McCutcheon JR, Rahaman MS (2017) A high flux polyvinyl acetate-coated electrospun nylon 6/SiO2 composite microfiltration membrane for the separation of oil-in-water emulsion with improved antifouling performance. J Membr Sci 537:297–309

135. Jabur AR, Abbas LK, Moosa SA (2016) Fabrication of electrospun chitosan/nylon 6 nanofibrous membrane toward metal ions removal and antibacterial effect. Adv Mater Sci Eng 1–11
136. Parlayici S, Avci A, Pehlivan E (2019) Electrospinning of polymeric nanofiber (nylon 6,6/graphene oxide) for removal of Cr (VI): Synthesis and adsorption studies. J Anal Sci Technol 10(13):1–13
137. Chen S-L, Huang X-J, Xu Z-K (2011) Functionalization of cellulose nanofiber mats with phthalocyanine for decoloration of reactive dye wastewater. Cellulose 18:1295–1303
138. Li S, Yue X, Jing Y, Bai S, Dai Z (2011) Fabrication of zonal thiol-functionalized silica nanofibers for removal of heavy metal ions from wastewater. Colloids Surf A 380:229–233
139. Sathishkumar P, Kamala-Kannan S, Cho M, Kim JS, Hadibarata T, Salim MR, Oh B-T (2014) Laccase immobilization on cellulose nanofiber: The catalytic efficiency and recyclic application for simulated dye effluent treatment. J Mol Catal B Enzym 100:111–120
140. Cai Z, Song X, Zhang Q, Zhai T (2017) Electrospun polyindole nanofibers as a nano-adsorbent for heavy metal ions adsorption for wastewater treatment. Fibers Polym 18(3):502–513
141. An AK, Guo J, Lee E-J, Jeong S, Zhao Y, Wang Z, Leiknes T (2017) PDMS/PVDF hybrid electrospun membrane with superhydrophobic property and drop impact dynamics for dyeing wastewater treatment using membrane distillation. J Membr Sci 525:57–67
142. Martin DM, Ahmed MM, Rodriguez M, Garcia MA, Faccini M (2017) Aminated polyethylene terephthalate (PET) nanofibers for the selective removal of Pb(II) from polluted water. Materials 10:1352
143. Surgutskaia NS, Martino AD, Zednik J, Ozaltin J, Lovecka L, Bergerova ED, Kimmer D, Svoboda J, Sedlarik V (2020) Efficient Cu2+, Pb2+ and Ni2+ ion removal from wastewater using electrospun DTPA-modified chitosan/polyethylene oxide nanofibers. Separat Purif Technol 247:116914
144. Singh P, Mondal K, Sharma A (2013) Reusable electrospun mesoporous ZnO nanofiber mats for photocatalytic degradation of polycyclic aromatic hydrocarbon dyes in wastewater. J Colloid Interface Sci 394:208–215
145. Chen H, Lin J, Zhang N, Chen L, Zhong S, Wang Y, Zhang W, Ling Q (2018) Preparation of MgAl-EDTA-LDH based electrospun nanofiber membrane and its adsorption properties of copper(II) from wastewater. J Hazard Mater 345:1–9
146. Chen Y, Ma Y, Lu W, Guo Y, Zhu Y, Lu H, Song Y (2018) Environmentally friendly Gelatin/β-Cyclodextrin composite fiber adsorbents for the efficient removal of dyes from wastewater. Molecules 23:2473
147. Zhao R, Wang Y, Li X, Sun B, Wang C (2015) Synthesis of β-Cyclodextrin-Based electrospun nanofiber membranes for highly efficient adsorption and separation of methylene blue. ACS Appl Mater Interfaces 26649–26657
148. Bai L, Jia L, Yan Z, Liu Z, Liu Y (2018) Plasma-etched electrospun nanofiber membrane as adsorbent for dye removal. Chem Eng Res Des 132:445–451
149. Naseri A, Samadi M, Mahmoodi NM, Pourjavadi A, Mehdipour H, Moshfegh AZ (2017) Tuning Composition of Electrospun ZnO/CuO Nanofibers: Towards Controllable and Efficient Solar Photocatalytic Degradation of Organic Pollutants. J Phys Chem C 121:3327–3338
150. Satilmis B, Budd PM, Uyar T (2017) Systematic hydrolysis of PIM-1 and electrospinning of hydrolyzed PIM-1 ultrafine fibers for an efficient removal of dye from water. React Funct Polym 121:65–75
151. Zheng F, Zhu Z (2018) Flexible, Freestanding, and functional SiO2 Nanofibrous Mat for DyeSensitized Solar Cell and Photocatalytic Dye Degradation. ACS Appl Nano Mater 1:1141–1149
152. Ma FF, Zhang D, Zhang N, Huang T, Wang Y (2018) Polydopamine-assisted deposition of polypyrrole on electrospun poly(vinylidene fluoride) nanofibers for bidirectional removal of cation and anion dyes. Chem Eng J 354:432–444
153. Lee J, Yoon J, Kim JH, Lee T, Byun H (2018) Electrospun PAN–GO composite nanofibers as water purification membranes. Inc J Appl Polym Sci 135:45858

154. Wang L, Ali J, Zhang C, Maiohot G, Pan G (2020) Simultaneously enhanced photocatalytic and antibacterial activities of TiO2/Ag composite nanofibers for wastewater purification. J Environ Chem Eng 8:102104

155. San ON, Celebioglu A, Tumtas Y, Uyar T, Tekinay T (2013) Reusable bacteria immobilized electrospun nanofibrous web for decolorization of methylene blue Dye in wastewater treatment. RSC Adv 1:1–8

156. Sethuraman RG, Venkatachalam T, Kirupha SD (2018) Fabrication and characterization of Zn doped CuO nanofiber using newly designed nanofiber generator for the photodegradation of methylene blue from textile effluent. Mater Sci Poland 36(3):520–529

157. Liu Y, Hou C, Jiao T, Song J, Zhang X, Xing R, Zhou J, Zhang L, Peng Q (2018) Self-assembled AgNPs containing nanocomposites constructed by electrospinning as efficient dye photocatalyst materials for wastewater treatment. Nanomaterials 8:35

158. Lee SS, Bai H, Liu Z, Sun DD, Novel-structured electrospun TiO2/CuO composite nanofibers for high efficient photocatalytic cogeneration of clean water and energy from dye wastewater. Water Res 1–46

159. Kanjwal MA, Chronakis IS, Barakat NAM (2015) Electrospun NiO, ZnO and composite NiO–ZnO nanofibers/photocatalytic degradation of dairy effluent. Ceram Int 41:12229–12236

160. Aquino RR, Tolentino MS, Crisogono BMZ, Salvacion SKV (2018) Adsorption of methylene blue (MB) Dye in wastewater by electrospun polysulfone (PSF)/organo-montmorillonite (O-MMT) nanostructured membrane. Mater Sci Forum 916:120–124

161. Tijing LD, Choi J-S, Lee S, Kim S-H, Shon HK (2014) Recent progress of membrane distillation using electrospun nanofibrous membrane. J Membr Sci 453:435–462

162. Tijing LD, Woo YC, Johir MAH, Choi J-S, Shon HK (2014) A novel dual-layer bicomponent electrospun nanofibrous membrane for desalination by direct contact membrane distillation. Chem Eng J 256:155–159

163. Yao M, Woo YC, Tijing LD, Cesarini C, Shon HK (2017) Improving nanofiber membrane characteristics and membrane distillation performance of heat-pressed membranes via annealing post-treatment. Appl Sci 7(78):1–11

164. Ma H, Hsiao BS, Chu B (2014) Functionalized electrospun nanofibrous microfiltration membranes for removal of bacteria and viruses. J Membr Sci 452:446–452

165. Yoon K, Kim K, Wang X, Fang D, Hsiao BS, Chu B (2006) High flux ultrafiltration membranes based on electrospun nanofibrous PAN scaffolds and chitosan coating. Polymer 47:2434–2441

166. Bhuyan MS, Bakar MA, Akhtar A, Hossain MB, Ali MM, Islam MS (2017) Heavy metal contamination in surface water and sediment of the Meghna River, Bangladesh. Environ Nanotechnol Monit Manag 8:273–279

167. Li X, Wang M, Wang C, Cheng C, Wang X (2014) Facile Immobilization of Ag Nanocluster on Nanofibrous membrane for oil/water separation. ACS Appl Mater Interfaces 6:15272–15282

168. Aliabadi M, Irani M, Ismaeili J, Piri H, Parnian MJ (2013) Electrospun nanofiber membrane of PEO/Chitosan for the adsorption of nickel, cadmium, lead and copper ions from aqueous solution. Chem Eng J 220:237–243

169. Bozorgi M, Abbasizadeh S, Samani F, Mousavi SE (2018) Performance of synthesized cast and electrospun PVA/chitosan/ZnO-NH2 nano-adsorbents in single and simultaneous adsorption of cadmium and nickel ions from wastewater. Environ Sci Pollut Res 25:17457–17472

170. Manea LR, Chirita M, Hristian L, Popa A, Sandu I (2016) Researches on the realization of wool-type yarns with elastomer core on classical spinning technology. Mater Plast 53:361–366

171. Scarlet R, Manea LR, Sandu I, Martinova L, Cramariuc O, Sandu IG (2012) Study on the solubility of polyetherimide for nanostructural electrospinning. 63:688–692

172. Manea LR, Scarlet R, Leon AL, Sandu I (2015) Control of Nanofibers production process through electrospinning. Rev Chim (Bucharest) 66(5):640–644

173. Shirazi MM, Kargari A, Ramakrishna S, Doyle J, Rajendrian M, Babu PR (2017) Electrospun membranes for desalination and water/wastewater treatment: A comprehensive review. 3:209–227

174. Keshtkar AR, Irani M, Moosavian MA (2013) Removal of uranium (VI) from aqueous solutions by adsorption using a novel electrospun PVA/TEOS/APTES hybrid nanofiber membrane: Comparison with casting PVA/TEOS/APTES hybrid membrane. J Radioanal Nucl Chem 295:563–571

175. Zhou J, Lu Q-F, Luo J-J (2017) Efficient removal of organic dyes from aqueous solution by rapid adsorption onto polypyrrole–based composites. J Clean Prod 167:739–748

176. Cho IH, Zoh K-D (2007) Photocatalytic degradation of azo dye (Reactive Red 120) in TiO2/UV system: Optimization and modeling using a response surface methodology (RSM) based on the central composite design. Dyes Pigm 75:533–543

177. Nairat M, Shahwan T, Eroglu AE, Fuchs H (2014) Incorporation of iron nanoparticles into clinoptilolite and its application for the removal of cationic and anionic dyes. J Ind Eng Chem 1–9

178. Punzi M, Anbalagan A, Borner RA, Svensson B-M, Jonstrup M, Mattiasson B (2015) Degradation of a textile azo dye using biological treatment followed by photo-Fenton oxidation: Evaluation of toxicity and microbial community structure. Chem Eng J 270:290–299

179. Wen T, Zhang D-X, Zhang J (2013) Two-Dimensional Copper(I) Coordination Polymer Materials as Photocatalysts for the Degradation of Organic Dyes. Inorg Chem 52:12–14

180. Cai T, Li H, Yang R, Wang Y, Li R, Yang H, Li A, Cheng R (2015) Efficient flocculation of an anionic dye from aqueous solutions using a cellulose-based flocculant. Cellulose 22:1439–1449

181. Turhan K, Turgut Z (2009) Decolorization of direct dye in textile wastewater by ozonization in a semi-batch bubble column reactor. Desalination 242:256–263

182. Yagub MT, Sen TK, Afroze S, Ang HM (2014) Dye and its removal from aqueous solution by adsorption: a review. Adv Coll Interface Sci 209:172–184

Electrospun Nanofibers for Coating and Corrosion

Subhash B. Kondawar, Hemlata J. Sharma, Sushama M. Giripunje, and Pravin S. More

Abstract Metallic and alloy materials are needed special surface treatments and coatings to be acted as environmentally anticorrosion. Corrosion is a typical critical issue in the performance of metallic components used in industrial applications. Surface engineering techniques have been well developed with the advent of nanotechnology for the improvement of corrosion resistance. Traditional techniques of coating used chromates which are environmentally creating human health problems. To resolve this issue, surface modification by using nanostructure coating is the hot topic of research in the field of anticorrosion properties of nanomaterials. Among the various nanoscale deposition on the metallic or alloy materials such as sputter deposition, physical vapor deposition, chemical vapor deposition, laser-assisted ablation, electrodeposition, atomic layer deposition, spin-coating, chemical bath deposition, etc., the deposition by polymeric fibrous materials at the nanoscale has found to be well anticorrosive. The polymeric nanofibers coating on metal by electrospinning is an effective technique used for the reduction of corrosion rate. This chapter is mainly focused on the fabrication technique of electrospun nanofibers which can be efficiently used as corrosion inhibitors as well as self-healing agents to reduce the corrosion rate and compared with other coating techniques.

Keywords Polymers · Nanofibers · Electrospinning · Coating · Corrosion

S. B. Kondawar (✉) · H. J. Sharma
Department of Physics, Rashtrasant Tukadoji Maharaj Nagpur University, Nagpur, India

S. M. Giripunje
Department of Physics, Visvesvaraya National Institute of Technology, Nagpur, India

P. S. More
Department of Physics, Institute of Science, Mumbai, India
e-mail: pravin.more@iscm.ac.in

© The Author(s), under exclusive license to Springer Nature Switzerland AG 2021
S. K. Tiwari et al. (eds.), *Electrospun Nanofibers*, Springer Series on Polymer and Composite Materials, https://doi.org/10.1007/978-3-030-79979-3_5

1 Introduction

The corrosion can be well defined as the electrochemical oxidation of metal in reaction with oxygen that converts the refined metal to a more chemically stable form. Corrosion is not limited to metal. It can be found in polymer, ceramic, and steel. Due to chemical or physical interactions of the material with the environment, there is a gradual destruction of metal. Corrosion of metals is one of the most serious problems and several techniques have been used to protect metals from corrosion. To control and prevent the corrosion of metals, it needs to be properly coated. The durability of the metal used as current collector electrodes in most of the applications including solar power generation panels and electrodes for capacitors and supercapacitors is greatly affected by environmental corrosion. Therefore, there is a demand to protect the metals by using suitable self-healing anticorrosion coatings. Metallic structures in the modern industry rely on organic coatings for corrosion protection. To prevent corrosion on the surface of metals exposed to a corrosive environment, polymer coatings are widely applied as such coatings are cost-effective and environmentally friendly. These coatings have the ability to self-heal after suffering mechanical damage. The ability to self-heal can be obtained by adding corrosion inhibition substances to the coating system. Nanocomposites containing active protective components that have self-healing properties became the top priority in organic coating systems for corrosion protection. Due to the synergistic effect of components forming polymeric nanocomposites, more attention has been devoted to organic–inorganic coatings. The organic component provides excellent conductivity, high mechanical flexibility, and improved adhesion on the metallic surface, whereas the inorganic nano-filler provides a high aspect ratio and improves the electrical and electrochemical performances by creating the active sites for reaction. The corrosion resistance of metallic materials can be well improved by the surface engineering techniques by using chromates. But these techniques are not suitable as the use of chromates is found to be dangerous for human health [1]. With the advent of nanotechnology, the materials behave differently at nanoscale as nanomaterials posses very large surface area to volume ratio and very large number of surface atoms in comparison with the interior or bulk atoms. And therefore, coating of such nanomaterials over the metallic surface will definitely improve the corrosion resistance.

The main purpose of coatings for corrosion inhibition is to create a functional barrier in environments to protect against corrosion of the materials. Generally, coatings can be categorized as metallic, inorganic, organic, and hybrid coatings depending on the materials which are to be used for coating. Since the materials behave differently at the nanoscale, the coating by nanostructure can improve coating properties, and therefore, nanomaterials are studied for nanocoating with improved strength, hardness, and corrosion behavior.

The nanofibers are found to be very good resistive to prevent corrosion due to their remarkable properties such as large surface area, high porosity with good elasticity, and mechanical strength. They are lightweight with small diameters and controllable pore structures making them ideal for a wide range of applications [1]. In recent years, a number of processing techniques have been used to prepare polymeric nanofibers such as drawing, template synthesis, phase separation, self-assembly, and electrospinning [2]. Among them, electrospinning is an advancement in the field of nanotechnology and is a popular technique that allows the deposition of synthesized polymeric nanofibers onto metallic surfaces. The deposited electrospun nanofibers can act as an effective barrier coating to reduce the rate of corrosion of metals. Zhao et al. reported the improvement of corrosion resistance when the healing agent is mixed in core–shell nanofibers [3]. Harb et al. suggested that nanocomposites have high-efficiency protective coatings. Inorganic oxides modified polymeric materials coatings have proved to be very good corrosion protection [4]. Aldabbagh et al. mentioned that polyamide coatings can be a good corrosion resistance for aluminum [5]. Electrospun polyvinyl alcohol (PVA) and polyvinyl chloride (PVC) fibers coating works as corrosion passivation application under chloride solution [6].

The selected coating material should be environmentally friendly to which it will be exposed to and should be pollution-free. By electrospinning technique, polyvinyl alcohol (PVA) fibrous material was successfully deposited on aluminum alloy and the corrosion behavior was studied in 3 wt % NaCl solution by means of electrochemical impedance spectroscopy (EIS). It is not only a cost-effective approach to provide a protective layer onto metallic surfaces but also provides new coating methods in which polymeric mats that are corrosion inhibitors developed self-healing functionalities [7]. Recently, it has been reported that ZnO-polymeric composite materials bilayer coating with a p-n heterojunction produce various effects of corrosion inhibition through various mechanisms on carbon steel substrate [8]. Superhydrophobic carbon nanofibers coatings have been reported as potential candidates for protecting metal and alloy surfaces in various engineering and environmental applications [9]. PANI microfibers blended with PMMA doped with DBSA and CSA can also be used as anticorrosion coatings of carbon steel. The coating is said to be reliable if it has low moisture penetration and chemically resistive [10, 11].

Corrosion mostly occurs in aqueous environments. In that condition, a metal can experience a number of degradations in the environment. The best coating or deposition technique which shields metallic surfaces against corrosion is electrospinning.

This chapter is mainly focused on various types of coating techniques in general and particularly the electrospun nanofibers coating along with the fabrication technique of electrospun nanofibers used as self-healing agents to prevent corrosion and for encapsulation of corrosion inhibitors.

2 Techniques of Nanostructure Coating

To prevent the corrosion of metals, the covering to the metallic surface by a certain substance is termed to be coating. There are various techniques for coating on the surface of metals. In most of the available techniques, chromates are used which create human health and environmental problems. To resolve this issue, various techniques of coating have been developed in accordance with the understanding of methods of nanostructure materials and the emergence of instrumentation to characterize the materials at the nanoscale. The development of preparation of nanostructure materials has proved to be very successful for deposition over the surface of the substrate which was later found to be useful for coating the metallic substrates against corrosion. Nanostructure deposition over the surface of metal has proved to be an anticorrosive coating which supports the metal against corrosion. The nanostructure coating over the metallic substrates can be done by various processes such as atomic layer deposition, laser thermal spray-coating, laser ablation coating, sol–gel coating, sputter deposition, chemical vapor deposition (CVD), physical vapor deposition (PVD), and electrospun nanofibers coating.

2.1 Atomic Layer Deposition

The atomic layer deposition (ALD) technique is a homogeneous and conformal technique that has been recently accepted compared to other deposition techniques with precise control over thickness and composition. It considers self-limiting surface reactions and adsorbing molecular monolayer in sequence, i.e., the growth rate of a thin film can be precisely limited to only one molecular layer at one time. Also, surfaces can be uniformly coated by adjusting the number of ALD cycles and only one monolayer of the precursor is adsorbed on the surface during each pulse, regardless of whether there is an excess of precursor on the substrate [12]. We can introduce ALD coatings on metallic surfaces for the purpose of enhancing the chemical resistance against corrosion and oxidation and to improve the chemical stabilities of other nanostructure metals [13]. The pinhole-free films produced by ALD technique have better protection to the metallic surfaces because they can entirely prevent the direct contact of the metal and the corrosive media [14]. Shan et al. reported the coating of TiO_2 by using the ALD technique on stainless steel to show its lower corrosion rate [15]. Gao et al. investigated the anticorrosion effect of coating of various metal oxides such as Al_2O_3, ZrO_2, HfO_2, and TiO_2 deposited by the ALD technique on the surface of silver nanoparticles and found that all of these metal oxide coatings exhibited a significant reduction of corrosion current density by a magnitude of tens of orders in a neutral NaCl solution [16]. Jeong et al. reported atomic layer deposition of TiO_2-Al_2O_3 core–shell nanoparticles uniformly on silicon wafer used as a substrate (Fig. 1) [17]. This ALD technique can be extended for the study of anticorrosion properties of the metals coated with core–shell nanoparticles.

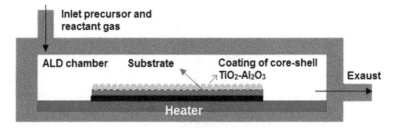

Fig. 1 Atomic Layer Deposition process (Reproduced with permission from Ref. [17])

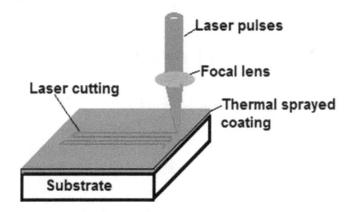

Fig. 2 Schematic diagram of laser-assisted thermal spray-coating

2.2 Laser Thermal Spray-Coating

The laser-assisted thermal spraying process is a technique of deposition of nanostructure material on the metal substrate by using laser and spray mechanism. Amorphous nanostructure can be coated by using laser thermal spray-coating [18]. Figure 2 shows the schematic diagram of laser-assisted thermal spray-coating. Zhao et al. used laser thermal spray-coating for deposition of AlFeSi coating on steel and studied the effect of laser power on corrosion performance of coating [19].

2.3 Sol–Gel Coating

Sol represents the solute colloidal particles dispersed in liquid and gel represents the network of sol filled with liquid. Nanomaterials of 0D (quantum dots), 1D (quantum wires), and 2D (quantum well) can be easily prepared. Preparation of 2D thin film by sol–gel is used for coating the surface of the metallic substrate. In the case of sol–gel coating, the nanoparticles are prepared by sol–gel technique followed by either

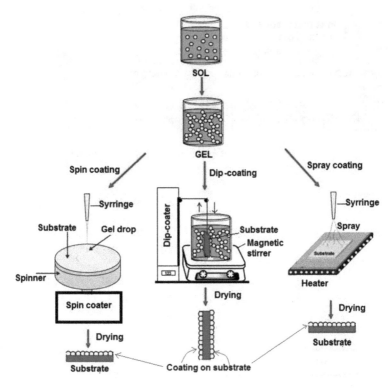

Fig. 3 Schematic diagram of sol–gel spin-coating, dip-coating, and spray-coating of nanostructure thin film on substrate

spin-coating or dip-coating or spray-coating to deposit the nanostructure material on the substrate as shown in Fig. 3. In the case of sol–gel dip-coating process, the substrate on which the coating is required is dipped in solution and taken out periodically for multilayer coating. In the case of sol–gel spin-coating process, the syringe is used to drop the sol on the surface of the substrate which is kept under spinning and heating for uniform coating on the substrate. In the case of sol–gel spray-coating, the sol is sprayed on the substrate and the subsequent heating by the heater gives the uniform coating of nanostructure thin film on the substrate. Zhang et al. reported the preparation of TiO_2-PTFE nanocomposites coating on stainless steel substrate by using a sol–gel dip-coating technique and studied anticorrosion properties [20]. Conceicao et al. showed the corrosion behavior of magnesium alloy (AZ31) sheets coated with PEI by sol–gel spin-coating process and the corrosion protection of the coatings was evaluated by impedance analyses. According to them, all tests were done under N_2 atmosphere in NaCl solution, using 15 wt.% solution in N'N'- dimethylacetamide (DMAc), and 2 μm thickness of film has been achieved. It has also been noted that value of impedance also decreases as the time increased during exposure of NaCl solution [21].

Sol–gel process is used for increasing the corrosion resistance of the metal, and it has good surface protection, good chemical stability, and good oxidation control for metal substrates [22]. But the drawback with metal oxide-based coatings is that they are brittle and thicker and often works at high temperature; so to overcome with these limitations, the sol–gel coatings by hybrids are of an interesting area which works at room temperature and have good thermal as well as environmental stability due to synergistic effect of organic and inorganic materials. Copper and bronze are oftenly used metals in kitchen utensils but in the wet environment, they form hydroxides and harmful complexes and accelerate corrosion process [23]. The corrosion inhibition efficiency was found to increase with increasing 3-mercaptopropyltrimethoxysilane (MPTS) concentration which was used as a precursor in the formation of sol–gel coating over aluminum and copper surfaces. In a similar way, nanofibers synthesized by electrospinning technique and calcination procedure dissolved in a solvent can be coated on metal surfaces to protect it from corrosion in the neutral environment at room temperature by sol–gel technique [24]. The particles of hydroxylated nanodiamond (HNDs) contained in the sol–gel coating diminished the corrosion effect on magnesium surface [25]. Because sol–gel itself contains a large number of defects and cracks in the coating but as we inculcate nanoparticle, then it overcomes defects in the material and enhanced corrosion resistance of the synthesized coated film. It has been reported that with the heat treatment, the stabilized polyacrylonitrile (PAN) nanofibers synthesized by sol–gel via electrospinning were uniform and continuous to improve the oxidation resistance of the carbon nanofibers [26].

2.4 Laser Ablation Coating

In the laser ablation process, the material is removed from the surface by irradiation of the laser beam. In this method, the ablation or vaporization of the bulk material by the laser beam of high power in the UV range of wavelength is used to produce the nanoparticles film deposited on the substrate. The temperature of the irradiated spot increases when the laser beam is irradiated on the target which then evaporates the atoms from the target. The interaction of evaporated atoms with the atoms of inert gas creates a plasma plume near to the target. Sufficient pressure of gas carries the vaporized atoms toward the cool substrate for condensation on the surface of the substrate to produce film of nanoparticles. Particle size distribution is determined by gas pressure and laser pulse. Usually, the gas pressure in the range of 0.1 to 1.0 torr is used for narrow particle size distribution. Shortening the laser pulse 20–30 ms gives a smaller particle size. Generally, CO_2 laser, Nd-YAG laser, ArF excimer laser, or XeCl excimer laser of shorter wavelength of the UV spectrum of the order of 200 nm are used in the laser ablation process for the preparation of nanoparticles. Figure 4 shows the schematic of laser-assisted ablation coating of nanostructure thin film on the substrate. Claries et al. demonstrated the laser ablation technique for coatings of amorphous calcium phosphate and crystalline hydroxyapatite with different morphologies [27].

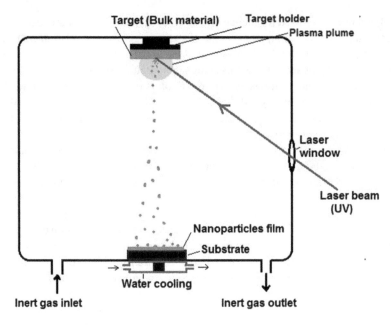

Fig. 4 Schematic of laser-assisted ablation coating of nanostructure thin film on substrate

2.5 *Sputter Deposition*

Interaction of an ion with the target causes the formation of plasma which is the mixture of photons, electrons, atoms, ions, and molecules in the space between the target and substrate (two electrodes) with the application of DC or AC. The density of plasma depends on the pressure of the inert gas. More the plasma density, sputtering of atoms will be more and hence more deposition of nanoparticles on the substrate takes place. Electrons are emitted from target due to high potential of DC source applied between target and substrate. These electrons ionize the inert gas which causes the formation of plasma between target and substrate. The ionized gas then strikes on the target to sputter the atoms to be deposited on the substrate. DC sputtering is not suitable for sputtering of insulating or even semiconducting target. In this case, electron oscillates with the radio frequency (13.56 MHz) of RF generator between the target and substrate causing more ionization of inert gas, even the target is an insulator thereby increasing the efficiency of the sputtering of target atoms as compared to that of DC sputtering. Electron is subjected to E and B simultaneously. Electron moves in a helical or spiral path and is able to ionize more atoms of the gas, thereby increasing the efficiency of the sputtering of target atoms for enhancing deposition of nanoparticles on the substrate. Baldwin et al. highlighted the versatility of the magnetron sputtering process to produce novel multilayer pyrotechnic coatings and corrosion-resistant supersaturated Al/Mg alloy coatings [28]. Figure 5 shows the schematic of magnetron sputter deposition. Further, the plasma density can be

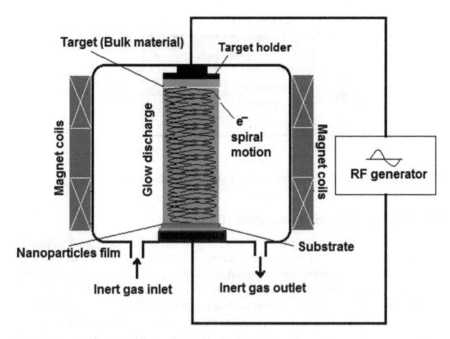

Fig. 5 Schematic of magnetron sputter deposition

increased by using microwaves of frequency 2.45 GHz in a direction such that the plasma density can be enhanced due to more radius of spiral motion of electron at resonance when microwave frequency coincides with the natural frequency of rotation of electrons in magnetic fields. Therefore, this technique is known to be a microwave ECR (Electron Cyclotron Resonance) excited plasma deposition.

2.6 Physical Vapor Deposition (PVD)

Physical vapor deposition is a vaporization coating technique that involves the transfer of material at the atomic level. The process can be described by the following steps; (i) the material to be deposited is converted into vapor by physical means (high-temperature tube furnace), (ii) the vapor is carried by the inert gas to a region of low pressure toward substrate, and (iii) the vapor condenses on the cool substrate to form a thin film. PVD processes are used to form multilayer coatings, composition deposits, very thick deposits, and freestanding structures. Physical vapor deposition is used to produce thin, corrosion-resistant coatings. The metallic behavior can be modified by the deposition of PVD layers to improve surface hardness, wear resistance, and corrosion resistance of PVD-coated materials when deposited on magnesium and their alloys; their application has extended to the biomedical field. A typical PVD process is shown in Fig. 6.

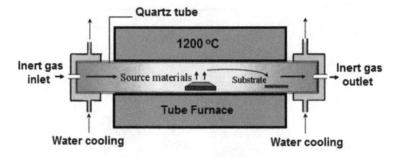

Fig. 6 Schematic of PVD process for coating nanoparticles on the substrate

2.7 Chemical Vapor Deposition (CVD)

To improve the quality of coating, the aerosol-assisted CVD method can be used to create ordered nanostructured material and deposit a thin layer of hydrophobic compound on the surface [29]. CVD is used to deposit various types of materials that may be nanocomposites onto metallic surfaces. The production of uniform thickness of films with fine coating and complicated shapes can be enabled to protect against oxidation and corrosion due to its excellent throwing power and low porosity. Its uniqueness lies in its ability to control and check the quality of the coating at various stages of processing. The schematic diagram of chemical vapor deposition is shown in Fig. 7. Chemical vapor deposition involves the deposition of solid material at the nanoscale from the gaseous phase. In this method, precursor gases are diluted with a carrier gas (inert gas) and delivered to the reaction chamber (quartz tube) at ambient temperature and they react with processing gas or decompose to form a solid phase to be deposited on the heated substrate place on a quartz boat at the middle of the tube furnace. In CVD, the volatile metal compound from the gas mixture is introduced into the reaction chamber deposits in the form of thin film on the surface of substrate placed at the center in the chamber.

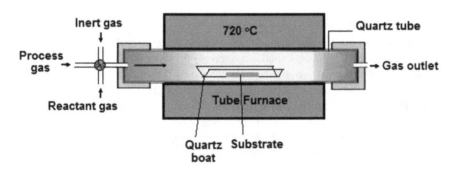

Fig. 7 Schematic diagram of chemical vapor deposition

Fig. 8 Chemical bath
deposition (CBD) technique

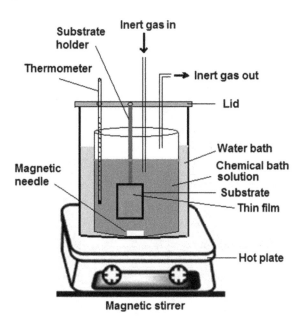

2.8   Chemical Bath Deposition

The chemical bath deposition (CBD) technique as shown in Fig. 8 involves the
controlled precipitation from theTsolution of a compound on a suitable substrate
inserted in a chemical bath solution. The reaction and the deposition occur simultaneously inside the chemical bath. Hence, it is named as CBD. It offers many
advantages of preparing semiconducting nanoparticle thin films over CVD and PVD
due to control of film thickness and deposition rate by varying pH, temperature, and
reagent concentration of the solution.

2.9   Ionized Cluster Beam Deposition

Nanoparticles thin film can also be deposited on the surface of the substrate using
ionized cluster beam deposition. Figure 9 shows the schematic of the deposition of
nanoparticles thin film by ionized cluster beam deposition technique. It consists of an
evacuated chamber which has the source of evaporation, a crucible with a fine nozzle
through which material evaporates, the beam of clusters through which the nozzle
gets ionized by electrons, and the accelerated ionized beam of cluster deposited on
the substrate to form nanoparticles film.

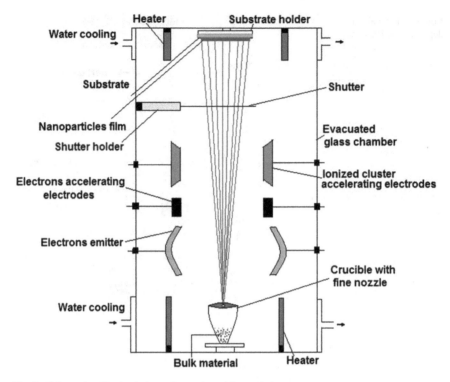

Fig. 9 Schematic of ionized cluster beam deposition technique

3 Electrospun Nanofibers Coating

There are various techniques available for the fabrication of nanofibers, but electrospinning is the simplest and cost-effective technique with facile control of the
diameter of fiber [30]. Electrospun nanofibers coating is a versatile technique of
nanostructure coating on the metallic substrate. Ultrafine electrospun polymeric or
composite nanofibers can be coated on the metal surface by electrospinning technique, which is an efficient technique to synthesize uniform, conductive nanofibers
with an average diameter between tens to hundreds of nanometer range by using
electrostatic forces [31]. Figure 10 shows the electrospinning setup for aligned and
randomized nanofibers coating on metal. The basic principle of this electrospinning
technique is "electrostatic interactions". The peristaltic pump, where a high voltage
power supply is connected, pushes the solution from the needle. Single droplet ejected
at the tip of needle will be collected in a counter electrode called collector. Electrospinning instrument consists of three major parts; first and foremost influencing part
is high voltage power supply, the second one is the syringe and needle assembly
(collectively called as spinneret in needleless type), and the last one is the collector.
Many types of electrospinning instruments are available in the market nowadays.
They differ only in the type of spinneret and collector. Some instruments employ

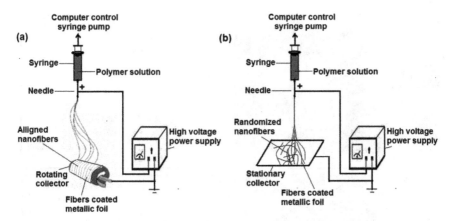

Fig. 10 Electrospinning setup for **a** aligned nanofibers and **b** randomized nanofibers coating on metallic foil

an electrode material as spinneret while few others contain needleless spinneret. Many parameters influence the electrospinning of fibers. The selection of a suitable polymer for electrospinning depends on the enduse of nanofibers. The molecular weight (M. wt.) of the chosen polymer is an important factor that alters the characteristics of nanofibers. The same polymer of different molecular weights produces fibers of different diameters. The diameter of the nanofibers prepared by electrospinning technique is dependent on the physical (distance, voltage, flow rate, and collector), chemical (concentration, conductivity, molecular weight, viscosity, solvent volatility, and molecular structure), and environmental (humidity and temperature) parameters. It consists of three main parts namely high voltage power supply, syringe pump, and metallic collector. The polymeric solution in the syringe generates an electrically charged jet when flowing through the capillary needle under a high electric field.

Bhute et al. reported the synthesis of polyvinylidene fluoride/cellulose acetate (PVdF/CA) and PVdF/CA-AgTiO₂ polymer nanofibers coating on aluminum substrate by electrospinning [32]. SEM images and histograms of electrospun PVdF/CA and PVdF/CA-AgTiO₂ polymer nanofibers membranes are shown in Fig. 11. It is concluded from the SEM images histograms that the diameter of the nanofibers coated on the metallic substrate can be easily controlled by changing the electrospinning parameter. The smaller fiber diameter due to the increase in viscosity of the electrospinning solution possess a large surface area beneficial for increasing adsorptive properties toward increasing the corrosion resistance of the metals on which it has been coated.

It has been found that polyamide (PA) coated on the aluminum surface by electrospinning technique is found to be good corrosion resistance [5]. Coating of nanofibers is used for improving the corrosion resistance of the coated metallic surfaces and used in various industries such as coating of pipes and tanks [33]. Metal oxide polymer nanocomposite, PCL/ZnO-NiO-CuO (polycarprolactone/ zinc

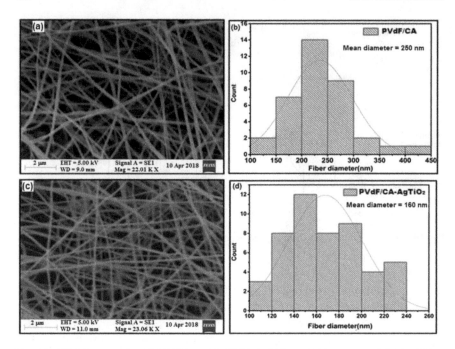

Fig. 11 SEM images of **a** PVdF/CA, **c** PVdF/CA-AgTiO$_2$ and histograms of **b** PVdF/CA, **d** PVdF/CA-Ag-TiO$_2$ nanofibers membranes (Reproduced with permission from Ref [32], Bhute MV, Kondawar SB, *Solid State Ionics* 333, 38–44 (2019) Copyright (2019) Elsevier)

oxide–nickel oxide–copper oxide) have been successfully synthesized through electrospinning technique and deposited on mild steel and found excellent corrosion resistance in HCl solution. Nanofibers of other polymers such as PCL, PAN-Al$_2$O$_3$, and polyaniline/poly(methylmethacrylate) PANI/PMMA could also be deposited by electrospinning technique to increase the corrosion resistance of metal surface [12]. Nanofibers coatings have good electrical and thermal conductivity, good environmental or thermal stability, good surface appearance, good chemical resistance, and better corrosion resistance. Self-healing coatings have been recently proposed. The principle of anticorrosion coatings on the metallic surface is to protect the coated sample from corrosive agents present in the corrosive environment. Coating allowed the flow of corrosive agents to the metal surface, but coating materials must have high electric resistance and high adhesion to the substrate. The nanocomposites coating, epoxy/polyaniline-ZnO nanorods, polyaniline-TiO$_2$ composite, organic–inorganic hybrid coatings, and sandwiched polydopamine for TiO$_2$ create a synergistic effect and have enhanced the corrosion resistance.

4 Corrosion

Natural conversion of a refined metal to a more stable in the form of oxide, hydroxide, or sulfide state which causes the metal deterioration is often called as corrosion. Corrosion is an irreversible interfacial reaction of a material with its environment, resulting in the loss of material [34]. Corrosion is the surface wastage that occurs when metals are exposed to reactive environments due to interaction between a metal and environments which results in its gradual destruction. The decay of materials by chemical or biological agents is also done by corrosion. Corrosion is a threat to the environment. For instance, water can become contaminated by corrosion products and unsuitable for consumption. Corrosion prevention is integral to stop contamination of air, water, and soil. The most common type of iron corrosion occurs when it is exposed to oxygen and in the presence of water, which creates a red iron oxide commonly called rust. Rust can also affect iron alloys such as steel. The term corrosion can be applied to all materials, including non-metals. But in practice, the word corrosion is mainly used in conjunction with metallic materials. Mechanical properties of metals reduce due to corrosion and therefore, it is necessary to study the anticorrosion coating and corrosion protection [35]. Based on corrosion mechanism, corrosion of metals can be divided into four categories; chemical, physical, electrochemical, and biological corrosion as shown in Fig. 12.

Due to oxidation–reduction reactions, chemical corrosion takes place when metal meets corrosive media directly. The entire reaction surface is covered by the corrosion product and protective film is formed in the process of chemical corrosion. Physical corrosion changes the shape, size, or phase of a substance due to the metal melting process. The electrochemical reaction between metal and electrolyte solution (mostly aqueous solution) is responsible for electrochemical corrosion. Biological

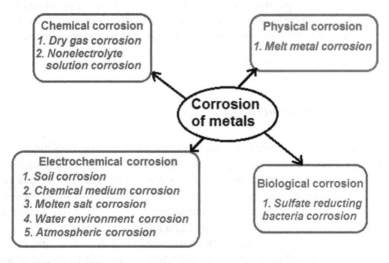

Fig. 12 Categories of metal corrosion based on corrosion mechanism

corrosion is the corrosion affected by the presence or activity of microorganisms such as bacteria on the surface of the metal.

4.1 Corrosion Resistance

Corrosion resistance is the ability to prevent environmental deterioration by chemical or electrochemical reactions. Desirable characteristics of corrosion-resistant alloys, therefore, include high resistance to overall reactions within the specific environment. As far as corrosion resistance of stainless steel is concerned, it is highly resistive for corrosion in many environments. On the other hand, carbon and low alloy tool steels will corrode in the same environment. The very thin oxide layer of the order of nanometer on stainless steel plays the role of its corrosion resistance. In the corrosive environment, the oxide layer act as a passive layer which forms because of the chromium added to stainless steel. For better corrosion resistance, there should be more than 10.5% chromium added to have a more stable passive layer in stainless steel. To further enhance the corrosion resistance of the stainless steel, the elements such as nickel, molybdenum, manganese, etc. can be added. Similarly, if the passive layer of the stainless steel is exposed to oxygen, corrosion resistance can be maintained. By boldly exposing the steel surface, the corrosion resistance will be greatest. During well exposing the surface of stainless steel, the passive layer on the surface at some localized spots may be broken under certain circumstances. The corrosion under this condition at localized spots is said to be pitting corrosion. Under some aqueous environments containing chloride, pitting corrosion is common for this situation. There are various examples of pitting-type corrosion which includes coastal atmospheres and road salt combined with rainwater or tap water which contains chloride mostly. Stainless steels are very popular for various applications but the elements it contains mostly chromium and nickel in high concentration may be very harmful to human beings as concerned health due to their carcinogenic and toxic effects [36].

4.2 Corrosion Measurement

By using electrochemical impedance spectroscopy (EIS), the important electrochemical parameters such as coating capacitance (Qc), coating resistance (Rc), double-layer capacitance, and charge transfer resistance are considered for the evaluation of corrosion resistance of organic coatings. For the evaluation of these parameters for the coating, an electrochemical cell consisting of three electrodes namely platinum wire as a counter electrode, Ag/AgCl as a reference electrode, and a working electrode made of coating is used in an instrument called as potentiostat. [7]. The impedance of the coated materials can be determined using this technique in which the response of the coated material can be detected in terms of a small amplitude *ac* perturbation as a function of frequency. A physical method is required to interpret the impedance–frequency response for metal–coating–electrolyte interface by an electrical equivalent circuit which provides a clear description of the different contributions to the impedance. Reproducibility of the measurement of the impedance of coating in this technique is an issue to obtain the results with high accuracy for which statistical methods to interpret the data can be used.

The type of methods for expressing the degree of metal corrosion depends on corrosion types. Generally, for general corrosion, it can be expressed by the average corrosion rate. The corrosion rate can be calculated by expressing the average corrosion rate in general for any type of corrosion using Eq. (1) [5].

$$\text{Corrosion rate} = \frac{I_{corr} K E_w}{D\,a} \tag{1}$$

where I_{corr} is the current of corrosion, K is a constant according to units of corrosion rate, E_w is the equivalent weight, D = density, and a is the area of the sample.

There are several ways to prevent and control corrosion which are summarized in Fig. 13.

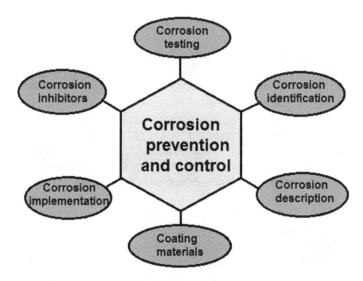

Fig. 13 Schematic of the ways for corrosion prevention and control

4.3 Corrosion Inhibitor

The best way of corrosion protective coatings is the self-healing of the damage caused by corrosion. As a coating suffers mechanical damage, which is automatically repaired by a chemical component that acts as a corrosion inhibitor in the coating. Corrosion inhibitor is a chemical mixed with an environment that contains liquid or gas which decreases the corrosion rate of the metal when it is exposed to that environment. To prevent or control corrosion, the effective way is the corrosion inhibitor as it allows one to use less expensive metals for a corrosive environment. Though there are several corrosion inhibitors, but the process of selecting a particular corrosion inhibitor is quite difficult when it is the first time selected for the material to prevent its corrosion as the effectiveness of a corrosion inhibitor depends on the fluid composition. Formation of coating is a common mechanism for inhibiting corrosion which prevents the access of the corrosive substance to the metal [37–39].

The corrosion rate can be slowed down in the presence of an inhibitor. The inhibition efficiency can be determined by finding the difference between the rate of corrosion in the absence and presence of the inhibitor to reduce the rate of corrosion expressed by Eq. (2);

$$R_i = \frac{\vartheta_0 - \vartheta}{\vartheta_0} \tag{2}$$

where ϑ_0 and ϑ are the corrosion rate measured in the absence of inhibiter and in the presence of inhibitor, respectively. The inhibition efficiency mostly depend on an important parameter inhibitor concentration. In many ways, the inhibitors can be applied to prevent corrosion. The most common corrosion inhibitors are acid pickling baths which are used to chemically deposit the iron oxide scale on steel which adsorb onto the metal surface thereby blocking the electrochemical attack of the steel. The effect of the inhibitor on the corrosion process can be understood and classified by comparing polarization curves in environments with and without the inhibitor added. Figure 14 shows the schematic examples of the types of inhibition and their effects on the polarization curves. Inhibitors can be classified as an anodic, cathodic, or mixed inhibitor as shown in this figure. An anodic inhibitor reduces the rate of the anodic reaction with less effect on the cathodic reaction. A cathodic inhibitor reduces the rate of the cathodic reaction with less effect on the anodic reaction. A mixed inhibitor reduces both the rates of cathodic and anodic reactions equally. The corrosion potential (Ecorr) is different in a solution containing an inhibitor compared to that solution which does not have an inhibitor. Ecorr will be more for an anodic inhibitor, less for a cathodic inhibitor, but there is no change for a mixed inhibitor. The change in Ecorr is used as an diagnostic for the type of inhibitor as long as the corrosion rate is reduced from the metal.

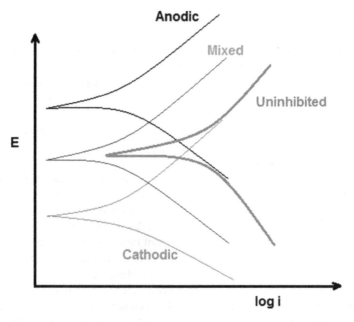

Fig. 14 Effect of anodic, cathodic, and mixed type of inhibition on the polarization curves

5 Electrospun Nanofibers Coating for Corrosion Protection

A new and innovative way of fighting against corrosion is electrospun nanofibers-based coatings. Electrospinning is a versatile and flexible low-cost process for producing nanostructure ultra-thin fibers from a wide range of materials (polymers, composites, and ceramic) onto any type of surface with arbitrary geometries. Recently, with the advent of nanotechnology, the fabrication of nanofibers has been progressed with excellent control of their morphology for the applications toward the control of the rate of corrosion. The corrosion rate of metal (steel, aluminum, and magnesium alloys) can be reduced by the effective barrier coating produced by these electrospun nanofibers which have a large surface area, good mechanical strength, and high elasticity can improve the mechanical properties of the coating. As compared to polymeric coating, the electrospun composite coating layers on magnesium alloy via electrospinning have proved to enhance the corrosion resistance. Electrochemical evaluation and salt-spray test are the tests that can be conducted to evaluate the corrosion resistance of different samples [3, 40]. Pure PANI and PANI/PMMA nanofibers synthesized by the electrospinning technique displayed superior anticorrosion properties. Polyaniline (PANI) is one of the most interesting conducting polymers due to the ease of chemistry, redox behavior, and excellent anticorrosion properties. The mechanism behind this is the enhanced anticorrosion behavior of PANI as it has the ability to form a protective layer due to its redox catalytic properties and to reduce the amount of H^+ at the metal/PANI interface and due to their doping–dedoping structural morphology [11].

The degradation of Mg alloy covered with poly(lactic acid) was studied to find the influence of electrospinning and dip-coating approaches for enhancement of the corrosion resistance. The electrospinning technique to fabricate nanofibrous materials further found to be beneficial in improvement for prolonged immersion periods [41]. Es-saheb et al. investigated the fabrication of nanofiber coating of polyvinyl alcohol (PVA) and polyvinyl chloride (PVC) on the pure aluminum surface using the electrospinning deposition technique. They have prepared NaCl solution and studied the effect of coating of PVA and PVC nanofibers on the corrosion resistance of aluminum in this solution. The surface analysis of the fabricated nanofibers was studied by using scanning electron microscopy. The corrosion test was studied using cyclic potentiodynamic polarization by electrochemical impedance spectroscopy measurements. Actually, chloride solution corrodes the aluminum when it is placed in NaCl solution, but the results obtained from corrosion tests confirmed that the presence of PVA and PVC nanofibers coatings on top of aluminum surface acted as corrosion resistance and hence it was greatly precluded this corrosion. It was also confirmed from this test that the polarization resistance of aluminum in NaCl solution was found to be increased due to PVA and PVC coating which decreases the rate of corrosion drastically. This study concluded that PVA and PVC nanofibers coatings can protect the aluminum surface against corrosion [42]. Muthirulan et al. reported the preparation of poly-o-phenylenediamine (PoPD) nanofibers by the electrospinning technique and studied corrosion protection properties of synthesized nanofibers-coated stainless steel in NaCl solution. The comparison of uncoated stainless steel and coated stainless steel was studied for the detection of corrosion resistance and found the strong adherent and inhibition effect of PoPD nanofiber film-coated stainless steel substrate which exhibited good corrosion resistance. Thus, for chloride-containing solution, PoPD nanofiber coatings by electrospinning were found to be a potential coating material for stainless steel against corrosion [43]. Figure 15 shows the potentiodynamic polarization curves for uncoated and PoPD nanofiber-coated 316L SS steel in 3.5% NaCl solution. The corrosion potential (Ecorr) and current density (Icorr) were found to be higher for PoPD nanofiber-coated 316L SS compared to uncoated 316L SS. From the potentiodynamic polarization curves, it was observed that the value of Ecorr shifted toward a more positive value when the steel surface was coated by the PoPD nanofiber film. This confirmed that the best protection against corrosion was provided by the coating of PoPD nanofiber on the surface of stainless steel. The protective layer on the stainless steel by PoPD nanofiber coating may be to the presence of p electrons in the aromatic ring and the lone pair of electrons in the nitrogen atom responsible for the surface modification of stainless steel against corrosion.

Grignard et al. reported the fabrication of poly-(heptadecafluorodecylacrylate-co-acrylic acid)-bpoly(acrylonitrile) nanofibers coated on the aluminum surface by using electrospinning technique for the study of superhydrophobicity and corrosion resistance [44]. Firouzi et al. reported the fabrication of polyvinyl alcohol (PVA) nanofibers coating on aluminum alloy by using electrospinning technique and studied

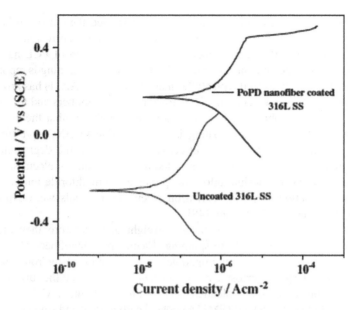

Fig. 15 Potentiodynamic polarization curves for uncoated and PoPD nanofiber-coated 316L SS steel in 3.5% NaCl solution (Reproduced with permission from Ref [43], Muthirulan P, Kannan N, Meenakshisundaram M, *J. Adv. Res.* 4, 385–392 (2013), Copyright (2013) Elsevier)

the anticorrosion performance of NaCl solution by using electrochemical impedance spectroscopy (EIS) [7]. Electrochemical tests revealed that the corrosion resistance of electrospun PVA nanofibers-coated aluminum in NaCl solution was found to be 26 kΩ after 20 h as compared to the corrosion resistance of uncoated aluminum in the same solution (3.8 kΩ).

Aldabbagh et al. prepared polyamide (PA-6) nanofiber coatings on aluminum surface using the electrospinning technique under two different voltages (24 kV and 34 kV). They studied the coating morphology roughness, 3D structural properties, and hydrophobic behavior. The electrochemical corrosion of aluminum without and with PA-6 nanofibers coating in NaCl solution was investigated. It was found that the PA coating decreased the corrosion current and corrosion rate as well as increased the corrosion resistance for aluminum in the NaCl solution. PA-6 electrospun nanofibers are very efficient for protecting the metal surface from corrosion, also the high voltage prepared electrospun nanofibers to lead to an increase in the corrosion resistance of the metal surface [5]. Castro et al. reported the fabrication of electrospun PHBV nanofibers-coated magnesium alloy by using electrospinning technique and studied for biodegradable implant applications. Their study concluded that the electrospun-coated PHBV nanofibers-coated magnesium alloy showed better corrosion resistance than uncoated magnesium alloy and also verified that the porous structure of PHBV nanofibers was obtained by electrospinning technique provided extracellular matrix which promoted cell growth for tissue healing [45]. Nafi et al. prepared electrospun PLLA nanofibers coating on pure magnesium and magnesium alloy (AZ91) using electrospinning technique and studied the corrosion behavior in Hanks' solution. The results confirmed the reduction in the corrosion rate of the materials coated with PLLA nanofibers [46]. The preparation of PCL nanofibers coated on HNO_3 treated magnesium alloy (AZ31) by electrospinning is reported by Hanas et al. Previously treated with HNO_3 magnesium alloy (AZ31) has provided a good adhesion between the coating of electrospun PCL nanofibers and the metallic substrate. The results obtained from corrosion tests confirmed that the bioactivity of electrospun PCL nanofibers coating is enhanced due to HNO_3 pre-treatment which was further found to be very effective in controlling the degradation rate [47]. In another study, they have demonstrated that the coating of electrospun PCL nanofibers on AZ31 magnesium alloy was protected from chloride ion attack and avoided the formation of pitting corrosion when the materials were exposed to supersaturated simulated body fluid [48].

Polymers possess very high molecular weight and therefore they are used to prepare nanofibers using electrospinning. Electrospun nanofibers of polymers including conducting polymers, hydrophobic polymers, and superhydrophobic polymers are found to be good corrosion resistant when coated on metallic surfaces. Zhao et al. reported the fabrication of blended electrospun polyaniline (PANI) microfibers with poly(methyl methacrylate) (PMMA)-coated carbon steel and studied for corrosion behavior in 0.1 M H_2SO_4 solution. They have well compared the corrosion protection efficiency of blended electrospun polyaniline (PANI) microfibers with poly(methyl methacrylate) (PMMA)-coated carbon steel with that of traditional drop-casting PANI/PMMA coating on carbon steel and confirmed that nearly 500 times higher corrosion protection efficiency for blended electrospun nanofibers coating. The efficiency of the corrosion protection for blended electrospun nanofibers coating was found to be 99.96 even after 20 days of immersion in the acidic solution. The increased anticorrosion behavior of blended electrospun nanofibers coating is due to cathodic protection by releasing H_2 gas occupying holes of the PANI/PMMA coating because of blocking H^+ diffusion on the surface of carbon steel. Trapping of H_2 in the holes of PANI/PMMA coating limits the diffusion of the anodic reaction produced by Fe^{2+} ions which also retard the anodic reaction [11]. Tafel polarization curves for carbon steel, PANI/PMMA-coated carbon steel, and blended electrospun PANI/ PMMA microfibers with 25 wt% PANI-coated carbon steel in 0.1 M H_2SO_4 aqueous solution are shown in Fig. 16. Due to extraordinary compact microstructure of blended electrospun PANI/ PMMA microfibers coatings with 25 wt% PANI, the superior corrosion protection property was observed. This is also responsible for effective anticorrosion protection.

The intrinsically ultrahigh hydrophobic or superhydrophobic polymers used to fabricate nanofibers by electrospinning provide long-term anticorrosive protection to the metallic surfaces. Polyvinylidenefluoride (PVDF) has high hydrophobicity with excellent antioxidation and anticorrosion properties and therefore, it is favorable to have electrospun for coating over the metallic surfaces [49]. Cui et al. could fabricate electrospun blended polyvinylidenefluoride (PVDF)/stearic acid (SA) nanofibers coated on aluminum (Al) and studied its long-term anticorrosion property. Even after 30 days, they found that PVDF/SA nanofibers-coated aluminum exhibited lower

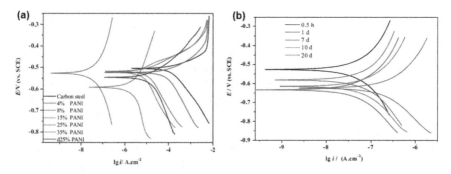

Fig. 16 Tafel polarization curves **a** for carbon steel, coated carbon steel with PANI/PMMA coatings in 0.1 M H_2SO_4 aqueous solution and **b** for coated carbon steel with electrospun PANI/ PMMA coating at 25 wt% PANI after different days of immersion (Reproduced with permission from Ref [11], Zhao Y, Zhang Z, Yu L, Tang Q, *Synth. Met.* 212, 84–90 (2016) Copyright (2016) Elsevier)

corrosion current density in comparison with PVDF nanofibers-coated aluminum in NaCl solution. From the Tafel polarization curves as shown in Fig. 17 for bare Al sheet after immersion in NaCl aqueous solution for 2 h, pure PVDF nanofibers-coated Al sheet after immersion in NaCl aqueous solution for 6 h, and PVDF/SA nanofibers-coated Al sheet after immersion in NaCl aqueous solution for 6 h as well as 30 days, it is concluded that a superior anticorrosion performance for long-term anticorrosion of aluminum sheet was obtained from PVDF/SA nanofibers coating [50].

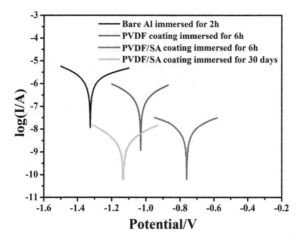

Fig. 17 Tafel polarization curves of a bare Al sheet after immersion in NaCl aqueous solution for 2 h, pure PVDF nanofibers-coated Al sheet after immersion in NaCl aqueous solutions for 6 h, and PVDF/SA nanofibers-coated Al sheet after immersion in NaCl aqueous solutions for 6 h and 30 days (Reproduced with permission from Ref [50], Cui M, Xu C, Shen Y, Tian H, Feng H, Li J, *Thin Solid Film.* 657, 88–94 (2018) Copyright (2018) Elsevier)

Polyvinyl chloride (PVC) or polystyrene (PS) nanofiber coatings fabricated by electrospinning with superhydrophobic behavior on brass surfaces confirmed good anticorrosion property. The coating of electrospun polymer nanofibers made of PVC is found to be a corrosion inhibitor when they are deposited on metallic substrates such as aluminum, steel, and brass surfaces [51, 52]. Yabuki et al. reported pH-controlled self-healing polymer coatings with cellulose nanofibers as corrosion inhibitor to protect corrosion of carbon steel by varying the pH values of the polymer coatings to control the adsorption and desorption of the corrosion inhibitor from the cellulose nanofibers. By scratching the coatings, the polarization resistance was measured in a sodium chloride solution for evaluation of the self-healing ability of polymer coatings with cellulose nanofibers. The resistance of the scratched coatings was largely dependent on the pH of the polymer. A drastic increase in polarization resistance was observed when the scratched polymer coating was prepared at pH 11.4 after 24 h corrosion test. The polarization curves of the scratched polymer coating were measured in sodium chloride solution using the same system as that used for the measurement of electrochemical impedance [53]. Doan et al. reported the fabrication of electrospun core–shell nanofibers as a self-healing coating material for the protection of steel against corrosion. They have prepared two types of core–shell nanofibers. In A type core–shell nanofibers, the corrosion healing agent core contains a stoichiometric mixture of poly(dimethylsiloxane) and crosslinker poly(diethoxysiloxane) (PDES) with PVA as shell, while in B type core–shell nanofibers, the corrosion healing agent core contains the crosslinking catalyst dibutyltindilaurate (DBTL) with the same PVA as shell. The schematic of fabrication of core–shell electrospun nanofibers of PVA as shell and healing agent as core along with SEM images is shown in Fig. 18.

The technique of coating of core–shell nanofibers on steel is shown in Fig. 19. The technique of coating consisting of core–shell nanofibers was prepared on clean sandblasted steel by electrospinning and then silicone binder was deposited to affix fibers on steel followed by scribing the coating with a corrocutter and allowing the coating to heal for 24 h. The linear polarization experiment was performed to find the effectiveness of the healed coating toward preventing corrosion by observing the corrosion current increased with applied potential. The corrosion inhibition efficiency of the healed coating was found to be 88% indicating the effective healing behavior against corrosion [54].

6 Conclusion

With the advent of nanotechnology, the nanostructure coating on metallic and alloy materials significantly improved the corrosion resistance. The control of the coating structure at the nanoscale level allows improving the intrinsic properties of the surface compared to bulk materials. A nanostructure deposition technique with increasing popularity in the field of nanotechnology is electrospinning. Electrospinning is a popular and cost-effective technique used for the deposition of synthesized polymeric

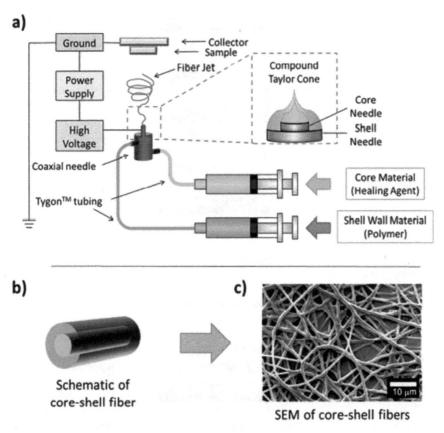

Fig. 18 Schematic of fabrication of core–shell electrospun nanofibers: **a** schematic of coaxial electrospinning setup, **b** schematic of core–shell fibers of PVA (shell) and healing agent (core), and **c** SEM image of randomly deposited core–shell fibers (Reproduced with permission from Ref [54], Doan TQ, Leslie LS, Kim SY, Bhargava R, White SR, Sottos NR, *Polymer* 107, 263–272 (2016), Copyright (2016) Elsevier)

nanofibers onto metallic surfaces. The deposited electrospun nanofibers reduce the corrosion rate of the metal by acting as an effective barrier coating. Corrosion protection of metallic surfaces can be well controlled by depositing electrospun nanofibers. The deposited electrospun nanofibers reduce the corrosion rate of the metal by acting as an effective barrier coating. Some of the coating or deposition techniques which shield metallic surfaces against corrosion are electrospinning, atomic layer deposition, sol–gel coating, sputters coating, laser-assisted ablation coating, chemical vapor deposition (CVD), and physical vapor deposition (PVD). Electrospun polymeric nanofibers coating have demonstrated the potential deposition technique useful for superior anticorrosion performances for long term metal substrate preservation.

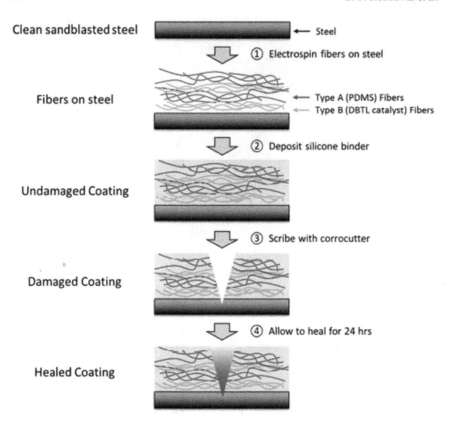

Clean sandblasted steel   ← Steel

                                                                          ① Electrospin fibers on steel

Fibers on steel   ← Type A (PDMS) Fibers
                                                                          ← Type B (DBTL catalyst) Fibers

                                                                          ② Deposit silicone binder

Undamaged Coating

                                                                          ③ Scribe with corrocutter

Damaged Coating

                                                                          ④ Allow to heal for 24 hrs

Healed Coating

Fig. 19 Schematic of coating fabrication process (Reproduced with permission from Ref [54], Doan TQ, Leslie LS, Kim SY, Bhargava R, White SR, Sottos NR, *Polymer* 107, 263–272 (2016), Copyright (2016) Elsevier)

References

1. Rivero P, Redin D, Rodríguez R (2020) Met 10:1
2. Saheb M, Elzatahry A, Sayed E, Sherif M, Alkaraki A, Kenawy E (2012) Int J Electrochem Sci 7:5962
3. Zhao X, Yuan S, Jin Z, Zhu Q, Zheng M, Jiang Q, Song H, Duan J (2020) Prog Org Coat 149:105893
4. Loto RT, Joseph OO, Akanji O (2015) J Mater Environ Sci 6:2409
5. Aldabbagh B, Alshimary H, Eng J (2017) Technol 35:987
6. Sherif E, Saheb M, Zatahry A, Kenawyand E, Alkaraki A (2012) Int J Electrochem Sci 7:6154
7. Firouzi A, Gaudio, Iamastra F, Montesperelli G, Bianco A (2015) J Appl Polym Sci 41250:1
8. Hu C, Li Y, Li T, Qing Y, Tang J, Yin H, Hu L, Zhang L, Xie Y, Ren K (2020) Colloids Surf A 585:1
9. Siddiquia A, Maurya R, Katiyar P, Balania K (2020) Surf Coat Tech 404:1
10. Ulaeto S, Rajan R, Pancrecious J, Rajan T, Pai B (2017) Prog Org Coat 111:294
11. Zhao Y, Zhang Z, Yu L, Tang Q (2016) Synth Met 212:84
12. AlFalah M, Kamberli E, Abbar A, Kandemirli F, Saracoglu M (2020) J Surf Interfac 21:100760

13. Hsu P, Wu H, Carney T, McDowell M, Yang Y, Garnett E, Li M, Hu L, Cui Y (2012) ACS Nano 6:5150
14. Pereira, Escrig J, Palma J, Dicastillo C, Patiño C, Galotto M (2018) J Vac Sci Technol B 36:1
15. Shan C, Hou X, Choy K (2008) Surf Coat Tech 202:2399
16. Gao Y, Walsh M, Liang X (2020) Appl Surf Sci 532:147374
17. Jeong H, Yoo J, Park S, Lee J (2020) Vacuum 179:1
18. Dai X, Zhou S, Wang M, Lei J, Wang C, Wang T (2017) Surf Coat Technol 324:518
19. Zhao W, Kong D (2019) Appl Surf Sci 481:161
20. Zhang S, Liang X, Gadd GM, Zhao Q (2019) Appl Surf Sci 490:231
21. Conceicao T, Scharnagl N, Blawert C, Dietzel W, Kainer K (2010) Corros Sci 52:2066
22. Wang D, Bierwagen G (2009) Prog Org Coat 64:327
23. Castro Y, Ferrari B, Moreno R, Duran A (2003) J Sol-Gel Sci Technol 26:735
24. Vignesh R, Edison T, Sethuraman M (2014) J Mater Sci Technol 30:814
25. Nezamdoust S, Seifzadeh D, Yangjeh A (2020) Trans Nonferrous Met Soc China 30:1535
26. Kim B, Kim C, Yang K, Kim K, Lee Y (2010) Appl Surf Sci 257:1607
27. Cleries L, MartmHnez E, FernaHndez-Pradas JM, Sardin G, Esteve J, Morenza JL (2000) Biomaterials 21:967
28. Baldwin KR, Bates RI, Arnell RD, Smith CJE (1996) Corrosion Sci 38:155–170
29. Jehn H (2000) Surf Coat Technol 125:212
30. Zheng J, He A, Li J, Xu C, Han C (2006) Polymer 47:7095
31. Alam A, Sherif E, Al-Zahrani S (2013) Int J Electrochem Sci 8:8388
32. Bhute MV, Kondawar SB (2019) Solid State Ion 333:38
33. Tjong S, Chen H (2004) Mater Sci Eng C 45:1
34. Heusler KE, Landolt D, Trasatti S (1989) Pure and Appl Chem 61:19
35. Galvele JR (1999) Corrosion 55:723
36. Bordji K, Jouzeau JY, Mainard D (1996) Biomaterials 17:491
37. Balbo A, Chiavari C, Martini C, Monticelli C (2012) Corros Sci 59:204
38. Mansfeld F, Smith T, Parry EP (1971) Corrosion 28:289
39. Kosec T, C′urkovic′ HO, Legat A (2010) Electrochimica Acta 56:722.
40. Deyab M, Awadallah A (2020) Prog Org Coat 139:105423
41. Abdal-hay A, Barakat NAM, Lim JK (2013) Colloid Surf A 420:37
42. Es-saheb M, Elzatahry AA, Sherif ESM, Alkaraki A, Kenawy ER (2012) Int J Electrochem Sci 7:5962
43. Muthirulan P, Kannan N, Meenakshisundaram M (2013) J Adv Res 4:385
44. Grignard B, Vaillant A, Coninck JD, Piens M, Jonas AM, Detrembleur C, Jerome C (2011) Langmuir 27:335
45. Castro J, Krishnan KG, Jamaludeen S, Venkataragavan P, Gnanavel S (2017) J Bio Tribo Corros 3:52
46. Nafi AW, Afifi A, Abidin NIZ, Aziz HA, Kalantari NK (2018) Sains Malays 47:169
47. Hanas T, Kumar TSS, Perumal G, Doble M (2016) Mater Sci Eng C 65:43
48. Hanas T, Kumar TSS (2017) Mater Today Proc 4:6697
49. Liu F, Hashim NA, Liu Y, Abed MRM, Li K (2011) J Memb Sci 375:1
50. Cui M, Xu C, Shen Y, Tian H, Feng H, Li J (2018) Thin Solid Film 657:88
51. Es-Saheb M, Sherif EM, El-zatahry A, Rayes MME, Khalil KA (2012) Int J Electrochem Sci 7:10442
52. Sherif ESM, Es-Saheb M, El-Zatahry A, Alkaraki AS (2012) Int J Electrochem Sci 7:6154
53. Yabuki A, Shiraiwa T, Fathona IW (2016) Corrosion Sci 103:117
54. Doan TQ, Leslie LS, Kim SY, Bhargava R, White SR, Sottos NR (2016) Polym 107:263

Polymer Nanofibrous and Their Application for Batteries

Ahmed Ali Nada

Abstract Rechargeable batteries have been rapidly developed to meet the continuous demand for pursuing sustainable and renewable energy resources in order to reduce the fossil fuel consumption and CO_2 emission. In such batteries, electric energy is produced a via reduction–oxidation reactions between two electrodes and electrolyte. In order to obtain such batteries with higher energy and power densities, nanoscaled functional electrodes have been an approach. In this review, electrospinning basics and challenges to develop bead-free and smooth electrospun fiber are briefly provided. Advanced techniques to produce functional electrospun fibers such as coaxial and emulsion electrospinning are described. Basics of voltaic cells were extensively illustrated by highlighting the role of each component in the cell. Differences between primary and secondary cells are explained. Applications of such functional nanofibers in the battery electrodes and electrolytes to enhance the overall performance are discussed. The discussion was focused on Metal-ion and metal–air batteries including the rule of the electrospun substrates on enhancing anodes, cathodes, and electrolytes performances. Electrochemical characteristics in terms of ionic conductivity, battery density, and capacity were highlighted. The add-on value in the performance of such advanced nanofibers in metal–air and metal-ion batteries is broadly highlighted.

Keywords Energy storage · Rechargeable batteries · Metal–air batteries · Metal-ion batteries · Power density

A. A. Nada (✉)
Pre-treatment and Finishing of Cellulosic Textiles Dept., Textile Research Division, National Research Centre (Scopus Affiliation ID 60014618), Dokki, Giza, Egypt
e-mail: aanada@ncsu.edu

Centre for Advanced Materials Application, Slovak Academy of Sciences, Dúbravská cesta 9, 845 11 Bratislava, Slovakia

1 Introduction

1.1 Basics of Battery Electrochemistry

It has been a long time for climate activists calling for effective, efficient, and clean alternative energy sources to fossil fuel with limited carbon dioxide emission to save the planet earth [1, 2]. Concurrently, many scientists have been working globally on developing alternative clean energy resources such as wind energy [3, 4], solar energy [5, 6], and biogas energy [7, 8], etc. Such different sustainable resources of energy require energy storage systems. Basically, dry cells that have been developed since the 1880s are considered as a turning point for storing electricity and operating portable electronic devices [9]. Dry cells have been divided into primary (non-rechargeable) cells and secondary (rechargeable) cells. Both cell types follow the same operating mechanism by converting chemical energy to electric energy [10, 11]. However, rechargeable cells can convert back electricity to chemical energy as well [12].

Simply, electricity can be generated when electrons move from one point to another through a wire. Such phenomenon can be achieved when electron-weak puller metal is connected with electron-strong puller metal through a wire in which electrons move from anodes to cathodes spontaneously [13]. A traditional example of dry battery can be illustrated when zinc metal (weak pull for electrons) and copper (strong pull for electrons) rods are connected to each other through electrolyte. In such a process, one side (zinc) will lose electrons through oxidation reaction at anodes and the other side (copper) will gain electrons through reduction reaction at cathodes. This process is known as galvanic or voltaic cells where chemical reactions, specifically an oxidation/reduction reaction, create electricity [14].

The basic structure of the voltaic cells is illustrated in Fig. 1 in which the main components are represented, and the main idea of moving electrons is clarified [15].

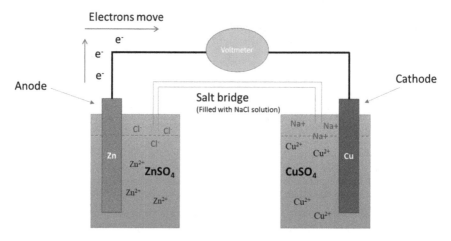

Fig. 1 Voltaic cell

A typical voltaic cell is composed of two half-cells, each half-cell contains a metal that is immersed in a solution [16]. Initially, Zn metal is oxidized by losing electrons that move through a wire to a cathode, made of copper (Cu), resulting in releasing Zn^{2+} ions in the $ZnSO_4$ solution. On the other half-cell, Cu is gaining electrons from the Zn half-cell resulting in converting Cu^{2+} ions in the $CuSO_4$ solution into Cu metal atoms. As a result, Zn rod in the first half-cell dissolves as Zn metal convert to Zn^{2+} ions. However, Cu rod, in the second half-cell, increases in mass as Cu^{2+} ions converts to Cu metal atoms. More importantly, a salt bridge made of NaCl helps to balance these charges that built up in the two half-cells by releasing Na^+ ions in the Cu half-cell and releasing Cl^- ions in the Zn half-cell.

Accordingly, the oxidation reaction can be summarized in the following equation (Eq. 1) in which sold Zn metal losing two electrons, as they move out, to turn to Zn^{2+} ion which is dissolved into the solution of the oxidation half-cell.

$$Zn(s) \rightarrow Zn^{2+}(aq) + 2e^- \qquad (1)$$

The reduction reaction can also be summarized in the following equation (Eq. 2) in which Cu^{2+} ion which is in the solution of the reduction half-cell gains 2 electrons and turns into solid Cu that is attacked by the copper rod (cathode).

$$Cu^{2+}(aq) + 2e^- \rightarrow Cu(s) \qquad (2)$$

Finally, the overall reactions that took place into the two half-cells can be put together in the following equation (Eq. 3):

$$Zn(s) \rightarrow Cu^{2+}(aq) + 2e^- \rightarrow Zn^{2+}(aq) + 2e^- \rightarrow Cu(s) \qquad (3)$$

Based on the reduction potential, E^0_{red} of the reduction half-cell and the reduction potential of the reverse oxidation half-cell (E^0_{oxi}) as mentioned in Eq. 4, total cell potential can be calculated by the addition of the two half-cells potential. In the case of Zn/Cu dry cells, the an overall cell potential is the sum of $E^0_{red} = 0.339$ V and $E^0_{oxi} = 0.762$ V to end up with overall standard cell potential equal to 1.101 V [17].

$$E^0_{oxi} = -E^0_{red} \qquad (4)$$

The calculation of the overall cell potential is carried out based on the standard reduction potential (Table 1) that occurred in the standard condition state (concentrations 1 mol/L, pressures 1 atm and temperature 25 °C) [18].

If the reduction/oxidation reaction process takes place in an unlike the standard condition, the following equation (Eq. 5) is applied to calculate the standard cell potential:

$$E_{cell} = -E^0_{cell} - \frac{RT}{nF} \ln Q \qquad (5)$$

Table 1 Standard Reduction
Potentials

Half reaction	Potential
Pb^{4+}	+1.67 V
Cu^{2+}	+0.34 V
Zn^{2+}	−0.76 V
Al^{3+}	−1.66 V
Li^+	−3.05

where E_{cell} = The cell potential at non-standard condition; E^0_{oxi} = cell potential at standard condition; R (gas constant) = 8.314 $J \cdot K^{-1} mol^{-1}$; T = Kelvin temperature; F = Faraday's constant; n = number of moles of electrons; Q = reaction quotient [19].

1.2   Factors Affecting the Cell Potentials and Current Density

In light of oxidation/reduction mechanism, it has been reported many factors that influence the cell electrochemical properties such as the recharging cell mechanism, the cell voltage, and the cell current density [20].

In rechargeable batteries, oxidation/reduction reactions, occurred in the dry cells, have to be reversible in which negative charges are compelled to move toward the anode (in the charging phase) instead of the cathode. For instance, lead–acid battery [21] is one of the most popular rechargeable batteries found in the automobiles which have lead metal as an anode and lead oxide as a cathode [22]. The cell potential of one single lead-acid cell equal to 2.02 V, while a six-pack of lead-acid cells packed together in a parallel position can obtain ~12 V battery [23]. Accordingly, cell potential can be increased by using multiple packs of cells connected in barrel position.

However, the cell current density can be significantly increased by increasing the surface area of anode by using, for instance, electrodes in powder form instead of metal form [24]. Based on this phenomenon, the electrospinning technique has been recently employed to obtain nanofibers as a platform for advanced rechargeable/chargeable batteries in terms of improving the power density and cyclability [25]. Therefore, nanofibers have been immensely involved in secondary battery composition in anodes [26], cathodes [27], separators [28], and as catalytic materials [29], as it will be discussed in the next subsections.

1.3   Basics of Electrospinning Technique

Electrospinning technique has been discovered in the last decades in order to produce fibers in nanoscaled diameter by using high-power supply [30–33]. Basically, the

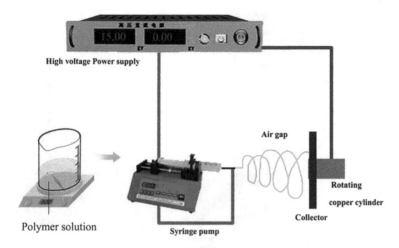

Fig. 2 Basic diagram of the electrospinning apparatus

average of the human hair diameter is 17–171,000 nm, while the average diameter of electrospun fibers can range from few microscales to 30–50 nm or less [34]. Electrospinning technology has several benefits such as high surface area of the electrospun fibers; producing fibers made of unconventional substrates and using dry process with limited hazard solvents [35].

The main components of the electrospinning technique can be summarized as high-power supply (0–60 kV or more); feeding system (syringe pump in nozzle or drum in needless systems) and receiver plate, drum, or conveyor. Figure 2 shows the basic diagram of the house-made electrospinning apparatus in which single nozzle is used to inject the polymer solution through a syringe needle connected to a high-power supply. The ejected fibers are passing through the airgap in which solvents get evaporated and dry nanofibers are deposited onto a plate receiver.

However, electrospinning apparatus has been developed by replacing the plate collector by drum collector (Fig. 3) with speed controller [36, 37]. Such a drum collector has been used to enhance the fiber alignment and control the fiber diameter [38, 39].

In order to increase the productivity of the electrospun fibers, the single nozzle has been changed to multi-nozzle [40, 41] or shower-like extrusion systems (Fig. 4), and the ejected fiber are received either on drum collectors or conveyors [42, 43]. Recently, needleless electrospinning [44, 45] has been emerged to increase the mass production of nanofibers. In such technique, ground-charged wire is adjustable, in the horizontal direction, to control the airgap distance. The receiving non-woven substrate is fixed on rolling cylinders, in the vertical direction, with a speed controller to adjust the thickness of the electrospun mats (substrate speed range 0–5000 mm/min). Power supply can be monitored and provide a wide range span from 0 to 80 kV. The polymer cartridge boat carries solutions range from 20 mL to 50 mL [46].

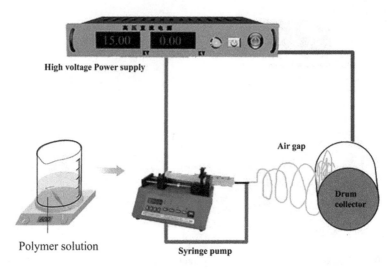

Fig. 3 Advanced electrospinning apparatus with drum collector

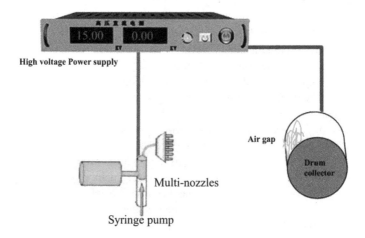

Fig. 4 Multi-nozzle electrospinning apparatus with drum collector

2 Applications of Electrospun Fibers in Batteries

2.1 Electrospun Fibers in Metal–Air Batteries

Principally, metal–air batteries are based on pure metal anodes made of lithium, zinc, or aluminum and external cathode of ambient air with aqueous electrolytes. Typically, as electrons move from anodes during the discharging process, metal anode is getting oxidized. The specific capacity and energy density of such metal–air batteries are

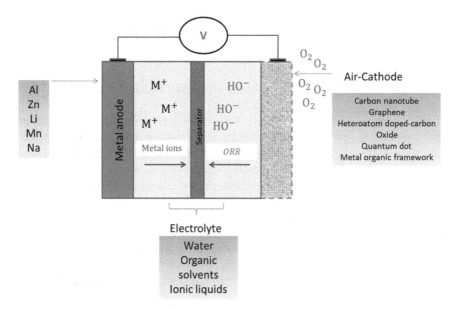

Fig. 5 Schematic diagram of metal–air batteries

much higher than that of metal-ion batteries. For example, lead–acid batteries can produce 30 watt-hours per kilogram (kg) of storage unit, while lithium-ion batteries can produce 100–200 watt-hour per kg of storage unit. However, aluminum–air batteries can produce 8100 watt-hour per kg of storage unit [47].

Typically, the generated electricity in such type of batteries, in discharging process, is carried out by the oxidation reaction of the metal to superoxides or peroxides (Fig. 5) by which electrons move out to the air-cathodes reacting with oxygen to produce water (H_2O) and hydroxides (HO^-) in a process called oxygen reduction reaction (ORR). Metals such as lithium, sodium and potassium which are very sensitive to water, are oxidized in aprotic solvents instead [48].

The reliability of such batteries can be precisely measured by calculating different parameters such as the polarization performance, the round-trip efficiency, and the Coulombic efficiency. The polarization performance can be calculated by measuring the current density at a specific discharge voltage, while round-trip efficiency gives an information about the ratio between the energy released in the discharging process and the required energy in the charging process. The Coulombic efficiency is the most important parameter for rechargeable batteries which is defined as the ratio between the charge capacity and the discharge capacity at full cycle of discharge–charge process. In optimum conditions for batteries, Coulombic efficiency must be %99.98 to show initial capacity higher than 80% after 1000 charging cycles.

Metal–air batteries [49] have some limitations with the irreversible consumption of the metal electrodes resulting in low Coulombic efficiency. Also, the reduction reaction of oxygen in the air-cathode is very sluggish in kinetic. Much has been done

to mitigate such drawbacks and to enhance the life-cycle to metal–air batteries that
will be addressed in the following subsections [50].

2.1.1 Electrospun-Based Electro-Catalyst for Metal–Air Batteries

In order to enhance the rechargeability of metal–air batteries, bifunctional catalysts
have been used to regulate the reduction reaction of oxygen and lower the overpotentials of the discharging and charging processes. Carbon nanotube-based bifunctional
catalysts have been used [51] immensely to accelerate the oxygen reduction reaction (ORR) in the discharging process (Fig. 6) and the oxygen evolution reaction
(OER) in the charging process. Heteroatom-doped graphene-based electrocatalysts
[52], Perovskite oxides [48], and many others have been used to accelerate the same
process. However, other research groups have been worked on changing the surface
area of such electrocatalysts by using electrospinning technique to boost their activities. Also, hydrogels [53–55] with high ionic conductivity have been used to enhance
the reversibility and stable cycle [56], as flexible supercapacitor electrodes [57] and,
gel electrolyte [58].

Xu et al. [59] developed porous Perovskite oxide-based catalyst nanotube by
using the electrospinning technique to produce nanofiber forms that were exposed
to a calcination process at 650 °C for 3 h. Final porous catalyst showed a significant
acceleration to the ORR and OER processes, lower the overppotentials and resulting
in improving the round-trip efficiency in lithium–air batteries. Such high catalytic
activity of the porous Perovskite-based catalyst showed high specific capacity and

Fig. 6 Oxygen reduction
reaction mechanism in the
metal–air cathodes

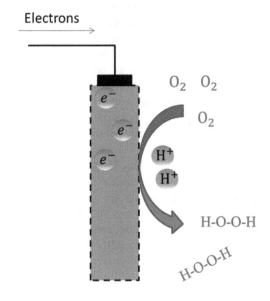

good cycle stability for batteries. Park et al. [60] reported a new class of bifunctional electrocatalyst to ORR and OER based on porous nanorods of $La_{0.5}Sr_{0.5}Co_{0.8}Fe_{0.2}O_3$ and reduced graphene oxide/nitrogen-doped. Authors blended the inorganic catalysts with fluoro-based polymer, Nafion, as a carrier to obtain electrospun fibers which was calcinated at 700 °C for 3 h. Nanoscaled composite was mixed with N-doped reduced graphene oxide to end up with very active bifunctional electrocatalyst for rechargeable metal–air batteries by accelerating ORR and OER in alkaline electrolyte. This is due to the large surface area of the porous structure of the prepared electrocatalyst.

Peng et al. [61] fabricated air-cathode for zinc–air battery by using the coaxial electrospinning technique in different approaches. Author used polyacrylonitrile polymer as a shell layer which impregnated with cobalt acetate tetrahydrate and thiourea. While poly(methyl methacrylate) was used as a core substrate. Mixture was electrospun at 18 kV with shell diameter 750 nm and core diameter 420 nm. After thermal treatments of the electrospun fibers, poly(methyl methacrylate) was vanished to gases and the shell substrate (poly-acrylonitrile) was carbonized via stabilization thermal treatment at 280 °C for 0.5 h and at 800 °C for 2 h under argon atmosphere. This process produced hollow fibers made of carbon nanofibers decorated by cobalt/sulfur-doped. Data revealed that the hollow and porous structure of the doping of nitrogen and sulfur showed a better electrical conductivity, bifunctional catalysts, and exhibited preferable performance toward ORR.

2.1.2 Electrospun-Based Composites of Anodes for Metal–Air Batteries

Anodes in metal–air batteries, during the discharging and charging processes, is subjected either to shape change or corrosion especially in alkaline electrolytes which lead to shortening of cycle life [50]. To solve this issue, it has been developed different methods to decrease or prevent the corrosion process during the discharging/charging process. It has been reported that alloying pure metals such as aluminum [62] with different metals such as copper [63], magnesium [64], titanium [65], etc., has shown a significant impact to reduce the corrosion tendency in the alkaline electrolytes. Also, it has been investigated that adding either organic or inorganic additives [66, 67] as inhibitors can reduce the corrosive effect of electrolytes. Recently, coating anode surface by protecting interlayer has been reported to control the anode corrosion. Sol–gel method has been used to fabricate Al_2O_3 thin film on Zn-electrode in order to mitigate the electrode corrosion [68].

Zuo et al. [69] electrospun polyacrylonitrile incorporated with Al_2O_3, using DMF as a solvent, to achieve a 4 µm electrospun mat on an aluminum foil sheet. Fiber diameters were around 450 nm and thermally stabilized at 300 °C for 2 h. The produced interlayer helped to suppress the corrosion of the aluminum anode and resulted in a significantly high capacity (1255 mAh/g at 5 mA/cm^2) and a remarkable stability.

2.1.3   Electrospun-Based Composites of Cathodes for Metal–Air Batteries

Metal–air batteries are distinguished by air-cathode in which the main reactant, oxygen, is obtained from air. Therefore, cathode is composed of electrocatalyst (to reduce the electrode overpotential) and a porous layer. The latter layer acts as a gas-diffuser to regulate the oxygen diffusion through the electrocatalyst. The main concern in the air-cathode is the accumulation of reaction products such as Li_2O_2, and Li_2O, in case of lithium–air batteries, on the cathode. Therefore, air-cathode has to be fabricated of high electroconductive substrates with large surface area and porous structure in order to facilitate the electrons pathways. As a result, the fabrication methods of the air-cathodes have been widely investigated to overcome the above-mentioned issues.

Casting is the most considerable method to fabricate the air-cathode in which conductive paste was cast on conductive metals in a foam form. However, such method leads to side reactions during the discharging and charging processes [15]. Therefore, other routes have been discovered to produce self-standing electrospun composites as air-cathodes.

Song et al. [70] fabricated such self-standing air-cathode based on cobalt ions linked to benzimidazolate ligands that has been blended with carbonization-capable polymers such as polyacrylonitrile. The latter slurry was electrospun followed by two-post thermal treatments to end up with Co_3O_4/nanotube composites that used as an air-cathode without the need to a binder or metal foam. The results demonstrated much higher discharging capacities (760 mA h/g), lower charging overpotential and enhanced cycle performance.

Nitrogen-doped carbon nanofibers containing iron carbide has been electrospun by Ma et al. group [71] and was utilized as an air-cathode in flexible aluminum–air battery. Authors synthetized Fe_3C nanophases that were encapsulated in nitrogen-doped carbon nanofibers. Typically, polyacrylonitrile polymer was used as a carrier to contain the iron metal-organic frameworks which was electrospun with an average fiber diameter of 300 nm. The produced fibers were exposed to a thermal treatment in nitrogen atmosphere to obtain N-doped porous carbon nanofibers decorated with Fe_3C nanoparticles. This study showed outstanding catalytic activity and stability toward oxygen reduction reaction and showed a stable discharge voltage (1.61 V) for 8 h, giving a capacity of 1287.3 mA h/g.

Bending-resistant cathode for aluminum–air battery has been fabricated based on carbon nanofibers incorporated with magnesium oxide (Mn_3O_4). It has been reported [72] that magnesium oxide was mixed with polyacrylonitrile to be electrospun in single needle at 17 kV. Electrospun fibers, with average fiber diameter of 400 nm, were calcinated at 900 °C for 1 h. Data showed that the fabricated Al–air battery can be discharged over 1.2 V under 2 mA/cm^2 at dynamic bending state, and a specific capacity up to 1021 mA h/cm^2.

Bui et al. [73] reported the electrochemical performance of non-woven mats made of carbon nanofibers incorporated with different metals (platinum, cobalt, and palladium) as air-cathodes for lithium-oxygen battery (in organic solvent as electrolyte). Authors fabricated the air-cathodes by using a typical carbonization-capable polymer, polyacrylonitrile, as a core substrate and poly(vinylpyrrolidone) [46] as a shell substrate. Metal precursors (platinum acetylacetonate, palladium acetate and cobalt acetate tetrahydrate) were added to the core substrate and were electrospun at 12 kV. Electrospun fibers with fiber diameter 720 nm (390 shells and 165 nm core) were subjected to thermal treatments started by stabilization in air at 300 °C for 4 h, and then under nitrogen at 300–750 °C for 1 h and at 750–1200 °C for 2 h. The electrochemical performance of the pure carbon nanofiber cathode was compared with those decorated with the metals. Results stated that the discharge/charge profiles are looked the same. However, the specific capacity of platinum-decorated carbon nanofibers was much higher (5133 mAh/g at 1000 mAh/g) than that of the reference cathode (1533 mAh/g at 1000 mAh/g). Also, data showed overpotentials of both on discharging and charging processes reduced with platinum-decorated carbon nanofibers and remarkably prolonged cycle life (163 cycles).

2.2 Electrospun Fibers in Metal-Ion Batteries

Metal-ion rechargeable batteries are considered as fast-growing technologies for energy storage devices, in the last few decades. This is due to their high density of power and energy associated with long life cycle [74]. Basically, metal-ion cell as a rechargeable cell, metal ions move from negative electrodes to positive electrodes in discharging process and vice versa in charging process. The flow of such ions leads to a flow of electrons that move in a circuit to generate energy. Accordingly, Metal-ion batteries consist of four main components namely: anode, cathode, electrolyte, and separator. The most common metals used in such batteries are aluminum, lithium, sodium, and potassium. In the following subsections, utilization of electrospinning technique in metal-ion batteries will be limited and discussed for lithium-ion batteries as a role model.

2.2.1 Electrospun Materials for Lithium-Ion Batteries

Although lithium-ion batteries have been used globally in different applications, researchers are still urged to develop low cost, high power density, and high energy lithium-ion batteries. Typically, lithium-ion batteries have been made of graphite cathodes and lithium metal-based anodes in the presence of lithium salt mixed in organic solvents [75]. As a result of many incidents regarding to lithium-operated portable devices, lithium oxides and different alloys have been developed to provide safe and reliable lithium batteries. The main measurements used to reflect the batteries' performances are the specific capacity and the operation current densities. These two parameters are determined by the electrochemical performance of the electrode materials used in batteries [76].

Recently, nanostructured electrodes have been utilized to enhance the electro-chemical performance in metal-ion batteries. This is due to the tremendous increase in the surface area which leads to a decrease in the mass and charge diffusion, shortens the transporting path of ions, increases the electron transfer, and finally improves the intercalation kinetics [77].

Many different approaches have been tackled to prepare one-dimensional nano-materials for electrodes such as chemical vapor deposition, self-assembly, solvothermal method, solution-growth and electrospinning technique. The latter approach, electrospinning technique, is very simple way to produce one-dimensional nano-sized materials for electrodes with wide diversity in morphological character-istics.

2.2.2   Electrospun-Based Cathodes in Lithium-Ion Batteries

In lithium-ion batteries, much has been studied to improve the energy density and the operating cell voltage. Lithium-based transition metal oxides showed very promising approach to obtain high energy density for such type of batteries. Of these composi-tions, lithium/iron/phosphate composite has been used due to it thermal stability and large energy density (170 mAh/g). However, this alloy showed lower cell operating voltage (3.5 V). Kang et al. [78] studied the effect of introducing manganese to the previous composite in electrospun fibers. The authors prepared electrospinning solu-tion based on poly(vinyl pyrrolidone) with lithium, manganese, iron, and phosphate salts and ejected it under certain condition for electrospinning. Solutions were elec-trospun at 10–15 kV with average fiber diameter of 100–500 nm. The electrospun fibers were air-dried at 100 °C, calcinated at 500 °C for 10 h, and then post-calcinated at 800 °C in nitrogen atmosphere. Hagen et al. [79] reported that the cyclic voltam-metry analysis showed the cell operating voltage increased from 3.5 to 4.1 V in the presence of manganese element in lithium cathode. Study showed that by increasing the manganese content in the electrospun fibers, cell voltage increased over 4.6 V with maximum discharge capacities 125 mAh/g.

Many different polymers have been used in order to provide self-supporting cathodes made of lithium-based transition metal oxides by utilizing electrospin-ning technique. Polyacrylonitrile has been used by Toprakci et al. [80] with $LiFePO_4$ composite to provide carbon nanotube-supported Li-cathode with total high capacity 166 mAh/g.

Lithium vanadium phosphate ($Li_2V_2(PO_4)_3$) composite is another Li-based transi-tion metal oxide that has been addressed in order to obtain better energy density. This cathode composition provides 190 mAh/g capacities with stable three-dimensional framework and poor electrochemical output. To solve this problem, the ionic conduc-tivity has been improved by developing nano-scaled platform of $Li_3V_2(PO_4)_3$. For example, Chen et al. [81] used poly (4-vinyl) pyridine as a carrier, mixed with NH_4VO_3, $NH_4H_2PO_4$, and citric acid for electrospinning. Mixture has been elec-trospun at 32 kV to obtain nanofibers with average diameters of 170–440 nm. These fibers have been calcinated at 800 °C for 4 h to retain the fibrous structure with average diameters 90–220 nm. $Li_3V_2(PO_4)_3$) carbon nanofibers composite exhibited good cycle performance, capability in average voltage 3.0–4.8 V, and high discharge capacity of 190 mAh/g.

Another research group prepared two different layers of $LiFePO_4$ and $Li_4Ti_5O_{12}$ carbon nanofibers to study the electrochemical performance [82]. Different metal salts were mixed with polyacrylonitrile and polyvinylpyrrolidone, electrospun at 25 kV, pre-oxidized at 260 °C for 2 h and then calcinated at 800 °C for 10 h in N_2 atmosphere. The two-layer cathode showed very promising capacity in terms of charged/discharged processes up to 800 cycles at 1C with a retention capacity of more than 100 mAh/g and a Coulombic efficiency close to 100%.

2.2.3 Electrospun-Based Anodes in Lithium-Ion Batteries

Anodes for lithium-ion batteries have a great impact on the overall performance in terms of charge/discharge rate capability, cyclability, and energy storage capacities. Developed anodes for Li-ion batteries showed several principles that have to be addressed such as cost and environmental concerns, hosting large numbers of Li-ions during the charging process, made of materials that are insoluble in the electrolyte solvents, and have a reduction potential as high as lithium metal (–3.05 V).

Over the past decade, graphite has been chosen as a promising candidate for lithium-ion anodes due to its abundance, reversibility with high Coulombic efficiency, and energy capacities (372 mAh/g). However, towards electric car applications require long cell lifetime and cell operating voltage, there are many challenges to anode materials that have to be addressed in the fabrication process. Again, nanostructured anodes have been an approach to solve the problem addressed above. This hypothesis has been firstly approved, in 1996, when Liu et al. [83] prepared hard carbon samples from calcinated epoxy resins with nanoporosity. Study proved that large specific capacity for lithium was recorded as a function of the number of single carbon layers in the heat-treated epoxides. Since then, increasing the surface area of electrodes is an approach to enhance the lithium-ion batteries.

Electrospinning technique was one of these approaches to fabricate Li-ion anode with very high surface area which is capable of hosting many lithium ions during the charging process. Kumbar et al. [84] prepared carbon nanotube using a typical procedure of the electrospinning of polyacrylonitrile followed by a calcination step in inert atmosphere (N_2) at 800 °C. Polyacrylonitrile was electrospun at 21 kV with gap distance 14 cm to obtain fibers with average diameters 200 nm. The electrochemical lithium storage properties showed discharge capacities of 826 and 370 mAh/g for the first and second cycles respectively, at a current density of 200 mA/ g with capacity 200 cycles.

Kim et al. [85] used 10% of polyacrylonitrile in dimethylformamide and electrospun the solution using 25 kV. Polyacrylonitrile electrospun fibers were obtained with average diameter 200–300 nm, thermally treated at 280 °C for 1 h and then calcinated at three different temperatures 700, 1000 and 2800 °C under inert atmosphere (argon). Data showed that the best results were recorded to the nanofibers that calcinated at 1000 °C to give a large reversibility of 450 mAh/g and high rate capability 100 mAh/g.

Other elements have been addressed to improve the graphite capacity (372 mAH/g) such as incorporating tin, tin oxide, and tin composites. Theoretical capacity of tin is 993 mAh/g and when it mixed with graphite anodes enhances the overall capacities. However, because its expansion during charge/discharge process, it cracks and results in rapid fading to the cell capacity. Nanostructured tin or tin alloy in different shapes such as nanoparticles, thin films and nano-wire have been reported to solve this problem. For this purpose, Zou et al. [86] prepared tin/carbon non-woven film via electrospinning technique. Author used polyvinyl alcohol (PVA), 10% wt/v of 80,000 MW, mixed with 10% wt/v of tin(II) chloride and 20% v/v distilled water. Solution was electrospun at 25 kV, air-gab distance (15 cm) and flow rate (1 mL/h). Electrospun fibers with tin nanoparticles in 0.33 nm was dedicated by high resolution transmission electron microscope (HRTEM) and was heated at 500 °C in argon/ H_2 atmosphere for 3 h. The electrochemical investigation of the non-woven film showed a reversible capacity after 20 cycle of a 382 mAh/g which is 96% of the capacity in the first cycle.

Nickel oxide is another element that showed a significant interest to enhance the lithium-ion batteries owing to its high theoretical capacity (718 mAh/g). However, in bulk form, nickel oxide showed very poor electrochemical performance due to its large volume change during charge/discharge process and low electronic conductivity. Porous nickel oxide anodes with nano-morphologies such as nano-sheet, nano-wall, and nano-spheres have been considered to improve the electrochemical properties. Wang et al. [87] prepared lithium-ion anode based on carbon nanofibers incorporated with nickel oxide. Authors used polyacrylonitrile (10% wt/v) mixed with nickel nitrate salt for electrospinning at 10 kV, air-gab distance 15 cm and flow rate 0.9 mL/h. Nanoweb was thermally treated at 300 °C for 1 h and at 600 °C for 5 h and then at 350 °C for 2 h in inert atmosphere (N_2). The porous anodes including nickel oxides showed high reversible capacity of 638 mAh/g over 50 cycles.

Another super promising candidate, silicon, has been reported for lithium-ion batteries owing to its very high gravimetric specific capacity (4200 mAh/g). Like other elements, silicone has four times volume expansion during charge/discharge process resulting in rapid fading of the battery capacity. Silicon nanoparticles (50 nm) have been mixed [88], in different ratios, with 7.5% wt/v solution of polyacrylonitrile (MW 86,000) in dimethylformamide to obtain electrospun composites. This nanoweb composite was pre-oxidized in air for 6 h at 240 °C to protect the fibrous morphology during following carbonization steps. Post-thermal treatment at 600 °C in inert atmosphere (argon) was conducted to the nanoweb composites for 8 h. At ratio C/Si (77/23 wt/wt), silicon composite carbon nanofibers showed a large reversible capacity up to 1240 mAh/g.

Agglomeration of silicone nanoparticles has been an issue to obtain well distributed electrospun precursor solutions for electrospinning. Xu et al. [89] worked on de-agglomeration of silicon nanoparticles by conducting different treatments in order to obtain well-dispersed solution for electrospinning. Silicon nanoparticles (50–100 nm) were initially stirred magnetically in piranha solution (H_2SO_4/H_2O_2=7:3

v/v) for 2 h at 80 °C, centrifuged in deionized (DI) water and diluted into 200 mL ethanol solution to obtain hydroxyl-terminated silicon. The latter was treated via 3-aminopropyl trimethoxysilane to obtain the amino-silane functionalized silicon nanoparticles. The modified silicon nanoparticles were mixed with 7% wt/v polyvinyl alcohol (MW 86000–124000) which is more compatible with the functionalized silicon nanoparticles. Nanoweb was stabilized at 200 °C for 2 h and then at 650 °C for 1 h in inert atmosphere (nitrogen gas). The obtained electrode exhibited an excellent electrochemical performance with a discharge capacity of 872 mAh/g (after 50 cycles) and capacity retention of 91%.

In the light of the above mentioned modifications, silicon nanoparticles had to be treated through sophisticated processes and hazard materials in order to provide well distributed nanoparticles. Therefore, it has been a challenge to find out an ecofriendly alternative to synthesis carbon-based composites with good electrochemical properties.

Cobalt oxide has been an example for experimental trials to obtain carbon-based electrode with large reversible capacity, excellent cyclic performance, and good rate capacity. Zhang et al. [90] prepared cobalt oxide-based carbon nanofibers via the electrospinning of polyacrylonitrile, Mw 150000, mixed with Cobalt acetate tetrahydrate. After thermal treatment at 650 °C for 2 h in nitrogen gas, data showed that cobalt compound was CoO rather than Co or Co_3O_4. The obtained cobalt-based electrode showed a good electrochemical performance of 633mAh/g after 52 cycles.

2.2.4 Electrospun-Based Separators and Electrolytes in Lithium-Ion Batteries

As described before, electricity is produced from lithium batteries as electrons move in a wire and lithium ions move in the electrolyte between cathodes and anodes back and forth during the charge/discharge process. Accordingly, electrolyte with very high ionic conductivity is highly demanded to the lithium-ion batteries as long as it meets the environmental, safety, and cost concerns. It has been using the organic solvents in such lithium-ion batteries until concerns about flammability, explosion, and volatilization have been raised [91].

On the other side, separators have very crucial role in secondary batteries and especially in lithium-ion batteries to ensure the safety of the batteries by preventing the direct contact between anodes and cathodes and allow ion transfer through microscopic holes. Accordingly, separators must satisfy all the physical and electrochemical conditions.

Much has been investigated that solid polymer electrolytes can be employed in lithium-ion batteries as electrolyte and separator, in the same time, provided to meet the basic requirements of the electrochemical conductivity and the chemical, thermal and mechanical stability [92]. In the very beginning, polyethylene oxide (PEO) has been investigated as a promising candidate to provide thinner and safer lithium-ion batteries. Different drawbacks in PEO that have been reported constrain its applications in lithium-ion batteries. PEO high crystallinity especially at room temperature leads to a huge constraint in the ionic conductivity. The deterioration in PEO mechanical strength at higher temperatures is another reason. These limitations of solid polymer electrolytes remain as main challenges to upscale the battery production.

Therefore, much has been done in this field to enhance the PEO crystallinity by different means such as chemical modifications, blending with other polymers, and mixing with conductive additives [91]. In this part, fabrication of such polymer solid electrolytes via electrospinning will be the main concern.

Samad et al. [93] investigated the effect of blending PEO solutions with a novel cellulosic reinforcement material, named GELPEO, on the mechanical properties of PEO electrospun fibers as solid polymer electrolyte. Author electrospun 10% wt/v of PEO (Mw= 300,000) associated with different concentrations of GELPEO (5, 10 and 20 wt/v) at 20 kV, flow rate 1 mL/h and the electrospun fibers were received on rotating drum at 200 rpm. Data showed that tensile strength values have improved by two-fold and the composite fibers are thermally stabilized up to 200 °C. Unlike the expected, addition of 5% wt/v of GELPEO did not show significant reduction in the ionic conductivity and gives very comparable measurements (4×10^{-4} S/cm) compared to PEO fibers (5×10^{-4} S/cm).

Another research group from Iran [94] studied the effect of ZnO and TiO_2 nanoparticles embedded on electrospun fibers of PEO associated with lithium perchlorate. Fibers were electrospun at 18.4 kV, flow rate 0.5 mL/h, distance 15 cm and fibers were collected on rotating drum at 100 rpm. Authors used a potentiostat/galvanostat instrument to evaluate the cycling stability of PEO with/without ZnO and TiO_2 nanoparticles. Data showed that highest values of the ion conductivities were recorded for of 0.21 wt% of the TiO_2 and ZnO to reach 0.045 mS/cm and 0.035 mS/cm, respectively. These values were considered much higher than that of the same composition used for casting thin films which recorded 0.0044 mS/cm and 0.0147 mS/cm for TiO_2 and ZnO respectively. However, it has been reported that such filler-filled electrospun solid electrolyte loss up to 40% of its capacity after 45 cycles.

The same research group [95] investigated the effect of ethylene and propylene carbonates (EC and PC), as plasticizers to the PEO electrospun fibers, on the electrolyte electrochemical performance. Authors recorded the highest value in ionic conductivity, at ratio 3:1 (EC:PC), 0.171 mS/cm. More PC to the PEO resulted in decreasing the cycle capacity significantly.

Zhu et al. [96] studied the effect of adding high ionic conductive ($Li_{0.33}La_{0.55}TiO_3$) nanowires to PEO electrospun fibers on the electrochemical properties of the final solid composite electrolyte. The ionic conductive nanowire was prepared by the electrospinning of PVP solutions mixed with lithium salt ($LiNO_3$), lanthanum salt ($La(NO_3)_3$) and Ti $(OC_4H_9)_4$, dissolved in DMF solutions and followed by thermal treatments. Authors obtained nanofibers of PEO incorporated with the ($Li_{0.33}La_{0.55}TiO_3$) nanowires along with propylene carbonate, plasticizer, as a solid composite electrolyte. Data showed that by adding 8% of the nanowires to PEO, the ionic conductivity of the composite has reached to the maximum value at 5.66×10^{-5} and 4.72×10^{-4} at 0 °C and 60 °C, respectively. The obtained composite electrolyte exhibited an initial reversible discharge capacity of 135 mAh/g and good cycling stability.

Poly(vinylidene fluoride), PVdF has been used as solid polymer electrolyte in lithium batteries due to its high mechanical stability, its polar nature owing to the fluorine atoms, and for being chemically inert. However, PVdF showed a low ionic conductivity owing to its crystallinity which leads to migration hindrance of lithium ions.

Gopalan et al. [97] prepared electrospun solid polymer electrolyte based on PVdF mixed with different amounts of poly(diphenylamine), PDPA (0.5, 1 and 2% w/w). Mixture was electrospun at 25 kV, 10 mL/h flow rate and 15 cm distance to obtain electrospun fibers with average diameter 200 nm. The final electrolyte composite of PVdF/PDPA electrospun fibers were soaked in a mixture of lithium salts (lithium perchlorate) and propylene carbonate. Electrochemical properties showed superior activity in terms of ionic conductivity, electrochemical stability and good interfacial behavior with electrode.

The same research group investigated the electrochemical properties of the composition of PVdF with polyacrylonitrile (PAN) prepared via electrospinning [98]. Mixture of PVdF and different proportions of PAN were dissolved in DMF:acetone (7:3 v/v) and electrospun at 25 kV, 10 mL/h flow rate, 20 cm distance and fibers were received on rotating drum. The electrolyte composite was obtained by soaking the electrospun mat in a mixture of lithium salts (lithium perchlorate) and propylene carbonate. PVdF/PAN composite electrolyte, prepared by 25% PAN, showed a high amount of lithium salt uptake of the electrolyte (300%) and a high ionic conductivity of 7.8 mS/cm.

The copolymer of PVdF with hexafluoropropylene (HFP) has been reported as a good solid polymer electrolyte owing to its good electrochemical stability and affinity to electrolyte solutions. Li et al. [99] reported the electrochemical properties of PVdF-co-HFP electrospun membrane. Polymer ($M_w = 4.77 \times 10^5$) at a concentration of 12–18 wt/v, dissolved in acetone/dimethylacetamide (7/3, wt/wt) was electrospun at 18 kV to obtain fibers of average diameter 1 μm. The electrolyte was prepared when PVdF-co-HFP was soaked in lithium salts/propylene carbonate solutions. The solid copolymer (PVdF-co-HFP) electrolyte showed a high electrolyte uptake and ionic conductivities of 10^{-3} S/cm.

Other research groups investigated the influence of the incorporation of ceramic fillers on the ionic conductivity of the solid polymer/composite electrolytes. Raghavan et al. [97] reported the electrochemical performance of the electrospun composite of PVdF-co- hexafluoropropylene and silica. The copolymer was mixed with in-situ prepared silica and ball mill prepared silica and their electrochemical characteristics were compared to PVdF-co- hexafluoropropylene fibers. Electrospun solutions were obtained at 20 kV, 0.1 mL/min flow rate, 16 cm distance, and received on rotating drum at 140 rpm. Obtained electrospun fibers were investigated on scan electron microscope (SEM) to show an average diameter of 1–2 μm. The final composite electrolytes were prepared by immobilizing lithium salt (lithium hexafluorophosphate) and ethylene carbonate/dimethyl carbonate in the electrospun mats. In general, the prepared composites exhibited high electrolyte uptake (550–600%), while the superior electrochemical performance recorded for the polymer electrolyte containing 6% in situ silica with ionic conductivity of 8.06 mS/cm at 20 °C.

Cui et al. [100] investigated more complicated approach by preparing solid composite electrolyte when PVdF was mixed with modified titanium dioxide (TiO_2). Initially, aminated TiO_2 was grafted by poly(methyl methacrylate) via atom transfer radical polymerization technique. The grafted TiO_2 was mixed with PVdF before electrospinning. Fibers were electrospun at 14 kV and 20 cm distance. Electrospun fibers of PVdF/grafted TiO_2 with an average diameter 0.333–0.336 um were soaked in lithium salt (1 M of lithium hexafluorophosphate) in ethylene carbonate/DMF. Data showed that the presence of grafted TiO_2 inhibits the crystallization of PVdF in the solidification process and enhances the ionic conductivity of the final solid composite electrolyte. The improved electrochemical performance was recorded for the composite electrolyte containing 6% wt (based on the weight of PVdF) grafted TiO_2 and showed ionic conductivity of 2.95 mS/cm at 20 °C compared to 2.51 mS/cm of PVdF electrolyte.

Thermoplastic polyurethane (TPU) electrospun fibers have been employed as mold for solid composite electrolyte when it soaked in fillers to enhance the ionic conductivity. In 2018, Gao et al. [101] electrospun 17% wt/v of TPU at 20 kV, at 40 °C and used rotating drum as a collector. The obtained mat was soaked in PEO solution containing nano-sized SiO_2 and lithium bis(trifluoromethanesulfonyl)imide (LiTFSI)salt. Data revealed that the final composite electrolyte of TPU-PEO with 5 wt% SiO_2 and 20 wt% LiTFSI showed an ionic conductivity of 6.1×10^{-4} S/cm at 60 °C with a high mechanical stability of 25.6 MPa. Battery made of this solid composite electrolyte and $LiFePO_4$ cathode showed a discharge capacity of 152, 150, 121, 75, 55 and 26 mA h/g at C-rates of 0.2C, 0.5C, 1C, 2C, 3C and 5C, respectively. The discharge capacity of this lithium-ion battery remains 110 mA h/g after 100 cycles at 1C at 60 °C with capacity retention of 91%.

However, Zainab et al. [102] used the electrospun fibers of polyurethane mixed with polyacrylonitrile to build up a composite used as separator in lithium-ion batteries. Fibers were obtained using electrospinning technique at 25 kV, at flow rate 1 mL/h, 15 cm distance and collected on rotating drum at 50 rpm. Data revealed that ionic conductivity has improved up to 2.07 S/cm, with high mechanical stability up to 10.38 MPa and good anodic stability up to 5.10 V were observed. The thermal stability of PU/PAN separator displayed only a 4% dimensional change after 0.5 h of long exposure at 170 °C.

Different combinations of several conductive polymers have been utilized as solid composite electrolyte for lithium-ion batteries. Peng et al. [103] obtained new electrospun electrolyte made of TPU and PVdF-*co*-HFP dissolved in DMF/acetone (1:1 wt/wt) and electrospun at 24.5 kV. The vacuum-dried electrospun mat was soaked in 1 M of lithium perchlorate/ethylene carbonate. Data showed that the ionic conductivity value has enhanced up to 6.62×10^{-3} S/cm. Composite showed very decent value of tensile strength (9.8±0.2 MPa) and elongation at break (121.5±0.2%). Battery, made of this composite electrolyte and Li/PE/$LiFePO_4$ cathode, provides a high initial discharge capacity of 163.49 mAh/g under 0.1 C rate.

Tan et al. [104] fabricated electrospun fibers made of a mixture of three polymers PAN, TPU and polystyrene (PS) in mass ratio 5:5:1. The three polymers were dissolved under vigorous stirring in DMF for 12 h at 60 °C before the spinning step.

Electrospun fibers of PAN/TPU/PS were obtained at 24 kV and extruded at flow rate 0.5 mL/h. The composite electrolyte was obtained by soaking these electrospun fibers in 1 M of a solution of lithium hexafluorophosphate/ethylene carbonate. The prepared solid composite electrolyte showed an ionic conductivity of 3.9×10 mS/cm at room temperature and an electrochemical stability of 5.8 V. However, battery made of this composite electrolyte and LiFePO$_4$ cathode, exhibited charge and discharge capacities of 161.70 mAh/g and 161.44 mAh/g, respectively, at a 0.1 C rate. Battery showed a stable cycle performance in the capacity retention of 94% after 50 cycles and high Coulombic efficiency.

Other research group investigated more complicated composites to provide better ionic conductivity, thermal and mechanical stabilities. Yang et al. [104] prepared solid composite electrolyte based on electrospun filaments and silicon-based conductive additive. The silicon-based additive was synthesized by refluxing γ-chloropropoyl trimethoxy silane in acid/anhydrous ethanol medium at 40 °C for 5 days. The dry substance was mixed, in different concentrations, to PVdF/PAN/PMMA polymers in solid ratio (2:2:1). The polymer solution (15% wt/v) was electrospun at 20 kV, flow rate 1.8 mL/h, air-gab distance 25 cm and fibers were collected on rotating drum at 50 rpm. Data showed that the average diameter of PVdF/PAN/PMMA electrospun fibers was 600 nm and with silicon-additive 2, 4, 6, 8, 10 and 12%wt/v diameters have increased to 740, 770, 820, 810, and 800 nm, respectively. The electrochemical characteristics of the composite electrolyte of 10 wt% silicon-additive exhibited a high electrolyte uptake of 660% and an excellent thermal stability. Also, the solid composite electrolyte showed ionic conductivity potent of 9.23 mS/cm at room temperature and electrochemical stability is up to 5.82 V.

Maurya et al. [105] prepared solid composite electrolyte based on electrospun membrane and hetero-nano particles of rare-earth elements. Different concentrations of lithium, lanthanum, barium, and zirconium salts were mixed together in the presence of citric acid at 120 °C until milky powder was obtained. Different concentrations of the calcinated powder (5, 10 and 15% wt/v) were added to a solution of PVdF-co-HFP (16% wt/v) dissolved in a mixture of solvents (dimethylacetamide and acetone 3:7). Solutions were electrospun at 18 kV, 12 cm distance and a flow rate of 0.5 mL/h. The obtained membranes were dried on vacuum at 60 °C, pressed to 25–45 μm thickness and soaked in 1 M of a liquid electrolyte of lithium hexafluorophosphate in ethylene carbonate/dimethyl carbonate. Data showed that the ionic conductivity has improved to 3.30 mS/cm at 25 °C with a working potential of 4.6 V. Battery assembled from this composite exhibited a superior specific capacitance of 123 F/g at a current density of 1 A/g with a capacity retention of 83% even after 1000 cycles.

Other research group used electrospinning technique to prepare sandwich-like composite electrolyte via layer by layer technique of different polymers. Qin et al. prepared a sandwich-like structure made of two layers of PVdF-co-HFP and a polyamide-6 layer in between. Electrospinning conditions were optimized to fabricate bead-free and uniform electrospun fibers. Both fibers were layer-by-layer received on the same target to produce the final membrane. The composite electrolyte was prepared by soaking the prepared sandwich-like membrane in lithium hexafluorophosphate. Sandwich mat picked up 270% of the liquid electrolyte and showed acceptable mechanical properties up to 17.11 megapascal. The composite electrolyte showed a high ionic conductivity of 4.2 mS/cm at room temperature and stable electrochemical window of a 4.8 V. In the assembled battery of this composite electrolyte and lithium anode and lithium iron phosphate cathode, high electrochemical stability, high discharge capacity and good cycle durability were observed.

3 Conclusion and Outlook

Rechargeable batteries hold a prestigious position in the map of the energy-storage research in order to decrease the usage of fossil fuel and decrease the emission of CO_2. Increasing the surface area of the main components of batteries showed a significant increase in the electrochemical performance of such batteries. The electrospinning technique, to fabricate nano-sized fibers, can be utilized in coating, decorating, or constructing the cell components which is considered as a turning point to overcome the drawbacks of the traditional way of fabrications.

Cathodes made of decorated carbon nanofibers with inorganic metals have been emerged recently to enhance the battery density and capacity. Electrospinning techniques can provide different shapes of electrospun fibers in which inorganic nanoparticles can be introduced either in the core or in the shell of carbon nanofibers. Calcinated electrospun fibers can enhance the specific capacity on metal–air battery as much as 5133 mAh/g at 1000 cycles compared to 1533 mAh/g at 1000 cycle of reference cathodes. Generally, electrospun fibers are made of carbonizable polymers that are capable to be carbonized by thermal treatments to provide carbon nanofibers or nanotubes. Such polymers usually have been mixed with many electroconductive enhancers made of nanoparticles of inorganic metals, heteroatom-doped metals, etc. Impregnated polymers have been electrospun either in form of single filaments, core/shell structures, or hollow fibers. Electrospun carbon nanofibers or nanotubes have very large surface area that facilitates the ionic transfer in much higher magnitude compared to regular composites.

Coating of the anode surface by protecting interlayer of electrospun fibers has been reported to control the anode corrosion and to provide high capacity and a remarkable stability.

Electrospun fibers have been used to fabricate solid composite electrolyte by which many drawbacks of liquid electrolyte have been covered. Such composites are made of the electrospun fibers of different polymers that showed good ionic conductivity and high liquid uptake. Such composites showed high mechanical stability and less liquid leakage.

For further development on the utilization of electrospun components in batteries, here are some personal perspectives.

- Utilization of new electrospinning apparatus that provides different fiber orientation and alignment has to be immensely investigated. Fiber direction may have a great impact on the electro-conductivity and the performance of the conversion processes from chemical energy to electric energy and vice versa.
- Synthesis of more carbonization-capable polymers possesses different functional groups may alter the final performance of nanofibers in the voltaic cells.
- The exposure of the electrospun fibers to plasma chamber in the presence of activated different gasses such as nitrogen, argon, fluorocarbon may lead to functionalized electrospun fibers resulting in more porous structure and more surface area.
- More studies are required to enhance the anode corrosion using the electrospun fibers in respect of the adhesion parameter of the electrospun fibers onto the anode surface.
- Monitoring the diameter of the electrospun fibers needs more focus to show the correlation between the fiber diameter and the battery performance.
- More integrated studies to enhance anodes, electrolytes and cathodes may show significant enhancement in the overall performance of batteries.
- Rechargeability is still behind any expectation for batteries and more research is required to enhance it.

Acknowledgements This work was performed during the implementation of the project Building-up Centre for advanced materials application of the Slovak Academy of Sciences, ITMS project code 313021T081 supported by the Integrated Infrastructure Operational Program funded by the ERDF. Author thanks the editor for the kind invitation.

References

1. Sarkar J, Bhattacharyya S (2012) Arch Thermodyn 33:23
2. Poizot P, Dolhem F (2011) Energy Environ Sci 4:2003
3. Kumar Y, Ringenberg J, Depuru SS, Devabhaktuni VK, Lee JW, Nikolaidis E, Andersen B, Afjeh A (2016) Renew Sustain Energy Rev 53:209
4. Aziz MS, Ahmed S, Saleem U, Mufti GM (2017) Int J Renew Energy Res 7:111
5. Jia T, Dai Y, Wang R (2018) Renew Sustain Energy Rev 88:278
6. Giwa A, Alabi A, Yusuf A, Olukan T (2017) Renew Sustain Energy Rev 69:620
7. Rosa AP, Chernicharo CAL, Lobato LCS, Silva RV, Padilha RF, Borges JM (2018) Renew Energy 124:21
8. Alexander S, Harris P, McCabe BK (2019) J Clean Prod 215:1025
9. Kuncoro CBD, Luo WJ, Kuan Y Der (2020) Int J Energy Res 1
10. Shepherd CM (1965) J Electrochem Soc 112:657
11. Patrício J, Kalmykova Y, Berg PEO, Rosado L, Åberg H (2015) Waste Manag 39:236
12. Tar B, Fayed A (2016) IEEE 59th Int Midwest Symp Circuits Syst (IEEE), pp 1–4
13. Weppner W (2000) Mater. Lithium-Ion Batter. (Springer Netherlands), pp 401–412
14. Qu D (2014) AIP Conf Proc, pp 14–25
15. Li M, Li YT, Li DW, Long YT (2012) Anal Chim Acta 734:31
16. Jorné J, Kim JT, Kralik D (1979) J Appl Electrochem 9:573
17. Sapkota P, Kim H (2009) J Ind Eng Chem 15:445

18. Koebel M (1974) Anal Chem 46:1559
19. Tti ST (1986) Pure Appl Chem 58:955
20. Tarhan L, Acar B (2007) Res Sci Technol Educ 25:351
21. Ballantyne AD, Hallett JP, Riley DJ, Shah N, Payne DJ (2018) R Soc Open Sci 5
22. Manwell JF, McGowan JG (1993) Sol Energy 50:399
23. Burzyński D, Kasprzyk L (2017) E3S Web Conf 14
24. Deng T, Lu Y, Zhang W, Sui M, Shi X, Wang D, Zheng W (2018) Adv Energy Mater 8
25. Jing M, Zhang X, Fan X, Zhao L, Liu J, Yan C (2016) Electrochim Acta 215:57
26. Li L, Ding Y, Yu D, Li L, Ramakrishna S, Peng S (2019) J Alloys Compd 777:1286
27. Singh A, Kalra V (2019) J Mater Chem A 7:11613
28. Wang L, Wang Z, Sun Y, Liang X, Xiang H (2019) J Memb Sci 572:512
29. Ma W, Xu Y, Ma K, Zhang H (2016) Appl Catal A Gen 526:147
30. Subbiah T, Bhat GS, Tock RW, Parameswaran S, Ramkumar SS (2004)
31. Frey MW (2008) Polym Rev 48:378
32. Bhattarai P (2014) Thapa KB. Sharma S, Basnet RB, p 3809
33. Yong K, Jeong L, Ok Y, Jin S, Ho W (2009) Adv Drug Deliv Rev 61:1020
34. Schiffman JD, Schauer CL (2008) Polym Rev 48:317
35. Kessick R, Fenn J, Tepper G (2004) Polymer (Guildf) 45:2981
36. Nada AA, Ali EA, Soliman AAF, Shen J, Abou-Zeid NY, Hudson SM (2020) Int J Biol Macromol
37. Al-Moghazy M, Mahmoud M, Nada AA (2020) Int J Biol Macromol 160:264
38. Katta P, Alessandro M, Ramsier RD, Chase GG (2004) Nano Lett 4:2215
39. Haider S, Al-Zeghayer Y, Ahmed Ali FA, Haider A, Mahmood A, Al-Masry WA, Imran M, Aijaz MO (2013) J Polym Res 20
40. Lee EJ, An AK, Hadi P, Lee S, Woo YC, Shon HK (2017) J Memb Sci 524:712
41. Kim IG, Lee JH, Unnithan AR, Park CH, Kim CS (2015) J Ind Eng Chem 31:251
42. Dos Santos AM, Dierck J, Troch M, Podevijn M, Schacht E (2011) Macromol Mater Eng 296:637
43. Wang S, Yang Y, Zhang Y, Fei X, Zhou C, Zhang Y, Li Y, Yang Q, Song Y (2014) J Appl Polym Sci 131:2
44. Yang Z, Peng H, Wang W, Liu T (2010) J Appl Polym Sci 116:2658
45. Yarin AL, Zussman E (2004) Polymer (Guildf) 45:2977
46. Zahran SME, Abdel-Halim AH, Mansour K, Nada AA (2020) Int J Biol Macromol 157:530
47. Qin J, Liu Z, Wu D, Yang J (2020) Appl Catal B Environ 278:
48. Han S, Hao Y, Guo Z, Yu D, Huang H, Hu F, Li L, Chen HY, Peng S (2020) Chem Eng J 401
49. Zhao H, Yuan ZY (2021) J Energy Chem 54:89
50. Wang HF, Xu Q (2019) Matter 1:565
51. Chen Z, Yu A, Higgins D, Li H, Wang H, Chen Z (2012) Nano Lett 12:1946
52. Li JC, Hou PX, Liu C (2017) Small 13:1
53. Ragab TIM, Nada AA, Ali EA, Shalaby ASG, Soliman AAF, Emam M, El Raey MA (2019) Int J Biol Macromol 135:407
54. Nada AA, Ali EA, Soliman AAF (2019) Int J Biol Macromol 131:624
55. Nada AA, Soliman AAF, Aly AA, Abou-Okeil A (2018) Starch Stärke 71:1800243
56. Han Q, Chi X, Zhang S, Liu Y, Zhou B, Yang J, Liu Y (2018) J Mater Chem A 6:23046
57. Shi Y, Pan L, Liu B, Wang Y, Cui Y, Bao Z, Yu G (2014) J Mater Chem A 2:6086
58. Wang SH, Hou SS, Kuo PL, Teng H (2013) ACS Appl Mater Interfaces 5:8477
59. Xu JJ, Xu D, Wang ZL, Wang HG, Zhang LL, Zhang XB (2013) Angew Chemie Int Ed 52:3887
60. Park HW, Lee DU, Zamani P, Seo MH, Nazar LF, Chen Z (2014) Nano Energy 10:192
61. Peng W, Wang Y, Yang X, Mao L, Jin J, Yang S, Fu K, Li G (2020) Appl Catal B Environ 268:
62. Birbilis N, Buchheit RG (2005) J Electrochem Soc 152:B140

63. Kowal K (1996) J Electrochem Soc 143:2471
64. Song G (2005) Adv Eng Mater 7:563
65. Wolfe RC, Shaw BA (2007) J Alloys Compd 437:157
66. Ibrahim NA, Nada AA, Hassabo AG, Eid BM, Noor El-Deen AM, Abou-Zeid NY (2017) Chem Pap 71:1365
67. Eid BM, Hassabo AG, Nada AA, Ibrahim NA, Abou-Zeid NY, Al-Moghazy M (2018) Adv Nat Sci Nanosci Nanotechnol 9:
68. Wongrujipairoj K, Poolnapol L, Arpornwichanop A, Suren S, Kheawhom S (2017) Phys Status Solidi Basic Res 254
69. Zuo Y, Yu Y, Liu H, Gu Z, Cao Q, Zuo C (2020) Batteries 6:1
70. Song MJ, Kim IT, Kim YB, Shin MW (2015) Electrochim Acta 182:289
71. Ma Y, Sumboja A, Zang W, Yin S, Wang S, Pennycook SJ, Kou Z, Liu Z, Li X, Wang J (2019) ACS Appl Mater Interfaces 11:1988
72. Yu Y, Zuo Y, Liu Y, Wu Y, Zhang Z, Cao Q, Zuo C (2020) Nanomaterials 10
73. Bui HT, Kim DY, Kim DW, Suk J, Kang Y (2018) Carbon N Y 130:94
74. Robert Ilango P, Peng S (2019) Curr Opin Electrochem 18:106
75. Kalluri S, Seng KH, Guo Z, Liu HK, Dou SX (2013) RSC Adv 3:25576
76. Li W, Zeng L, Wu Y, Yu Y (2016) Sci China Mater 59:287
77. Liu D, Cao G (2010) Energy Environ Sci 3:1218
78. Kang CS, Kim C, Kim JE, Lim JH, Son JT (2013) J Phys Chem Solids 74:536
79. Von Hagen R, Lorrmann H, Möller KC, Mathur S (2012) Adv Energy Mater 2:553
80. Toprakci O, Ji L, Lin Z, Toprakci HAK, Zhang X (2011) J Power Sources 196:7692
81. Chen Q, Zhang T, Qiao X, Li D, Yang J (2013) J Power Sources 234:197
82. Chen LL, Yang H, Jing MX, Han C, Chen F, Hu X yu, Yuan WY, Yao SS, Shen XQ (2019) Beilstein J. Nanotechnol 10:2229
83. Liu Y, Xue JS, Zheng T, Dahn JR (1996) Carbon NY 34:193
84. Suresh Kumar P, Sahay R, Aravindan V, Sundaramurthy J, Ling WC, Thavasi V, Mhaisalkar SG, Madhavi S, Ramakrishna S (2012) J Phys D Appl Phys 45:
85. Kim C, Yang KS, Kojima M, Yoshida K, Kim YJ, Kim YA, Endo M (2006) Adv Funct Mater 16:2393
86. Zou L, Gan L, Kang F, Wang M, Shen W, Huang Z (2010) J Power Sources 195:1216
87. Wang B, Cheng JL, Wu YP, Wang D, He DN (2012) Electrochem Commun 23:5
88. Wang L, Ding CX, Zhang LC, Xu HW, Zhang DW, Cheng T, Chen CH (2010) J Power Sources 195:5052
89. Xu ZL, Zhang B, Kim JK (2014) Nano Energy 6:27
90. Zhang M, Uchaker E, Hu S, Zhang Q, Wang T, Cao G, Li J (2013) Nanoscale 5:12342
91. Li L, Peng S, Lee JKY, Ji D, Srinivasan M, Ramakrishna S (2017) Nano Energy 39:111
92. Xue Z, He D, Xie X (2015) J Mater Chem A 3:19218
93. Samad YA, Asghar A, Hashaikeh R (2013) Renew Energy 56:90
94. Banitaba SN, Semnani D, Heydari-Soureshjani E, Rezaei B, Ensafi AA (2019), Mater. Res. Express 6
95. Banitaba SN, Semnani D, Heydari-Soureshjani E, Rezaei B, Ensafi AA (2020) Solid State Ionics 347
96. Zhu L, Zhu P, Yao S, Shen X, Tu F (2019) Int J Energy Res 43:4854
97. Gopalan AI, Lee KP, Manesh KM, Santhosh P (2008) J Memb Sci 318:422
98. Gopalan AI, Santhosh P, Manesh KM, Nho JH, Kim SH, Hwang CG, Lee KP (2008) J Memb Sci 325:683
99. Li X, Cheruvally G, Kim JK, Choi JW, Ahn JH, Kim KW, Ahn HJ (2007) J Power Sources 167:491
100. Cui WW, Tang DY, Gong ZL (2013) J Power Sources 223:206
101. Gao M, Wang C, Zhu L, Cheng Q, Xu X, Xu G, Huang Y, Bao J (2019) Polym Int 68:473
102. Zainab G, Wang X, Yu J, Zhai Y, Ahmed Babar A, Xiao K, Ding B (2016) Mater Chem Phys 182:308

103. Peng X, Zhou L, Jing B, Cao Q, Wang X, Tang X, Zeng J (2016) J Solid State Electrochem 20:255
104. Tan L, Deng Y, Cao Q (2019) Jing B. Liu Y, Wang X, p 3673
105. Maurya DK, Murugadoss V, Angaiah S (2019) J Phys Chem C 123:30145

Electrospinning of Lignin Nanofibers for Drug Delivery

Sandip K. Singh, Ajeet Singh, and Sasmita Mishra

Abstract In a biorefinery process, a key component of lignocellulosic biomass, lignin can be recovered at lab-to-pilot scales by using numerous methods. Lignin can be utilized for assorted applications as a renewable and sustainable source and also used as an excellent candidate to substitute and/or eliminate aromatic polymers derived from non-renewable petroleum sources. This chapter focuses on lignin's current state featuring assorted methods, including Kraft, organosolv, alkaline, and dilute acidic, to recover lignin from plant biomass. In addition, lignin types and their derivatives have been discussed and correlated with health and pharmacological activities. This chapter further discusses the lignin-derived carbon nanofibers (CNFs) and processes involving the production of CNFs by using an electrospinning method. The physico-chemical properties of lignin-derived CNFs were characterized using Raman, ^{13}C-^{1}H 2D heteronuclear single quantum coherence (HSQC) NMR, and more. Additionally, a proposed mechanism is presented to produce lignin bio-oils, and the formation of CNFs from this bio-oil is further discussed. The biomedical applications of nanofibers (NFs) and their role in drug delivery are also finally added and discussed. In the future prospective and concluding remarks, NFs derived from biopolymer can be a potential source as a substituent by elimination or substitution of non-renewable fossil feedstocks.

Keywords Electrospinning · Lignin · Nanofibers · Biomedical applications · Drug delivery · Plant biomass · Fossil feedstocks

S. K. Singh (✉)
Catalysis and Inorganic Chemistry Division, CSIR—National Chemical Laboratory, Pune 411008, India

A. Singh · S. Mishra
Discipline of Chemistry, Indian Institute of Technology, Indore 453552, India

© The Author(s), under exclusive license to Springer Nature Switzerland AG 2021
S. K. Tiwari et al. (eds.), *Electrospun Nanofibers*, Springer Series on Polymer and Composite Materials, https://doi.org/10.1007/978-3-030-79979-3_7

171

1 Introduction

The term electrospinning was conceived in the early sixteenth century by William Gilbert; he observed the formation of a cone-shaped water droplet in the presence of an electric field [1]. Thereafter, several systematic studies were done by Gray [2], Nollet and Stack [3], and Rayleigh [4] to, respectively, observe the electrohydro-dynamic atomization, electrospraying experiment to form an aerosol, and charged droplets behavior of water droplets. The coin differences between electrospraying and electrospinning, respectively, lied in the viscoelasticity and viscosity of the liquid used. In 1901, John Cooley and William Morton filed two patents and explained a prototype of setup for electrospinning [5, 6]. The Soviet Union in 1938 first implemented the electrospun nanofibers (NFs) to prepare "Petryanov filters" for capturing the aerosol particles. The strength of electric fields decides the shape of droplets, such as beyond a critical level, the spherical droplets turned to a cone, and is generally referred to as a Taylor cone "Geoffrey Taylor". Electrospinning has been enabled in several areas including catalytic applications, energy harvesting, conversion, and storage materials, and these materials are generally dominated by inorganic NFs. These NFs offered bundles of excellent physical, chemical, thermal, biological, and more substantial properties. Therefore, they are eventually used in various applications including environment and sustainability, air purification, catalysis, energy, photonics, electronics, biomedical, and more [7]. Nowadays, non-renewable fossil feedstocks are mostly applied to synthesize the NFs, in which large amounts of toxic gases are produced in environment, and the utilization of fossil feedstocks are the major threat of global warming through the greenhouse [8].

An alternate solution to save the Earth's atmosphere from Greenhouse gas is to use available renewable and sustainable biomass sources. Biomass is generally categorized into animal and plant-derived biomass. The plant-derived biomass is further categorized into edible and non-edible biomass; the edible biomass competes with food and can be unable to fulfill the market demands. Therefore, the utilization of non-edible plant biomass is the most durable solution for researchers working in the area of biomass utilization. Plant or lignocellulosic biomass is generally composed of three major biopolymer units; cellulose (homopolymer of glucose unit), hemicellulose (heteropolymer of C5 and C6 sugar units), and lignin (heteropolymer of generally, C3/C4/C5 (alkyls) + C6 (aryl)) [9, 10]. Cellulose is a homopolymer of β-linked D-glucose units, whereas C5 and C6 sugars formed the heteropolymer structure of hemicellulose. Covalent lignin–carbohydrate (LC) linkages are eventually present in layered plant cell walls of lignocellulosic biomass [11, 12]. Nowadays, these biopolymers are used widely in both academic and industries, including biofuels synthesis, polymers, materials, and more applications [13–16].

2 Lignin

Lignin is a natural occurring bio-copolymer, that is structured with mainly three phenolic moieties; coniferyl alcohol (G), *p*-coumaryl alcohol (H), and sinapyl alcohol (S) (Fig. 1) [17]. These moieties are present in a lignin structure through several types of linkages that are ether/ester (C–O–C)/ C(O)–O–C, (e.g., α-O-4, 5-O-4, and β-O-4)) and condensed carbon–carbon linkages, including alkyl–alkyl, alkyl–aryl, or aryl–aryl (C–C, (e.g., 5–5, β-β, β-5, and β-1 linkages)) (Figs. 1 and 2). Around 50–70% of these intramolecular linkages are ether bonds, predominantly β-O-4 linkages [18]. Several factors affected the concentrations of these linkages in plant species (e.g., hardwood, softwood, and grasses), including environmental and biological factors (Table 1) [19]. Lignin features are also linked with polysaccharides which are called as lignin–carbohydrates complex (LCC). LCC linkages in plants are linked through phenyl glycoside bonds, benzyl ethers bond, or esters bond [11, 12]. LCC linkages in plant species eventually provide mechanical strength. Lignin works as an antioxidant, antimicrobial, and antifungal agent. It provides more fighting strengths in plant biomass, to prevent the outer and inner damages of plant species from foreign attack, and also works to store the energy in a bundle of layered plant cell walls [17]. Lignin is an up-and-coming candidate to replace and/or eliminate the phenolic chemicals that are currently procured from non-renewable fossil feedstocks. Lignin

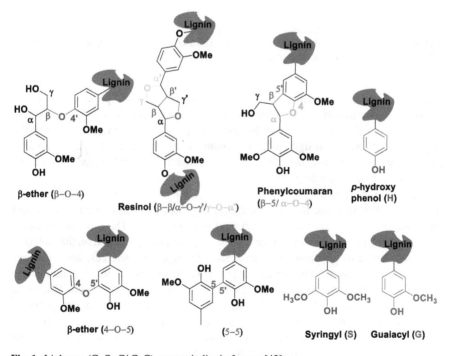

Fig. 1 Linkages (C–O–C/ C–C) present in lignin feature [42]

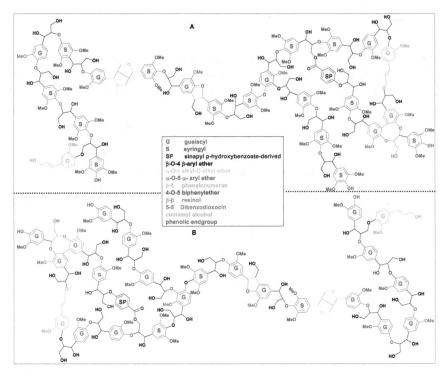

Fig. 2 Lignin copolymer structure for **a** syringyl moiety (hardwood lignin) with 24 units and **b** guaiacyl moiety (softwood lignin) with 24 units, redrawn from reports [30, 43]. For clarity of structure and bonding, see online or use color print for this figure

is produced in large quantities as a side product from pulp and paper industries. It is eventually burned to generate the energy and chemicals as a low-grade application. Lignin is also procured from biorefineries plant as a side product during the biofuel's synthesis from lignocellulosic biomass. Lignin is present in the layered plant cell walls in a native form and can be extracted from plant biomass by applying various treatments, including chemical (acidic, alkali, alkaline, oxidative, and more), solvent fractionation (organosolv fractionation along with small amounts of acids like sulphuric acid, ionic liquids, and more), physical (ball milling and more), and biological processes (enzymatic, fungi, and more) [20–22]. The physico-chemical properties and purity of recovered or technical lignin mainly depend on the isolation methods, degree of purification, fractionation history, plant sources, and measurement [23]. Table 2 shows the methods applied to recover lignin from plant biomass at academic and industrial scales [24–27]. These extraction methods are classified into two sub-categories. During the isolation process in the first sub-category, polysaccharide can be soluble, and lignin remains as a solid (e.g., Klason method and more), and in the second sub-category, lignin is soluble, and polysaccharide remains as an insoluble matter (e.g., organosolv, Kraft method, and more).

Table 1 Various types of linkages, relative abundance, and bond dissociation enthalpy in hardwood- and softwood-derived lignin

Linkage	Relative abundance of linkages in lignin feature (%) [32–34]		Bond dissociation enthalpy (kJ/mol) [35–39]
	Hardwood (Major guaiacyl lignin)	Softwood (Major syringyl lignin)	
Carbon–carbon linkage (30–40%)			
5–5'	4–11	10–25	490
β-β	3–7	2–4	335
β-1	5–7	3–7	270–289
α-1	–	–	360–390
β-5	4–6	9–12	–
Carbon–oxygen–carbon linkage (60–70%)			
β-O-4	46–65	43–50	290–335
α-O-4	4–8	6–8	215–270
4-O-5	3–7	2–5	330
α-O-γ'	–	–	270
– OCH₃	–	–	255–275
others	5–13	–	–
Functional group (per 100 parts unit). References [40, 41]			
Carboxyl	11–13	–	
Carbonyl	3–17	20	
Methoxy	132–146	92–96	
Aliphatic hydroxyl	–	120	
Benzyl hydroxyl	–	16	
Phenolic hydroxyl	9–20	20–28	

Unlike polysaccharides which have well-defined structure, lignin has a highly recalcitrant 3-D structure that provides the mechanical strength to plant cell walls [28]. Lignin composition and amounts within plant biomass depend upon plant species and vary with a set of various factors, including types of plant species, age of plants, location, weather, and more [23, 29]. Lignin contains, at a range of molar masses, various types of functional groups, including hydroxyl, carbonyl, ether, methoxy, phenolic, and more (Table 1) [30]. These functional groups played a crucial role in lignin's utilization for materials, polymers, chemicals, biological, and more applications [31].

Considering the earlier discussion impact of nanofibers, electrospinning, and renewable bioresource lignin, in this chapter, authors centered their attention on lignin features, and how lignin source and synthesis procedures have impacted their effect on healthcare. In addition, the impact of lignin or lignin derivatives over pharmacological activities has also been discussed. The synthesis of carbon nanofibers

Table 2 Various processes applied for lignin recovery [20, 25, 54, 55]

Process	Reaction conditions
Polysaccharides conversion	
Concentrated acid hydrolysis	Concentrated mineral acid (H_2SO_4, HCl, HF), 20–30 °C
Dilute acid hydrolysis	H_2O, H_2SO_4, HCl, HF, H_3PO_4, 170–300 °C,
γ-valerolactone-assisted acid hydrolysis	H_2O, GVL, H_2SO_4, 120–170 °C
Ionic liquid (IL)-assisted acid hydrolysis	IL: H_2O, H_2SO_4, HCl, 100–150 °C
Acid mechanocatalytic pretreatment	RT H_2SO_4, HCl, ball milling, post-hydrolysis, H_2O, 130 °C
Enzymatic hydrolysis	H_2O delignification, hot water, dilute acid, steam explosion, ammonia fiber explosion, deacetylation, and mechanical refining, 30–60 °C
Cellulolytic enzyme	H_2O, pH 4–5 (buffer), cellulolytic enzymes, ball-milled biomass, 40–60 °C
Enzymatic mild acidolysis	2-levels; level 1: cellulolytic enzymes (ball-milled biomass), level 2: mild acidolysis (dioxane/H_2O, HCl, 80–90 °C
Pyrolysis (fast)	Absence of O_2, acidic zeolite, 400–600 °C
Lignin conversion	
Kraft pulping	H_2O, NaOH, Na_2S, 140–170 °C
Sulfite pulping	H_2O, Na, NH_4, Mg or Ca salts of SO_3^{2-} or HSO_3^-, 140–170 °C
Soda pulping	H_2O, NaOH, (anthraquinone; AQ), 160–170 °C
Aqueous alkaline pretreatment	H_2O, NaOH, $Ca(OH)_2$, (AQ), 40–160 °C
Ammonia fiber explosion	H_2O, NH_3, fast decompression, 60–160 °C
Anhydrous ammonia pretreatment	Anhydrous NH_3, dry biomass, 100–130 °C
Ammonia recycle percolation	Aqueous NH_3, 150–210 °C
Flow-through dilute acid pretreatment	H_2O, H_2SO_4, HCl, 120–210 °C
Flow-through hot water pretreatment	H_2O, 160–240 °C
Steam explosion pretreatment + extraction	H_2O, (SO_2), fast decomposition, 100–210 °C
Organosolv pulping	Organic solvent/s H_2O, H_2SO_4, 100–220 °C
Formaldehyde-assisted fractionation	Dioxane, H_2O, formaldehyde, HCl, 80–100 °C
Carbon dioxide explosion	Supercritical carbon dioxide
Co-solvent enhanced lignocellulosic fractionation	THF-H_2O, H_2SO_4, $FeCl_3$, 160–180 °C
Reductive catalytic fractionation	Redox catalyst, H_2 (-donor), organic solvent (+H_2O), 180–250 °C
Wood-degrading microorganisms	Cultured, colonized, 24 °C

(continued)

Table 2 (continued)

Process	Reaction conditions
Fungal pretreatment	10–84 days, 60–75% moisture, 28–30 °C
Ionic liquid (IL) dissolution and ionosolv pulping	IL, H_2O, H_2SO_4, 90–170 °C
Mechanical pretreatment + extraction	Extensive ball milling, RT

Note GVL—gamma-valerolactone; RT—room temperature; THF—tetrahydrofuran

from lignin and other polymers by using electrospinning is also discussed. Finally, we presented the biomedical applications including drug delivery of CNFs derived from lignin and prospects of NFs productions from assorted types of lignin and their clinical applications, with brief conclusions.

2.1 Lignin Recovery

Pretreatment is an initial step to remove and/ or reduce the inhibitors including lignan, acetate, hemicellulose, lignin, inorganic contents, and more that strike to enzymes and desirable to provide the maximum accessibility for cellulose hydrolysis. It is also believed that pretreatment helped to increase the accessible surface area (i.e., porosity) for enzymatic hydrolysis to maximum amounts of sugar monomers yield [44]. Researchers also believed that the rate of enzymatic hydrolysis is a function of cellulose crystallinity, and crystallinity can be reduced, thereafter the pretreatment of lignocellulosic biomass or pure crystalline cellulose [44]. A series of pretreatment methods are developed, including physical, chemical, thermal, or biological, to recover lignin as a precipitate or solid and carbohydrates-rich pulp or in the solution phase. These methods are ranged from sophisticated to environment friendly and are used in various sets of conditions, including temperature, acids, base, ionic liquids (ILs), organic solvents, ammonia fiber expansion, enzymatic mild acidolysis, and more. Some of the pretreatment methods are briefly discussed below and details are shown in Table 2.

2.2 Kraft Process

Kraft pulping process is the most common chemical pretreatment process that is used to recover lignin from lignocellulosic biomass. It is a well-established industrial technology and currently covered more than 85% pulping of wood to produce delignified pulps. Sodium sulfide and sodium hydroxide as a "White Liquor" are used for the pulping process at elevated temperature and pressure. Lignin binds with polysaccharides, dissolved in white liquor, and separate as a precipitate by acidifying recovered black liquor [22, 45, 46]. The impact of delignification reaction conditions, including temperature and time in the presence of alkali, has been defined as H-factor [47] and the parameter is defined as below.

$$H = \int_0^t e\left(43.2 - \frac{16117}{T}\right)dt \tag{1}$$

where T(K) is temperature and t (h) is reaction time.

2.3 Dilute Acidic

Dilute acid pretreatment (DAP) is the most widely used deconstruction process of lignocellulosic biomass at a large scale to produce carbohydrates-rich pulp that pulp is mostly free and/or reduce from inhibitors content. DAP helps to cleave the covalent bonds in lignocellulosic biomass. In this process, mineral acids, including sulphuric acid, hydrochloric acid, or nitric acid, are used at a wide range of concentrations (≤ 4 wt.%) over a range of temperature (170–300 °C). Among all, sulphuric acid is mostly employed due to its economic process and high efficacy for lignin removal as well [48]. The combined severity factor (S) is used to correlate the acidic delignification process under different conditions, and it is defined by using the following equation [49, 50].

$$S = log\left(t\,exp\left(T - \frac{100}{14.75}\right)\right) - pH \tag{2}$$

where t is reaction time, T is temperature, and pH expressed is acid loading.

2.4 Alkaline Process

The alkaline pretreatment process is a widely used chemical pulping for delignification. This process uses several types of bases, including hydroxide of sodium, potassium, calcium, or ammonium over a range of reaction conditions, including time, temperature, and alkali loadings [51]. This process cleaved the ester and glycosidic side chains in layered plant cell walls and resulted in the lignin and hemicellulose solubilization in alkali solution. This process also helped for cellulose intercrystalline swelling. Additionally, the alkali process also removed and/ or reduced the acetate and uronic acid that linked as a hemicellulose substructure and helped to eliminate the inhibitors as well. This process also altered the treated biomass structure in terms of cellulose swelling that leads to maximum porosity and reduction in crystallinity or degree of polymerization [52]. The swelling of cellulose structure provided the maximum accessibility for enzymatic hydrolysis to produce maximum amounts of sugar monomer that is amenable for further biofuels production.

2.5 Organosolv Process

Organosolv process is used to delignify the lignocellulosic biomass using organic solvents, including aliphatic alcohols (i.e., ethylene glycol, methanol, ethanol, butanol, and more) and aromatic alcohol (i.e., phenol) in the presence or absence of acid catalysts. There are several parameters, including temperature (90–220 °C), time (25–100 min), aqueous organic solvent ratios (ethanol concentration 25–75% v/v), and acid amounts (0.5–2 wt%), involved to impact delignification process [53]. Organosolv lignin has eventually low molar mass and is mostly soluble in water. This lignin is considered as highly pure lignin. However, the amount of lignin recovery is low, and the recycling of organic solvent is another limitation of this process.

3 Types of Lignin and Their Effect on Health

Lignin can be obtained from a range of terrestrial plants and aquatic seaweeds, red algae, and more, by applying various processes (Table 2). Lignin is a remarkable renewable phenolic raw material and currently has limited uses. Although, specific potential applications of biorefinery lignin are in progress using the lignin directly or modified lignin. The pulp and paper industries produce a large quantity of lignin each year. Nearly 98% of recovered lignin is burned at the same factories to generate boiler heat and chemicals [56]. Approximately 2% of lignin, derived from lignosulfonate and Kraft lignin, is commercially exploited for about 1000,000 tons per annum and 100,000 tons per annum, respectively [57]. Nevertheless, commercial applications of biomaterials derived from lignin are promising in the coming years. Plant biomass sources and fractionation conditions of technical lignin must be specified prior to their use in potential applications because their physical, biological, thermal, and chemical properties are extremely variable due to developmental and environmental factors [58]. Similarly, their purity and physico-chemical properties are essential and need to be characterized. Table 3 shows the physico-chemical properties of some academic and industrial scales isolated lignin.

Lignin exhibits a very low level of toxicity. Generally, organic solvent-free lignin including Kraft and lignosulphate lignins has been extensively tested over humans and animals, and laboratory results showed that these lignins are essentially non-toxic for both humans and animals [59]. The LD50 values have been reported as more than 12 g/kg of body weight. Food and Drug Administration, United States of America, has approved both Kraft and lignosulphonate lignins to be used for manufacturing a

Table 3 Impact of various types of methods on physico-chemical properties of extracted lignin

Lignin	M_w(Da)	PDI	Carbohydrates (wt%)	Sulfur (wt%)	Ash (wt%)	Per phenyl propane unit				References
						Aliphatic -OH	Phenolic -OH	COOH	OCH$_3$	
Kraft	1500–200,000	2.5–7.3	2.3	1.0–3.0	0.5–3.0	~0.4	0.8–1.0	0.2–0.3	~0.6	[60–62]
Lignosulphonate	1000–400,000	6–8	–	3.5–8.0	4.0–8.0	–	0.2–0.3	0.2–0.3	~0.7	[63, 64]
Organosolv	500–17,000	1.7–4.5	1.0–3.0	0	0–2.0	~0.4	~0.4	~0.1	0.9–1.0	[9, 65–67]
Soda	1,000–15,000	2.5–3.5	1.5–3.0	0	0.7–2.3	~0.4	0.4–0.5	0.2–0.3	1.0–1.2	[23, 61, 63]
Hydrolysis	1,900–10,000	3.0–11.0	15.0–23.0	0–1.0	1.0–3.0	~0.5	0.2–0.3	~0.1	0.6–0.7	[64, 68, 69]
Ionic liquid	2,000–42,000	1.0–3.0	0.1–15.0	1.5	0.6–2.0	–	–	–	–	[70]

Note $\overline{M_w}$ and $\overline{M_n}$ = weight and number average molar mass; polydispersity index (PDI) = $(\overline{M_w}/\overline{M_n})$

S. K. Singh et al.

wide range of food-grade applications. As lab study shows lignin is almost present in substantial amounts in all vegetables and whole grains.

3.1 Impact of Lignin or Lignin Derivatives On Pharmacological Activities

Lignin is a heteropolymer structure of phenylpropanoid units. That is structurally linked with aliphatic alcohols and phenolic chains (Figs. 1 and 2). Several studies showed that these structures are associated with good thermal, chemical, physical, and biological properties. These properties of lignin or lignin-derived compounds worked out in different biological activities. These activities included the prevention of tumor growth that is demonstrated in rats [71], reduced serum cholesterol by binding with bile acids in the intestine [72], antioxidant activities, treatment of diabetes, obesity control, antiviral agents and immunomodulators, anticoagulant and anti-emphysema agents, and more (Table 4) [30, 73].

3.2 Lignin Carbon Fibers

Lignin is a low-grade biomaterial that is essentially used as a cement additive, burned to generate energy and chemicals, feed for cattle, and more. It can be a potential renewable, sustainable, and low-cost source to eliminate the synthetic polymers including polyacrylonitrile to produce carbon fibers. The production of carbon fibers from lignin can be improved by the raw material availability, reducing the cost and dependency of petroleum-derived products, and slash and/ or eliminating the greenhouse gas. That can be generated upon the synthesis and utilization of petroleum-derived products. Lignin can be a potential source for carbon fibers in contest of economic and environmental to the petroleum derived carbon fibers. Several factors need to be carefully implemented, including spinning conditions, treatment temperatures, ramping profiles, melt-spinning steps, and more, to obtain carbon fiber of superior strength from lignin [88]. Figure 3 shows the formation of low-cost carbon fibers from lignin (For more details on lignin isolation, see Table 2). The lignin-derived carbon fibers have been substantially applied in a wide range of technical applications, including energy storage (e.g., batteries, supercapacitors, dye-sensitized solar cells) and more [89]. The modified carbon fibers are used in sensing, diagnostics, aerospace, and more [90, 91].

Table 4 Effect of lignin or lignin derivatives on pharmacological activities

Lignin or lignin derivative	Source	Molar mass (Da)	Effect	Mechanism	References
Alkali lignin	*Acacia nilotica*	NA	Antidiabetic	Inhibited glucose movement and α-amylase	[74]
Lignosulfonic acid	NA	~8,000	Antidiabetic	Suppressed blood glucose levels via inhibition of α-glucosidase activity and intestinal glucose absorption	[75]
Lignophenols	*Cryptom-eria Japonica*	~2,000	Obesity control	Suppressed oleate-induced microsomal triglyceride transfer protein (MTTP) mRNA expression and cellular cholesterol	[76]
Lignophenols	Beech	1,500	Obesity control	Suppressed excess oxidative stress, infiltration and activation of macrophages, and glomerular expansion in STZ-induced diabetic kidneys	[77]
Lignosulfonic acid (LA)	NA	~8,000	Viral inhibition	LA inhibited the HIV and herpes simplex virus (HSV-2) replications	[78]
Lignin–carbohydrate Complexes (LCC)	*Pimpinella anisum*	7,000–8,000	Antiviral activities	Virus adsorption showed antiviral activities against HSV-1 and HSV-2, human cytomegalovirus, and measles virus	[79]
Cinnamic Acid-Based Lignins	NA	2,100–16,500	Antiviral activities	Inhibited HSV-1 entry to mammalian cells	[80]

(continued)

Table 4 (continued)

Lignin or lignin derivative	Source	Molar mass (Da)	Effect	Mechanism	References
Sulfated low-molecular-weight lignins	NA	3,300–4,100	Anticoagulants	Inhibited thrombin and factor Xa	[81]
Sulfated low-molecular-weight lignins	NA	3,300–4,100	Anticoagulants	Inhibited coagulation factor XIa and human leukocyte elastase, moderately inhibited cathepsin G	[82]
Sulfated, low-molecular-weight lignins	NA	3,300–4,100	Anticoagulants	Inhibited thrombin and factor Xa	[83]
Sulfated, low-molecular-weight lignins	NA	3,300–4,100	Antithrombin binding	Inhibited the free factor VIIa	[84]
Sulfated β-O-4 lignin	NA	NA	Anticoagulant and antiplatelet	Targeted exosite 2 of thrombin to reduce fibrinogen cleavage through allostery	[85]
Unsulfated or sulfated low-molecular-weight lignins	NA	NA	Anti-emphysema	Elastase, oxidation, and inflammation inhibition	[86]
Functionalized Kraft lignin	NA	NA	Antiproliferative	pH-responsive delivery of anticancer drugs	[87]

Note NA—not available

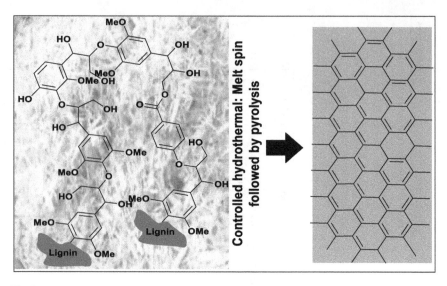

Fig. 3 Depiction of carbon fibers formation from lignin [92]

4 Electrospinning

Electrospinning is a very simple, highly productive, and emerging nanotechnology that is used to produce the thin natural micro- and nanometer scales (0.01–10 μm) fibers. These fibers are associated with substantial advantages that include the extremely lightweight, low manufacturing cost, highly flexible in surface functionality, reduced defects because of small diameter, high molecular orientation, and more. These fibers have been suggested for a wide range of applications due to their high specific surface area, small pore size, high porosity, and more. A "Taylor cone" of fibers is initiated at the tip when a high-voltage electric field is applied over the polymer solution. This method is applied to fabricate a range of NFs of polymers, composites, ceramics, metals, and more. Scaffolds are currently prepared by electrospinning the mixture of nanoparticles and polymers. The electrospinning process does not require any supplementary step including functionalization, heating, cooling, and more. This process requires a suitable solvent that can highly disperse the nanoparticles and could completely dissolve the polymers [93]. Tuning the electrospinning process parameters, including needle type, voltage, feed rate, tip-to-collector distance, and more, greatly affects the physico-chemical properties of electrospun NFs (Table 5) [94]. A capillary tube with pipette or needle with a small diameter, a high-voltage supplier, and a metal collecting screen, whichare the basic three main components, is used for electrospinning of polymers NFs [95].

Table 5 Process parameters for the generation of NFs using electrospinning [94]

Ambient condition	Process condition	Solution properties
• Temperature • Humidity • Environmental pressure • Time required for feeding • Air velocity in the chamber	• Needle diameter • Volume feed rate • Distance between needle and collector • Applied voltage • Motion and size of the target screen	• Elasticity • Surface tension • Electric conductivity • Molar mass of polymer • Polymer concentration • Viscosity • Types of solvent

5 Lignin-Derived Carbon Nanofibers

Lignin is a very low-grade biopolymer and produced in large quantities as a byproduct from pulp and paper industries, and through an integrated biorefinery to produce biofuels as well. Researchers have shown an attractive interest to convert lignin to high value-added commodities including chemicals, [25, 96, 97] materials, [98] polymers [99], and more. Lignin is the most approaching candidate to apply for manufacturing carbon nanofibers (CNFs) due to its meager cost, earth-friendly nature, aromatic and thermoplastics characteristics, and more. Several low-cost precursors have been reported elsewhere for the production of carbon fibers [88, 100]. Table 6 shows the total cost of CFs derived from lignin (US $ 6.3; precursor and production) which is lower as compared to conventional (US $ 34.6), textile-grade (US $ 16.6–38.6), and melt-spun (US $ 23.7) polyacrylonitrile (PAN) carbon fibers (CFs). This comparison study shows that lignin is a cost-effective precursor to produce CFs. There are several advantages including higher modulus, already oxidized, high carbon content, inexpensive and renewable, and more, and disadvantages including disordered and complex structure, diversity, may cross-link during melt spinning, and associated with lignin. Lignin is apparently inexpensive, renewable; therefore, the dependency over the non-renewable petroleum-derived resource could be reduced and/ or eliminated. Lignin is partially oxidized, has high carbon content ($\geq 55\%$), and associated with more fruitful properties categorize that lignin is a good-to-excellent

Table 6 Various sources of carbon fibers and their precursor and production costs [114–116]

Entry	Source	Precursor cost (per kg US $)	Production cost (per kg US $)	Total cost (per kg US $)
1	Lignin	1.1 (including spinning)	5.2	6.3
2	Conventional PAN	10.2	24.4	34.6
3	Textile-grade PAN	4.4–13.2	12.2–25.4	16.6–38.6
4	Melt-spun PAN	6.3	17.4	23.7
5	Polyolefin	1.57–2.36	NA	NA

Note PAN—polyacrylonitrile; NA—not available

candidate for producing low-cost CFs from manufacturing prospects by using a melt-spun and energy-efficient method [101]. Moreover, the generation of toxic gases including hydrogen cyanide (HCN), nitrous gases, and more can be avoided by using lignin as a precursor that occurred during the carbonization of petroleum-based precursors.

Two methods including vapor growth and electrospinning were eventually applied to produce CNFs. Additionally, few other methods, including template synthesis, phase separation, drawing, and self-assembly have been reported elsewhere [102, 103]. Lallave et al. reported the filled and hollow CNFs at room temperature and without polymer binders by applying co-axial electrospinning of Alcell lignin in ethanol [104]. Lignin nanofibers (L-NFs) were produced from the lignin/ethanol/platinum acetyl acetonate and lignin/ethanol solutions using electrospinning [105]. Lignin phosphoric acid mixtures were fabricated via electrospinning to obtain L-NFs [106]. To check the role of the metal complex with lignin and phosphoric reaction mixtures, platinum acetyl acetonate was used to get L-NFs via the electrospinning route [107]. Seven different types of softwood lignins were reported to produce lignin-derived beaded fibers in the varying concentration of DMF and water solution [108]. However, the production of uniform CNFs from recovered lignin is rarely seen, except the work done by Lallave et al. by using Alcell lignin and ethanol (1:1) solution [104, 105], and other work done by Dallmeyer et al. by using purified softwood Kraft lignin [109]. The low electrospinning ability of pure lignin occurred due to several points including low molar mass, nonlinear structure, and the formation of anion charge in solution [110]. The high contents of graphite and diamond-like carbons were obtained using electrospinning lignin/PAN solutions in the presence of catalysts including $Fe(NO_3)_3 \cdot 9H_2O$ or $PdCl_2$ [111]. The different sets of noble monometallic and bimetallic nanoparticles, including Pd, Pt, Au, Pd/Pt, Pd/Au, or Pt/Au, were applied to produce LCNFs mats by using the electrospun and thermostabilized routes [112]. An extension of the ice-segregation-induced self-assembly methodology was applied to make the LCNFs. In this method, a rapid freezing of aqueous lignin solution was done, and subsequent sublimation of resultant freeze dryer material was performed to obtain L-NFs at $-15\,°C, 100$ millitorr. Stabilization and carbonization processes were then performed under similar conditions reported elsewhere to get LCNFs [113].

Dalton et al. developed biobased CNFs that are derived from a mixture of PAN and lignin in different ratios (up to 70%). The impact of lignin addition is characterized regardless of the reduced diameter of CNFs from 450 to 250 nm, enhanced sample flexibility, and more [117]. Raman spectra of lignin-derived CNFs were done and provided two crystalline bands at $1350\,cm^{-1}$ and $1580\,cm^{-1}$ for D-band and G-band, respectively (Fig. 4) [118]. The electrical conductivity of biobased CNFs was determined and high amounts of lignin contents (i.e., 70%) resulted in higher electrical conductivity. This phenomenon was explained based on maximum inter-fiber fusion achieved with high lignin loadings in biobased CNFs. Table 7 shows the different lignin types and their respective fraction with PAN used to synthesize the biobased CNs. It was noted that the impact of lignin loading in between 50–70% was good to produce excellent NFs morphology.

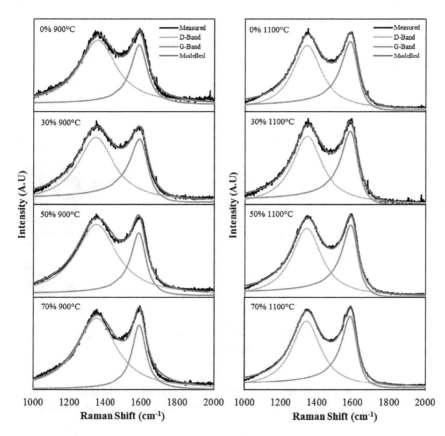

Fig. 4 Deconvoluted Raman spectra of biobased CNFs with different ratios of lignin. Figure adapted from Ref. [117] with Elsevier permission

Table 7 Types of lignin and their ratio with auxiliary polymer for lignin-based NFs synthesis

Entry	Type of lignin	Auxiliary polymer	Solution concentration (wt%)	Polymer-to-lignin ratio (wt%)	References
1	Kraft lignin	PAN	30	50:50	[119]
2	Kraft lignin	PAN	12	50:50	[120]
	Kraft lignin	PAN	18	80:20 and 50:50	[121]
3	Alkali lignin	PAN	20	50:50	[122]
4	Alkali lignin	PAN	15	50:50	[111]
5	Hardwood lignin	PAN	18	50:50	[123]
6	Lignin powder	PAN	25	70:30	[124]
7	Corn straw lignin	PAN	10	90:10	[125]
8	Enzymatic hydrolyzed lignin	PAN	10	60:40	[126]
9	Lignin bio-oil	PAN	20	15:85	[127]

Note PAN—polyacrylonitrile

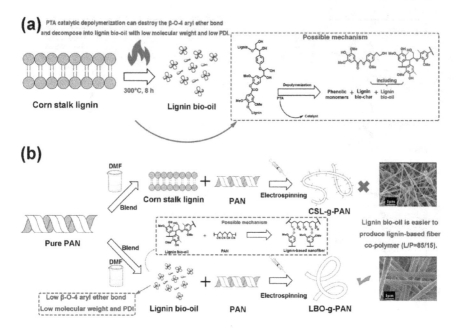

Fig. 5 Proposed mechanism for **a** lignin conversion to bio-oil and **b** lignin-based nanofibers. Figure adapted from Ref. [127] with Elsevier Permission

Figure 5 shows a possible mechanism for the formation of lignin or bio-oil-based CNFs. Corn stalk lignin was initially depolymerized to aromatic bio-oil that contained various types of functional groups, including hydroxyl, aldehyde, ester, acid, and more. These functional groups are highly susceptible to the formation of linkages with PNA. SEM images of lignin-derived CNFs and bio-oils-based CNFs were analyzed and average diameters of bio-oil-based CNFs were hanging between pure PAN and lignin-derived CNFs.

The ^{13}C-^{1}H 2D heteronuclear single quantum coherence (HSQC) NMR was recorded for PAN, lignin, and biobased CNFs, and the obtained spectra are shown in Fig. 6. New cross-signals were characterized in both lignin and biobased CNFs samples, besides the cross-signals presented in PAN (δ_H/δ_C 6.51/151.37 ppm). This study confirmed the formation of linkage between the PAN and lignin or bio-oils.

Alkali lignin was used to prepare the electrospun CNFs. The properties correlation suggested that lignin-derived CNFs have 10-times high absorption capacity, six-times high permeability, and two-time faster absorption kinetics relative to conventional CNFs [128]. These properties of lignin-derived CNFs relative to conventional CNFs were explained based on the high specific surface area, larger average pore diameter and high pore volume [128]. Kraft lignin is produced in maximum amounts by pulp and paper industries and used to prepare the CNFs. Acrylonitrile (AN), α,α'-azobisisobutyronitrile, and Kraft lignin were mixed to synthesize a copolymer, and the synthesized copolymer was transformed for electrospinning to prepare CNFs [129].

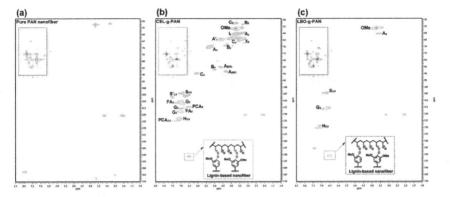

Fig. 6 2D ^{13}C-^{1}H HSQC NMR spectra of **a** pure PAN, **b** corn stalk lignin-g-PAN nanofiber, and **c** lignin-derived bio-oil-g-PAN nanofiber. Figure adapted from Ref. [127] with Elsevier Permission

6 Biomedical Application of Electrospun Nanofibers

In the field of biomedical applications, the advantages of using nanomaterials can be explored by using the NFs that are designed from a variety of polymers, either biocompatible or non-biodegradable. Due to the versatile and unique physical, chemical, thermal, mechanical, and biological properties of NFs, they have been used in a wide range of biomedical communities for their applications including drug delivery [130], gene delivery [131], skin tissue engineering [132], bone tissue engineering [133], cartilage tissue engineering [134], skeletal muscle tissue engineering [135], and more. For drug delivery applications, most of the drugs are introduced in the body by applying conventional methods, including oral tablets and/or intravenous injections. There are encounter problems and side effects including poor solubility, tissue damage during extravasation, cytotoxicity, undesirable biodistribution, and more. Considering these issues, advanced drug delivery systems can be implemented that aim to minimize toxicity, targeted drug release, extended drug release picture, enhanced bioavailability, and more. The CNFs scaffolds, fabricated by electrospinning, are the most attractive and modern competent drug delivery carriers. They are the transfigure agent in the nanomedicines field due to their structure and properties, including large surface area, diameters, porosity, and improved drug loading capacities, that enhance the drug permeation and retention effects and more [7].

6.1 Drug Delivery

The drug delivery system offers numerous selective forthcoming properties, including improving the therapeutic efficiency of several existing drugs. Several factors need to be considered prior to design an efficient drug delivery system (Fig. 7). The use of NFs for drug delivery systems is enormously developing in

Fig. 7 Factors need to be considered before designing an efficient drug delivery system [137, 138]

various applications including anticancer therapeutics, surgical implants, antibacterial sheets, wound dressings, tissue scaffolds, and more [136]. A various set of drug loading methods, including coating, embedding, encapsulating (co-axial and emulsion electrospinning), and more, are reported elsewhere [130]. These methods have been utilized to release the kinetics of drugs in a controlled manner. Over the past few decades, control drug delivery systems have found a lot of academic and industrial attention regarding biomedical applications for drug release. Both academic and industrial attention occurred because of various advantages offered compared to a conventional dosage form including a reduction in toxicity by delivering the drug at the targeted site of the body and controlled rate, improved therapeutic efficacy, and more.

Several types of NFs fabricated by electrospinning, that are derived from biocompatible or non-biodegradable materials, are used for a wide range of biomedical applications [139]. Similarly, lignin and lignin-derived biomaterials nanoparticles have also been applied in drug release, tissue engineering, and more applications [140, 141]. Kai et al. have synthesized a range of different concentrations of alkylated lignin (10–50%) with biodegradable poly(lactic acid) (PLA). These PLA-lignin copolymers further blended with PLA and proceeded to form the NFs composites by electrospinning. These NFs were evaluated for antioxidant activity over three different types of cells, including PC12, human dermal fibroblasts, and human mesenchymal stem cells [142]. Lignin was blended with other polymers, including poly(ε-caprolactone-co-lactide), polystyrene, and poly(ethylene glycol), and proceeded then for electrospinning to obtain NFs. These lignin-polymers-derived NFs were evaluated for various biomedical and/or healthcare applications [143, 144]. A highly stretchable lignin-based electrospun biomaterial was synthesized by using poly(methyl methacrylate) (PMMA) and poly(e-caprolactone) (PCL). Synthesized lignin-derived CNFs were processed for cell culture study. It was characterized that the synthesized biomaterial is biocompatible and showed human dermal fibroblasts interaction [145]. Electrospun poly (ε-caprolactone) (PCL)-grafted lignin (PCL-g-lignin) copolymer CNFs showed excellent antioxidant, anti-inflammatory, and low cytotoxicity properties. PCL-g-lignin NFs inhibited the reactive oxygen species generation and activated the antioxidant enzyme activity by an autophagic mechanism. The nanofibrous membrane of PCL-lignin can be implanted for arthroscopic and that provides an effective Osteoarthritis therapy [146]. Arginine-derived lignin NF was prepared by using an electrospun technique and obtained NF showed suitable viscosity that can be used for the spreadability for topical application. In *vivo*, the wound-healing assay was demonstrated in rats. The Arginine-based lignin NF accelerated wound healing and substantially increased re-epithelialization, collagen deposition, and angiogenesis relative to lignin NFs and arginine [147]. In another study, lignin copolymer was synthesized by using β-butyrolactone and/or ε-caprolactone. The synthesized lignin copolymer was further incorporated with poly(3-hydroxybutyrate) (PHB) to prepare PHB/lignin NFs by using electrospinning techniques. The obtained PHB/lignin NFs characterized using various techniques and results showed that the tensile strength and elongation of materials increased. The obtained materials were also screened for biodegradation and biocompatibility test and results showed that synthesized materials can be used for biomedical applications [148]. Lignin-polycaprolactone (PCL) copolymer was synthesized by ring-opening polymerization (ROP) in the presence of a tin (II) 2-ethylhexanoate catalyst. The obtained copolymer was further electrospunned with different PCL ratios (5:95, 10:90, and 15:85 wt/wt). The PCL/lignin-PCL nanofibers were cultured with Schwann cells and dorsal root ganglion (DRG) neurons to access nerve generation potential. The resulted cultures were characterized, and data suggested that lignin NFs promoted the cell proliferation of bone marrow mesenchymal stem cells and Schwann cells. The synthesized NFs also enhanced the myelin basic protein expressions of Schwann cells [149]. Poly(3-hydroxybutyrate) (PHB)-derived lignin copolymer was synthesized by the ROP mechanism. This copolymer was further reacted with PHB to produce composite NFs by using an electrospinning technique. The synthesized PHB/lignin NFs were demonstrated for an antioxidant activity to allow the excess free radicals neutralization. The nonirritating and biocompatible nature of PHB/lignin NFs was reported over the animal studies [150]. Cellulose-derived CNFs have also been studied for drug released kinetics for four drugs; naproxen (NAP), indomethacin (IND), ibuprofen (IBU), and sulindac (SUL). The liberation of these drugs from NFs was evaluated in solvent cast films, and the films followed the trend as NAP > IBU > IND > SUL, due to their higher interaction [151].

6.2 Electrospun Lignin-Based Nanocomposites for Drug Delivery

In the past few decades, nanocomposite scaffolds, derived from biopolymers, have received more attentions in a wide range of area, including tissue engineering, drug delivery, and more, from academia and industries [152]. A large number of nanocomposite interest can be defined due to its excellent physical, chemical, thermal, and biological properties, including, pore sizes, pore volumes, and more, that allow more readily cells differentiation and vascularization. These properties can be used for the substitution of bone and skin tissues [152]. Lignin and polycaprolactone (PCL) were used to synthesize a nanocomposite by using an electrospinning technique. A wide range of lignin concentration (i.e., 0, 5, 10, and 15 wt%) was mixed with PCL, and obtained lignin-PCL nanocomposite with 10 wt% of lignin showed the good properties, including porosity, fiber diameter, tensile strength, and Young's modulus, relative to pure PCL. These nanocomposites were tested for cell cytotoxicity, and characterized results showed that the 10 wt% lignin composite has been selected for the cell test [153].

7 Conclusions and Prospects

Nanofibers are unique and versatile class of nanomaterials with their excellent physical, -chemical, thermal, and biological properties. Moreover, their functional properties can be tuned as per need. NFs have attracted increasing attention in the past few decades and gained novel applications in assorted biomedical disciplines. Rapid and substantial advances in biology, chemistry, material science, and medical, micro-, and nanotechnologies have enabled us to produce renewable and sustainable sophisticated and biomimetic NFs for biomedical, drug delivery systems, and more applications. However, the clinical applications of renewable lignin-derived NFs are lesser relative to conventional non-renewable petroleum-derived NFs. Biomedical applications of L-NFs are still at their very early stage. This chapter offers a plethora of opportunities to design, synthesize, and study NFs from assorted lignins for biomedical applications.

Acknowledgements SKS would like to thank the Council of Scientific Industrial and Research for the Senior Research Fellowship. AS and SM acknowledge the Ministry of Human Resource Development (MHRD) and UGC, New Delhi, and IIT Indore.

References

1. De GW (1958) Magnete. Courier, New York
2. Gray S (1731) Philosophical transactions of the royal society of London 37:227.
3. Nollet JA, Stack T (1748) Philosophical transactions of the royal society of London 45:187.
4. Rayleigh L (1882) Lond Edinb Dublin Philos Mag J Sci 14:184
5. Cooley J, Morton W (1902) Apparatus for electrically dispersing fluids. U.S. Pat. US692631A
6. Morton WJ (1902) Method of dispersing fluids. U.S. Pat. US705691A
7. Xue J, Wu T, Dai Y, Xia Y (2019) Chem Rev 119:5298
8. Bennett SJ (2012) Implications of climate change for the petrochemical industry: mitigation measures and feedstock transitions. In: Chen W-Y, Seiner J, Suzuki T, Lackner M (eds) Handbook of climate change mitigation. Springer US, New York, NY, p 319
9. Singh SK, Dhepe PL (2016) Bioresour Technol 221:310
10. Singh SK, Dhepe PL (2018) Clean Technol Environ Policy 20:739
11. Koshijima T, Watanabe T (2003) Preparation and characterization of lignin-carbohydrate complexes. In: Timell TE (ed) Association between lignin and carbohydrates in wood and other plant tissues. Springer Berlin Heidelberg, Berlin, p 1
12. Jeffries TW (1990) Biodegradation 1:163
13. Werpy T, Petersen G (2004) Top value added chemicals from biomass: Volume I -results of screening for potential candidates from sugars and synthesis gas. National Renewable Energy Lab., Golden, CO (US), p 1
14. Holladay JE, White JF, Bozell JJ, Johnson D (2007) Top value added chemicals from biomass-volume ii, results of screening for potential candidates from biorefinery lignin: Pacific Northwest National Lab. (PNNL), Richland, WA (United States); National Renewable Energy Lab. (NREL), Golden, CO (United States)
15. Michael S (2008) Angew Chem. Int Ed 47:9200
16. Zhou C-H, Xia X, Lin C-X, Tong D-S, Beltramini J (2011) Chem Soc Rev 40:5588
17. Boerjan W, Ralph J, Baucher M (2003) Annu Rev Plant Biol 54:519
18. Alinejad M, Henry C, Nikafshar S, Gondaliya A, Bagheri S, Chen N et al (2019) Polymers 11:1202
19. Vanholme R, Demedts B, Morreel K, Ralph J, Boerjan W (2010) Plant Physiol 153:895
20. Mosier N, Wyman C, Dale B, Elander R, Lee YY, Holtzapple M et al (2005) Bioresour Technol 96:673
21. Lange H, Decina S, Crestini C (2013) Eur Polym J 49:1151
22. Zakzeski J, Bruijnincx PCA, Jongerius AL, Weckhuysen BM (2010) Chem Rev 110:3552
23. Dence CW, Lin SY. Introduction. In: Lin SY, Dence CW (eds) Methods in lignin chemistry. Springer Berlin Heidelberg, Berlin, Heidelberg, p 3
24. Zheng Y, Zhao J, Xu F, Li Y (2014) Prog Energy Combust Sci 42:35
25. Schutyser W, Renders T, Van den Bosch S, Koelewijn SF, Beckham GT, Sels BF (2018) Chem Soc Rev 47:852
26. van Kuijk SJA, Sonnenberg ASM, Baars JJP, Hendriks WH, Cone JW (2015) Biotechnol Adv 33:191
27. Cesaro A, Belgiorno V (2014) Chem Eng J 240:24
28. Gibson LJ (2012) J R Soc Interface 9:2749
29. Vanholme R, De Meester B, Ralph J, Boerjan W (2019) Curr Opin Biotechnol 56:230
30. Singh SK (2019) Int J Biol Macromol 132:265

31. Bajwa DS, Pourhashem G, Ullah AH, Bajwa SG (2019) Ind Crops Prod 139:111526.
32. Nimz H (1974) Angew Chem Int Ed 13:313
33. Pandey MP, Kim CS (2011) Chem Eng Technol 34:29
34. Li C, Zhao X, Wang A, Huber GW, Zhang T (2015) Chem Rev 115:11559
35. Wang X, Rinaldi R (2012) Chemsuschem 5:1455
36. Parthasarathi R, Romero RA, Redondo A, Gnanakaran S (2011) J Phys Chem Lett 2:2660
37. Younker JM, Beste A, Buchanan AC (2011) ChemPhysChem 12:3556
38. Dorrestijn E, Laarhoven LJJ, Arends IWCE, Mulder P (2000) J Anal Appl Pyrolysis 54:153
39. Rinaldi R (2015) Solvents and solvent effects in biomass conversion. In: Rinaldi R (ed) Catalytic hydrogenation for biomass valorization. The Royal Society of Chemistry, UK, p 74
40. Evtuguin DV, Neto CP, Silva AMS, Domingues PM, Amado FML, Robert D et al (2001) J Agric Food Chem 49:4252
41. Rodrigues Pinto PC, Borges da Silva EA, Rodrigues AE (2011) Ind Eng Chem Res 50:741.
42. Upton BM, Kasko AM (2016) Chem Rev 116:2275
43. Ralph J, Brunow G, Boerjan W (2007) Encyclopedia of lifescience:1.
44. McMillan JD (1994) Pretreatment of lignocellulosic biomass. American Chemical Society, Enzymatic conversion of biomass for fuels production, p 292
45. Gaspar AR, Gamelas JAF, Evtuguin DV, Pascoal Neto C (2007) Green Chem 9:717
46. da Costa SL, Chundawat SPS, Balan V, Dale BE (2009) Curr Opin Biotechnol 20:339
47. Vroom KE (1957) PPMC 58:228
48. Niju S, Swathika M, Balajii M (2020) Chapter 10 - Pretreatment of lignocellulosic sugarcane leaves and tops for bioethanol production. In: Yousuf A, Pirozzi D, Sannino F (eds). Academic Press, Lignocellulosic Biomass to Liquid Biofuels, p 301
49. Sievers DA, Kuhn EM, Tucker MP, McMillan JD (2017) Bioresour Technol 243:474
50. Chum HL, Johnson DK, Black SK, Overend RP (1990) Appl Biochem Biotechnol 24:1
51. Singh SK, Savoy AW, Yuan Z, Luo H, Stahl SS, Hegg EL et al (2019) Ind Eng Chem Res 58:15989
52. Behera S, Arora R, Nandhagopal N, Kumar S (2014) Renew Sustain Energy Rev 36:91
53. Chundawat SPS, Balan V, Sousa LDc, Dale BE (2010) 2 - Thermochemical pretreatment of lignocellulosic biomass. In: Waldron K (ed) Bioalcohol Production: Woodhead Publishing, p 24
54. Singh SK (2021) J Cleaner Prod 279:123546
55. Chundawat SPS, Beckham GT, Himmel ME, Dale BE (2011) Annu Rev Chem Biomol Eng 2:121
56. Thielemans W, Can E, Morye SS, Wool RP (2002) J Appl Polym Sci 83:323
57. Gosselink RJA, de Jong E, Guran B, Abächerli A (2004) Ind Crops Prod 20:121
58. Martone PT, Estevez JM, Lu F, Ruel K, Denny MW, Somerville C et al (2009) Curr Biol 19:169
59. Holladay JE, White J, Bozell JJ, Johnson D (2007) Biomass Fuels 2:PNNL.
60. Kai D, Tan MJ, Chee PL, Chua YK, Yap YL, Loh XJ (2016) Green Chem 18:1175
61. El Mansouri N-E, Salvadó J (2007) Ind Crops Prod 26:116
62. Chakar FS, Ragauskas AJ (2004) Ind Crops Prod 20:131
63. Norgren M, Edlund H (2014) Curr Opin Colloid Interface Sci 19:409
64. Vishtal AG, Kraslawski A (2011) BioResources 6:3547
65. Singh SK, Dhepe PL (2018) Ind Crops Prod 119:144
66. Alekhina M, Ershova O, Ebert A, Heikkinen S, Sixta H (2015) Ind Crops Prod 66:220
67. Tejado A, Peña C, Labidi J, Echeverria JM, Mondragon I (2007) Bioresour Technol 98:1655
68. Hatakeyama H, Tsujimoto Y, Zarubin M, Krutov S, Hatakeyama T (2010) J Therm Anal Calorim 101:289
69. Rabdsfovich ML (2014) Cellul Chem Technol 48:613
70. Bhalla A, Cai CM, Xu F, Singh SK, Bansal N, Phongpreecha T et al (2019) Biotechnol Biofuels 12:213
71. Reddy BS, Maeura Y, Wayman M (1983) J Natl Cancer Inst 71:419
72. Barnard DL, Heaton KW (1973) Gut 14:316

73. Vinardell M, Mitjans M (2017) Int J Mol Sci 18:1219
74. Barapatre A, Aadil KR, Tiwary BN, Jha H (2015) Int J Biol Macromol 75:81
75. Hasegawa Y, Kadota Y, Hasegawa C, Kawaminami S (2015) J Nutr Sci Vitaminol 61:449
76. Norikura T, Mukai Y, Fujita S, Mikame K, Funaoka M, Sato S (2010) Basic Clin Pharmacol Toxicol 107:813
77. Sato S, Mukai Y, Yamate J, Norikura T, Morinaga Y, Mikame K et al (2009) Free Radical Res 43:1205
78. Gordts SC, Férir G, D'huys T, Petrova MI, Lebeer S, Snoeck R et al (2015) PLOS ONE 10:e0131219.
79. Lee J-B, Yamagishi C, Hayashi K, Hayashi T (2011) Biosci Biotechnol Biochem 75:459
80. Thakkar JN, Tiwari V, Desai UR (2010) Biomacromol 11:1412
81. Henry BL, Desai UR (2014) Thromb Res 134:1123
82. Henry BL, Thakkar JN, Liang A, Desai UR (2012) Biochem Biophys Res Commun 417:382
83. Henry BL, Aziz MA, Zhou Q, Desai UR (2010) Thromb Haemost 103:507
84. Henry BL, Connell J, Liang A, Krishnasamy C, Desai UR (2009) J Biol Chem 284:20897
85. Mehta AY, Mohammed BM, Martin EJ, Brophy DF, Gailani D, Desai UR (2016) J Thromb Haemostasis 14:828
86. Saluja B, Thakkar JN, Li H, Desai UR, Sakagami M (2013) Pulm Pharmacol Ther 26:296
87. Figueiredo P, Ferro C, Kemell M, Liu Z, Kiriazis A, Lintinen K et al (2017) Nanomedicine 12:2581
88. Baker DA, Rials TG (2013) J Appl Polym Sci 130:713
89. Fang W, Yang S, Wang X-L, Yuan T-Q, Sun R-C (2017) Green Chem 19:1794
90. Huang J, Liu Y, You T (2010) Anal Methods 2:202
91. Wu X-F, Rahman A, Zhou Z, Pelot DD, Sinha-Ray S, Chen B et al (2013) J Appl Polym Sci 129:1383
92. Souto F, Calado V, Jr NP (2018) Mater Res Express 5:072001
93. Jose Varghese R, Sakho EhM, Parani S, Thomas S, Oluwafemi OS, Wu J. Chapter 3 - Introduction to nanomaterials: synthesis and applications. In: Thomas S, Sakho EHM, Kalarikkal N, Oluwafemi SO, Wu J (eds) Nanomaterials for solar cell applications: Elsevier, p 75
94. Haghi AK (2012) Electrospun nanofibers: an introduction. In: Haghi AK (ed) Electrospinning of nanofibers in textiles. 1st ed. New York Apple Academic Press, p 1
95. Kumar M, Hietala M, Oksman K (2019) Front Mater 6:1
96. Singh SK, Dhepe PL (2016) Green Chem 18:4098
97. Singh SK, Dhepe PL (2019) Ind Eng Chem Res 58:21273
98. Supanchaiyamat N, Jetsrisuparb K, Knijnenburg JTN, Tsang DCW, Hunt AJ (2019) Bioresour Technol 272:570
99. Grossman A, Vermerris W (2019) Curr Opin Biotechnol 56:112
100. Huang X (2009) Materials 2:2369
101. Liu HC, Chien A-T, Newcomb BA, Bakhtiary Davijani AA, Kumar S (2016) Carbon 101:382
102. Inagaki M, Yang Y, Kang F (2012) Adv Mater 24:2547
103. Zhang L, Aboagye A, Kelkar A, Lai C, Fong H (2014) J Mater Sci 49:463
104. Lallave M, Bedia J, Ruiz-Rosas R, Rodríguez-Mirasol J, Cordero T, Otero JC et al (2007) Adv Mater 19:4292
105. Ruiz-Rosas R, Bedia J, Lallave M, Loscertales IG, Barrero A, Rodríguez-Mirasol J et al (2010) Carbon 48:696
106. García-Mateos FJ, Berenguer R, Valero-Romero MJ, Rodríguez-Mirasol J, Cordero T (2018) J Mater Chem A 6:1219
107. García-Mateos FJ, Cordero-Lanzac T, Berenguer R, Morallón E, Cazorla-Amorós D, Rodríguez-Mirasol J et al (2017) Appl Catal B 211:18
108. Dallmeyer I, Ko F, Kadla JF (2010) J Wood Chem Technol 30:315
109. Dallmeyer I, Ko F, Kadla JF (2014) Ind Eng Chem Res 53:2697
110. Schreiber M, Vivekanandhan S, Mohanty AK, Misra M (2012) Adv Mater Lett 3:476
111. Xu X, Zhou J, Jiang L, Lubineau G, Payne SA, Gutschmidt D (2014) Carbon 80:91
112. Gao G, Ko F, Kadla JF (2015) Macromol Mater Eng 300:836

113. Spender J, Demers AL, Xie X, Cline AE, Earle MA, Ellis LD et al (2012) Nano Lett 12:3857
114. David WC (2011) Low cost carbon fiber overview, https://www.energy.gov/sites/prod/files/2014/03/f11/lm002_warren_2011_o.pdf 9 May,. 2011
115. Baker FS (2010) Low cost carbon fiber from renewable resources. https://www1.eere.energy.gov/vehiclesandfuels/pdfs/merit_review_2010/lightweight_materials/lm005_baker_2010_o.pdf
116. Paulauskas EFL (2010) High strength carbon fibers. https://www.hydrogen.energy.gov/pdfs/review10/st093_paulauskas_2010_p_web.pdf
117. Dalton N, Lynch RP, Collins MN, Culebras M (2019) Int J Biol Macromol 121:472
118. Ferrari AC, Robertson J (2001) Phys Rev B 64:075414.
119. Seo D, Jeun J, Kim H, Kang P (2011) Rev Adv Mater Sci 28:31
120. Choi DI, Lee J-N, Song J, Kang P-H, Park J-K, Lee YM (2013) J Solid State Electrochem 17:2471
121. Lei D, Li X-D, Seo M-K, Khil M-S, Kim H-Y, Kim B-S (2017) Polymer 132:31
122. Xu X, Zhou J, Jiang L, Lubineau G, Chen Y, Wu X-F et al (2013) Mater Lett 109:175
123. Oroumei A, Fox B, Naebe M (2015) ACS Sustain Chem Eng 3:758
124. Liu HC, Chien A-T, Newcomb BA, Liu Y, Kumar S (2015) ACS Sustain Chem Eng 3:1943
125. Ma A, Li C, Du W, Chang J (2014) J Nanosci Nanotechnol 14:7204
126. Li C, Ma A, Fu Y, Chang J (2014) J Bioprocess Eng Biorefinery 3:79
127. Du B, Sun Y, Liu B, Yang Y, Gao S, Zhang Z et al (2020) Polymer testing 81:106207
128. Beck RJ, Zhao Y, Fong H, Menkhaus TJ (2017) J Water Process Eng 16:240
129. Youe W-J, Lee S-M, Lee S-S, Lee S-H, Kim YS (2016) Int J Biol Macromol 82:497
130. Dhand C, Dwivedi N, Sriram H, Bairagi S, Rana D, Lakshminarayanan R et al (2017) Nanofiber composites in drug delivery. In: Ramalingam M, Ramakrishna S (eds). Nanofiber composites for biomedical applications. Woodhead Publishing, p 199
131. Lakshmi priya M, Rana D, Bhatt A, Ramalingam M (2017) Nanofiber composites in gene delivery. In: Ramalingam M, Ramakrishna S (eds) Nanofiber composites for biomedical applications. Woodhead Publishing, p 253
132. Naves LB, Almeida L, Rajamani L (2017) Nanofiber composites in skin tissue engineering. In: Ramalingam M, Ramakrishna S (eds). Nanofiber composites for biomedical applications. Woodhead Publishing, p 275
133. Liverani L, Roether JA, Boccaccini AR (2017) Nanofiber composites in bone tissue engineering. In: Ramalingam M, Ramakrishna S (eds) Nanofiber composites for biomedical applications. Woodhead Publishing, p 301
134. Rana D, Ratheesh G, Ramakrishna S, Ramalingam M (2017) Nanofiber composites in cartilage tissue engineering. In: Ramalingam M, Ramakrishna S (eds) Nanofiber composites for biomedical applications. Woodhead Publishing, p 325
135. Cai A, Horch RE, Beier JP (2017) Nanofiber composites in skeletal muscle tissue engineering. In: Ramalingam M, Ramakrishna S (eds) Nanofiber composites for biomedical applications. Woodhead Publishing, p 369
136. Nayak R, Padhye R, Kyratzis IL, Truong YB, Arnold L (2012) Text Res J 82:129
137. Supaphol P, Suwantong O, Sangsanoh P, Srinivasan S, Jayakumar R, Nair SV (2012) Electrospinning of biocompatible polymers and their potentials in biomedical applications. In: Jayakumar R, Nair S (eds) Biomedical applications of polymeric nanofibers. Springer Berlin Heidelberg, Berlin, Heidelberg, p 213
138. Prabaharan M, Jayakumar R, Nair SV (2012) Electrospun nanofibrous scaffolds-current status and prospects in drug delivery. In: Jayakumar R, Nair S (eds) Biomedical applications of polymeric nanofibers. Springer Berlin Heidelberg, Berlin, Heidelberg, p 241
139. Liu M, Duan X-P, Li Y-M, Yang D-P, Long Y-Z (2017) Mater Sci Eng, C 76:1413
140. Witzler M, Alzagameem A, Bergs M, Khaldi-Hansen BE, Klein SE, Hielscher D et al (2018) Molecules 23:1885
141. Beisl S, Friedl A, Miltner A (2017) Int J Mol Sci 18:2367
142. Kai D, Ren W, Tian L, Chee PL, Liu Y, Ramakrishna S et al (2016) ACS Sustainable Chem Eng 4:5268

143. Kai D, Zhang K, Jiang L, Wong HZ, Li Z, Zhang Z et al (2017) ACS Sustainable Chem Eng 5:6016
144. Kai D, Low ZW, Liow SS, Abdul Karim A, Ye H, Jin G et al (2015) ACS Sustainable Chem Eng 3:2160
145. Kai D, Jiang S, Low ZW, Loh XJ (2015) J Mater Chem B 3:6194
146. Liang R, Zhao J, Li B, Cai P, Loh XJ, Xu C, et al. (2020) Biomaterials 230:119601.
147. Reesi F, Minaiyan M, Taheri A (2018) Drug Delivery Transl Res 8:111
148. Kai D, Chong HM, Chow LP, Jiang L, Lin Q, Zhang K et al (2018) Compos Sci Technol 158:26
149. Wang J, Tian L, Luo B, Ramakrishna S, Kai D, Loh XJ et al (2018) Colloids Surf B 169:356
150. Kai D, Zhang K, Liow SS, Loh XJ (2019) ACS Appl Bio Mater 2:127
151. Kakoria A, Sinha-Ray S (2018) Fibers 6:45
152. Fernandes EM, Pires RA, Mano JF, Reis RL (2013) Prog Polym Sci 38:1415
153. Salami MA, Kaveian F, Rafienia M, Saber-Samandari S, Khandan A, Naeimi M (2017) J Med Signals Sens 7:228

1D Spinel Architectures via Electrospinning for Supercapacitors

Amrita De Adhikari

Abstract Supercapacitors have been publicized in current times as energy storage devices in regards to sustainability and renewability. Supercapacitors act as the conduit between the conventional capacitors and the batteries or fuel cells with superior power density, swift charge and discharge rates with long-term cycling performance. Hierarchical nanostructured materials with particular dimensions are reflected as electrode constituents for the high electrochemical performance of the supercapacitors. Researchers have focused on the utilization of fibrous templates having enlarged specific surface areas with multiple dynamic sites for the electrochemical reactions. The consequence of 1D nanostructures on the electrochemical behavior of the supercapacitors, which includes its specific capacitance, rate capability, cycling performance, has been summarized in this chapter. Electrospinning technique has proven itself as a simple, handy and greatly resourceful tactic to yield 1D architecture cross-breed nanomaterials having requisite dimensions. This chapter presents the electrospinning technique as an experimental approach by various researchers whereby tuning its important parameters such as deposition methods, solution flow rate for the electrospinning, applied electric potential, the needle tip and the collector distance, etc. in directive to obtain the desired electrochemical demonstration of the electrode materials. The chapter highlights the role of electrospinning technique for the formation of a wide variety of 1D spinels, which have strained attention as electroactive materials owing to their different metal centers and multiple redox states. The association among the configuration, construction and the electrochemistry associated is elaborated vividly. Finally, some concluding summaries with future outlook have been scripted predicting the fabrication of high-performance 1D electrospun electrode materials for supercapacitor applications.

Keywords Supercapacitors · 1D nanostructures · Electrospinning · Spinels · Electrochemical performance

A. De Adhikari (✉)
Department of Chemistry, Ben Gurion University of the Negev, 8410501 Beer Sheva, Israel

© The Author(s), under exclusive license to Springer Nature Switzerland AG 2021   199
S. K. Tiwari et al. (eds.), *Electrospun Nanofibers*, Springer Series on Polymer and Composite Materials, https://doi.org/10.1007/978-3-030-79979-3_8

1 Introduction

Energy has been the keystone of the ecosphere's uninterrupted wealth and its vandalism is delinquency and henceforth to preserve this is the responsibility of the human race. Over the decades, the extensive utilization of the energy from the earth's fossil fuel reserves gradually tended its extinction leading to ever-increasing pollution. Such consequences have compelled the world population to develop environmentally benevolent advanced energy storage devices. The global economic concerns and high operation cost of such devices are thus a challenge. Specifically, the electrochemical capacitors or supercapacitors (SCs) have drawn immense interest with extremely fast charge and discharge rates, high power density, long life cycle and reduced preservation cost and innocuous operations [1–3]. Depending on the mechanism of charge storage, SCs can be categorized into electrochemical double layer capacitors (EDLCs) and pseudocapacitors [1]. Compared with the EDLCs, pseudocapacitors offer higher energy density and specific capacitance ensuing from the fast reversible redox-active reactions occurring in electrochemical course [4]. Among different pseudocapacitive materials, low-cost, naturally abundant with multiple oxidation state binary transition metal oxides have been broadly considered in energy storage uses [5]. Compared with single transition metal-based oxides, the binary transition metal-based oxides are more potent from an electrical conductivity point of view [6]. In the need to inflate the performance of SCs, a variety of nanomaterials including 0D, 1D, 2D and 3D nanostructures have been contrived in modern times [7]. Most commonly, 1D nanostructured materials have drawn attention due to their extraordinary specific surface area, effectual transport of electrons and easy nanodevices construction [8]. In the meantime, the encroachment of stretchy and bendable electronic devices necessitates that the electrode ought to be free-standing without conductive agents and binders, but not compromising its electrochemical performance [9]. In the earlier few epochs, different synthetic stratagems like template-directed synthesis, vapor-phase approach, solution-phase method, melt-blown protocols, electrospinning technique, self-assembly method and electrodeposition have been engaged to organize 1D nanostructured materials [10, 11]. Among them, the electrospinning method is the utmost preferred route in terms of designed architecture, controlled dimensions, elasticity, adaptability and affluence of fiber creation [12, 13]. Consequently, electrospun nanofibers deliver appropriate features for use in SCs applications.

Various researches have been reported on the production of electrospun nanofibers, nanotubes, coaxial nanocables, core–shell structures and other hierarchical nanomaterials for SC application [14–16]. Electrospun materials as per electrodes for SCs are categorized into classes of material, which include carbon nanostructures, metal/metal oxides, nitrides, sulfides, organic–inorganic hybrid materials and conducting polymers [17–19]. Specifically, spinel transition metal oxide (AB_2O_4) is a varied choice of pseudocapacitive materials that are explored extensively for the SC electrodes [20, 21].

In this chapter, first, an ephemeral outline of the electrospinning methods intended for the formation of 1D architecture will be presented. The emphasis will then be directed to the 1D architecture compositions and morphologies. Further different discussions related to previous reports will be summarized, and in conclusion, open challenges and future projections of the electrospun 1D architecture will be addressed. Hence, cost-effective, simple and most importantly mass-productive electrospun devices can be tackled.

1. Electrospinning: Basics and techniques.

The fundamental of electrospinning (ES) technique deals with the electric forces to draw viscoelastic solution as charged filaments toward an oppositely charged collecting plate [22, 23]. Basically, the instrument comprises a syringe prepared with a needle or pipette tip, a pump in order to provide the viscoelastic solution, a high voltage power supply and a collector (see Fig. 1) [24, 25]. A little amount of the viscoelastic solution is propelled out of the needle, which is disposed to form a sphere-shaped drop due to the surface tension. Depending on the voltage power supply, a steady-charged liquefied jet is evicted leading to the formation of long, thin threads on the collector plate [26].

ES technique to generate nanofibers in the range of nanometers (nm) is controlled by different parameters such as viscidness of the pumped solution, voltage applied, surface tension, conductivity of the solution, feed rate, distance between the tip of the needle to the collector [28]. Industrial spinners are now also available depending on the specification required for designing the fibers. Spinners with different nozzle number (single, multi, uni or co-axial nozzle spinners) are available and also collectors of different shapes and sizes are also utilized as per requirement [25]. Currently, industrial production approaches have been planned and executed by various industries to produce electrospun nanofibers in bulk for commercialization. On the other hand, in the laboratories, electrospun nanofibers have been modified and tested with definite configurations and advanced assemblies, giving them superior properties for cutting-edge presentations. Adjusting electrospinning strictures such as polymer

Fig. 1 Pictorial depiction of the electrospinning procedure. Reproduced with permission from [27]

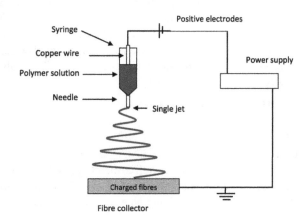

solution viscosity, types of solvent, voltage applied, distance between the needle and the collector, feeding rate, temperature and moisture, stimulates the structural morphology of the electrospun nanofibers; among which the thickness of the polymer solution is the furthermost significant factor, which controls the diameter and the formation of the nanofibers.

2. Electrospun fibers for supercapacitors.

SCs lies on the two electrical charge storage mechanisms: EDLCs and faradaic redox reactions for pseudocapacitors [29]. SCs counterparts other energy storage devices such as batteries or capacitors, which are used to transport high power throbs, load-leveling or storage of electricity [19]. Pseudocapacitors compared with that of the EDLCs are currently attracting much attention due to their high capacitances (200–1300 F/g) difference in contrast to EDLCs (50–150 F/g) [2]. Various pseudocapacitive materials are being investigated for the SCs application. Among which spinels especially 1D architecture has drawn immense attention. Lately, electrospinning technique has been studied as an adaptable and hassle-free manufacture method for porous, flexible, hollow nanofibrous electrode and for SCs. Fabrication of different 1D nanostructures like nanofibers and nanowires, electrospinning is of immense interest in supercapacitor-related applications [30]. Electrospinning has prodigious governance over the fiber morphology showing good uniformity, sponginess and architecture. Fabricating electrodes with 1D nanostructures with bi-continuous passage of electrons and ions will be advantageous for extraordinary energy density hybrid SCs that effectively combine the intercalation pseudocapacitance [31].

3. Advantages of 1D architecture.

1D nanostructure has attracted substantial consideration in the contemporary years for its application in energy storage devices owing to its intriguing features [32, 33]. Their exceptional structural features offer a large surface to bulk ratio, transportation directionally and short ionic conveyance distances, prerequisites for energy storage applications [26]. A widespread diversity of 1D materials have been reported including carbon, polymers, metal hydroxides, metals, chalcogenides, nitrides, etc. [34]. 1D nanostructure is categorized by its main function in flexible supercapacitors, conductive frameworks and active materials. 1D nanomaterial-based SCs such as sandwich-type, wire-shaped and chip-type SCs are also useful in energy storage devices [8].

a.   1D nanostructure for flexible SCs.

Metal current collectors such as Ni foil, Ni foams, stainless steel foils, Cu foils, Ti foils are the most commonly used till date. Even though such substrates have high conductivities, their rigidity and as well as their high molecular weight makes their flexibility limited. Therefore, a combination of 1D–1D nanostructure offers an extraordinary dimensional assembly giving an easy and facile electron pathway, fitting the needs of the flexible SCs [35, 36]. The rigid current collector is the drawback for developing the flexible SCs, so various 1D carbonaceous materials (carbon nanotubes, carbon nanofibers) are being widely used for the flexible SCs. Even the carbon-based materials exhibit low electrical-double layer capacitance, their high cyclic constancy makes it a good contestant for the SCs. There are also some natural materials that can be employed to construct 1D carbon-based nanostructures, like bacterial cellulose (BC), that can form a fine 1D structure by simple microbial formation [37]. Besides carbon-based frameworks, transition metal oxides and hydroxides have also been a good candidate for the flexible SCs including nanowire and nanotubes [38]. Mostly, MnO_2, V_2O_5, $NiCo_2O_4$, $Ni(OH)_2$, etc. are used as the cathodic materials and Fe_2O_3, MoO_{3-x} are used as the anodic materials [39–44]. Transition metal chalcogenides (TMCs) are another class of materials for electrodes that have short bandgap and advanced electrical conductivities than the transition metal oxides. The energy storage mechanism in the TMCs is also by the redox reactions, similar to that of the transition metal oxides [45]. Some commonly used TMCs are $NiCo_2S_4$, Ni_3S_2, Co_9S_8, $CoSe_2$ [46–49].

In electrochemical energy storage, 1D nanostructure possesses several advantages:

(i)   contract the electronic pathways and ease electrical conveyance in an axial direction (unlike their bulk counterparts) [50, 51];

(ii) in 1D nanostructure, the ion diffusion length will be shortened, benefiting the rate capability according to the relation $t = l^2/D$, where, t = time of diffusion of ions through the electrode, l = diffusion length, D = diffusion coefficient [7].

(iii) large electroactive surface area enhances the electrode kinetics facilitating ultrafast charge–discharge capability [52, 53];

(iv) they can be grown in-situ on the conductive substrates and thereby form uniform arrays resulting in a binder-free electrode and preventing conglomeration [54];

(v)   can be exploited to form numerous higher level multi-dimensional structures with amended surface area, good mechanical stabilities and improved cycling routine [53, 55].

The two major strategies of 1D nanoarchitecture synthesis are "bottom-up" and "top-down" approaches. The electrospinning technique is a "bottom-up" approach that has been utilized to formulate porous, tubular and core/shell 1D structures, which delivers momentous null space [31]. Till now, plenty of 1D nanowires/nanotubes/heterogeneous nanowires have been fabricated by electrospinning [16]. Electrospinning process parameters as well the annealing and calcination parameters play an important role in controlling the end electrospun product [56–58].

Various research groups have synthesized nanostructured spinel electrode materials, by tuning their surface area, microstructure and morphology. Recently, Yun et al. have reported the synthesis of ultra-long zinc–manganese oxide ($ZnMn_2O_4$) with one-dimensional hollow nanofiber morphology by electrospinning [59]. $Zn(CH_3COO)_2·2H_2O$ (2 mmol) and $Mn(NO_3)_2·4H_2O$ (4 mmol) were taken as the

Fig. 2 Images of ZMO-500
HNFs **a** TEM and **e** HRTEM;
and **b–d** elemental mapping.
Reproduced with permission
from [59]

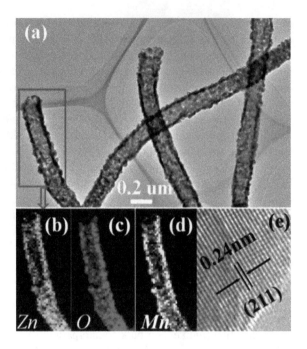

precursors, which were mixed with DMF and polyacrylonitrile (PAN) and stirred vigorously for 12 h. The solution mixture was then poured into the syringe fitted with a needle. The needle and the collector (aluminum foil) distance was kept 15 cm and a voltage of 6 kV was employed to assemble the 1D fibers. The obtained hollow fibers of $ZnMn_2O_4$ were calcined at different temperatures ranging from 500 to 800 °C. The optimum temperature was 500 °C and the sample obtained is designated as ZMO-500 HNFs. The 1D tubular structure comprising interconnected nanoparticles was explained and demonstrated by TEM and HRTEM study as shown in Fig. 2. Such nanostructured morphology offered a high surface area of 33.199 cm^2/g, which may provide an advantageous route for the efficient ion diffusion desirable for the supercapacitor application.

The electrochemical performance was also evaluated, which showed that ZMO-500 HNFs show superior performance compared with that of the other as created samples. The cyclic voltammetry (CV) and the galvanostatic charge–discharge (GCD) cycles also suggest the better performance of ZMO-500 HNFs.

The increased inner hollow area of the ZMO-500 HNFs endorses ion electrolyte storage capacity taking place at the time of the redox reaction, resulting to an enhanced specific capacitance of 1026 F g^{-1}. Also, the electrodes of the ZMO-500 HNFs displayed very decent cycle routine with a retention of capacitance to 100.8% after 5000 sequences at 6 A g^{-1}, suggesting its prospective use for numerous transportable microelectronic systems.

Yu et al. fabricated 1D $ZnCo_2O_4$/C composite nanofibers using a simple electrospinning method [60]. A consistent hybrid sol for electrospinning process was

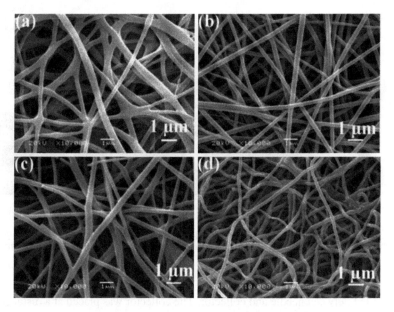

Fig. 3 SEM images of ZnCo$_2$O$_4$/C nanofibers synthesized at diverse carbonization temperatures: **a** 500 °C, **b** 600 °C, **c** 700 °C and **d** 800 °C. Reproduced with permission from [60]

prepared by dissolving Zn (CH$_3$COO)$_2$·4H$_2$O (0.439 g), Co(CH$_3$COO)$_2$·4H$_2$O (0.996 g), and polyvinylpyrrolidone (PVP; 3.610 g) in N,N-dimethylformamide (DMF; 20 mL for 10 h). The voltage given was 12 kV and the distance between the collector and needle tip collection was 140 mm. The as-obtained stabilized fibers were then carbonized at different temperatures viz. 500, 600, 700 and 800 °C. The final electrospun ZnCo$_2$O$_4$/C composite nanofibers were thereby obtained and were examined by SEM as in Fig. 3, which shows that the nanofibers experience crispation with the increase in temperature of carbonization.

The TEM images in Fig. 4 resemble the 1D nanostructure of the spinel composite. Furthermore, the specific surface area obtained for the synthesized ZnCo$_2$O$_4$ nanofibers was 199.0 m^2/g. Thus, a fine mesoporous structure was obtained, which can efficiently facilitate the transmission of electrolyte ions causing swift charge transfer, granting with a brilliant electrochemical outcome. The electrochemical features of the fabricated nanofiber electrodes were examined in 6 M KOH electrolyte solution. The calculated specific capacitances of the electrodes were found to be 327.5, 266.8, 213.0, 155.0 and 90.0 F/g at current densities of 0.5, 1.0, 2.0, 4.0 and 8.0 A/g, respectively.

Bhagwan et al. have reported porous, one-dimensional nanofibric network of nanofibers comprising CoMn$_2$O$_4$ (CMO) with cavities/pores that are produced by electrospinning and examined as electrode material for supercapacitors [61]. The viscoelastic precursor solution was prepared by mixing polymer and mixed metal precursor, which was fed into a 2.5 ml subcutaneous syringe having 0.337 mm internal diameter of stainless steel needle at a flow rate of 2 ml h^{-1}. Voltage of 18 kV

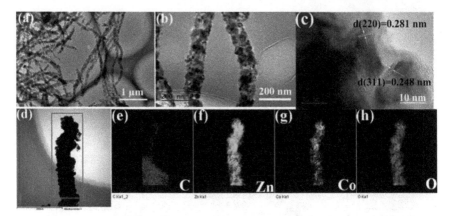

Fig. 4 Microscopic characterization of $ZnCo_2O_4$/C nanofiber obtained at 600 °C **a** TEM image, **b** TEM image magnified, **c** Image taken from HR-TEM analysis, **d–h** Elemental mapping images. Reproduced with permission from [60]

was applied between needle and collector. The collector was placed at a distance of 12 cm from the needle tip. The electrospun nanofibers were sintered at 400 °C for 4 h to obtain CMO nanofibers (Fig. 5).

The electrochemical studies on CMO nanofibers were performed in a three-electrode configuration. The voids/sites as observed from the morphological studies suggested that these voids can act as spot for the adsorption of ions, allowing more charge storage capacity that will act as the reaction sites. Therefore, the CMO nanofiber with high active surface area, proper arrangement and aspect ratio delivered an enhanced value of capacitance of $(320 (\pm 5)$ F g^{-1} at current density of 1 Ag^{-1} along with exceptional cyclability rate and capability up to 10,000 cycles. The same research group, Bhagwan et al., has also reported other 1D spinels like mesoporous

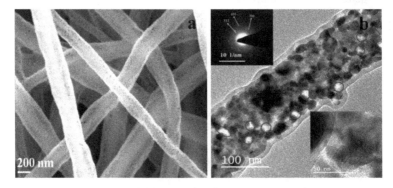

Fig. 5 a FESEM morphological of the cavities (black spots) of sintered nanofibers. **b** TEM image depicting the porosity of the nanofibers; inset: (upper) corresponding SAED and (lower) HRTEM patterns. Reproduced with permission from [61]

CdMn$_2$O$_4$ nanofiber [62] and hollow Mn$_3$O$_4$ [63] via electrospinning technique and used them in supercapacitor applications.

Tubular nanostructure, which is a special 1D structure, has also been studied because of possessing both hollow and 1D structures. Peng et al. fabricated the tube-in-tube structures of various spinels including NiCo$_2$O$_4$, CoMn$_2$O$_4$, NiMn$_2$O$_4$, CoFe$_2$O$_4$ and ZnMn$_2$O$_4$ and studied their electrochemical performance in supercapacitors [64]. They demonstrated a facile and simple approach to prepare multi-faceted tubular 1D nanostructure depending on an operative single-nozzle technique of electrospinning in combination with heat treatment. Figure 6 displays distinctive morphology of the tube-in-tube spinel structures as studied from SEM and TEM

Fig. 6 Tube-in-tube microstructures studied through FESEM and TEM as depicted in the images **a** and **b** NiCo$_2$O$_4$, **c** and **d** CoFe$_2$O$_4$, **e** and **f** NiMn$_2$O$_4$ and (g and h) ZnMn$_2$O$_4$; (The inset of each image shows the corresponding enlarged view). Reproduced with permission from [64]

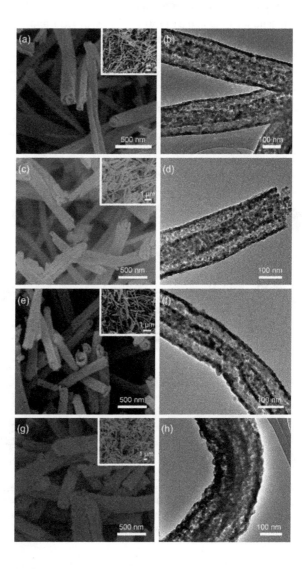

images. They pointed out that different spinels demonstrate a difference in the roughness of the, i.e. $NiCo_2O_4$ and $NiMn_2O_4$ show a greater number of pores whereas, $CoFe_2O_4$ and $ZnMn_2O_4$ have somewhat rough surface with a dense feature. The difference in the morphology can be suggested due to the difference in the nature of growth of the crystals and crystallization performance of the materials during the process of calcination. Herein, they have studied tube-in-tube structures of $NiCo_2O_4$ as an example to validate the preeminence of these nanostructures as electrodes for SCs applications. In order to study the electrochemical behavior of the electrodes, the characterizations were carried out in 2.0 M KOH electrolyte solution employing a conventional three-electrode configuration.

Figure 7 depicts the CV and GCD profiles of the $NiCo_2O_4$, revealing the specific capacitance obtained as 1756 1697, 1610, 1511 and 1457 F/g at a respective current density of 1, 2, 5, 10 and 20 A/g. This is explained as the comparatively extraordinary specific surface areas of the exclusive tube-in-tube hollow structures, which is found to be 47.3 m^2/g. It ensured improved interaction areas at the electrode/electrolyte edge and supplies more dynamic sites for fast faradaic redox reactions, thus leading to loftier electrochemical performance.

Coaxial electrospinning is another simple technology, which is used to arrange nanofibers in core–shell structures. Coaxial electrospinning results in production

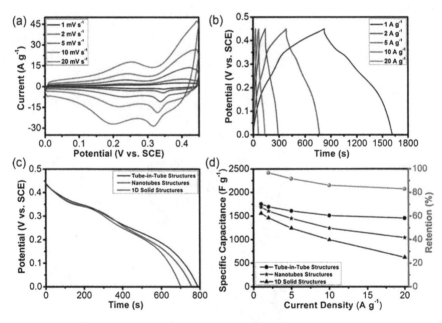

Fig. 7 Electrochemical studies of $NiCo_2O_4$ nanostructured electrodes: **a** CV curves at different scan rates extending from 1–20 mV/s; **b** galvanostatic charge/discharge profiles at different current densities; **c** discharge profiles at a current density of 1 A/g; **d** Specific capacitance and its retention variation with changing current density. Reproduced with permission from [64]

stability, which helps to control the fiber dimension as in this technique, there is particularly intended for the multi-channel spinneret, which consists of the core and shell networks [65]. Xiao et al. reported the well-regulated preparation of free-standing hybrid electrode with $NiCo_2S_4$ nanosheet interconnected arrays, which was allied on a double capillary carbon nanofiber ($NiCo_2S_4$@DCCNF) through the coaxial electrospinning method followed by hydrothermal treatment [9]. Such 1D hollow morphology is beneficial for reducing the ion transmission extent and lowering the ion movement struggle. The SEM, TEM and the HRTEM images in Fig. 8 show the smooth surface and the nanostructures are interconnected to each other showing an open 3D structure. EDX mapping images (Fig. 8j) define the dissemination of components in the DCCNF, verifying that the internal film is composed of carbon/$NiCo_2O_4$ and the external stratum is made of an amorphous carbonaceous structure, which functions as a conductor for electroactive materials. Such ordered construction of the DCCNF plays a noteworthy function in the accessibility of the electrolyte and better distribution of ions throughout the cycling processes, generating a constructive superior performance of the electrodes. Hence, the $NiCo_2S_4$@DCCNF hybrid electrode exhibited an extraordinary capacitance of 1275 F g^{-1} and 1474 F g^{-1} at a

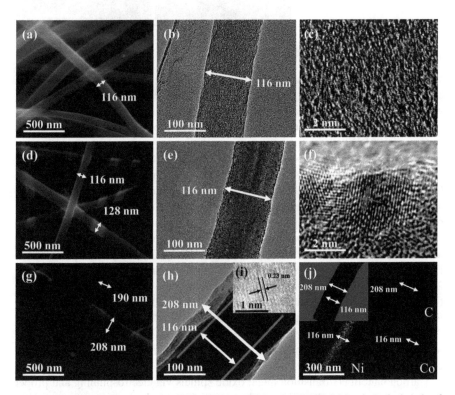

Fig. 8 Morphological study of pure CNF **a** SEM, **b** TEM and **c** HRTEM. Morphological study of CCNF **d** SEM, **e** TEM and **f** HRTEM. Morphological study of DCCNF **g** SEM, **h** TEM, **i** HRTEM, **j** TEM and EDX mapping. Reproduced with permission from [9]

Fig. 9 **a** Low magnification
before calcination and
b high-magnification after
calcination SEM image of
Ni(NO₃)₂/Co(NO₃)₂/PVP
composite electrospun
nanofibers with M:PVP =
0.61:1; **c** and
d high-magnification TEM
images (low magnification
image in inset (**c**);
e corresponding HRTEM
image; **f** SAED pattern [66]

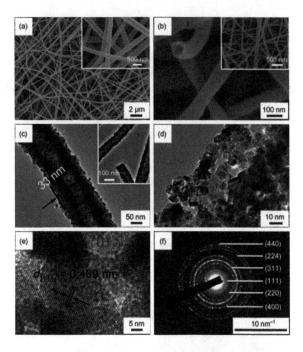

scan rate of 5 mV s^{-1} and at a current density of 1 A g^{-1}, respectively, in a three-electrode cell. The all-solid-state symmetric supercapacitor device was also tested, which attains a high capacitance of 166 F g^{-1} and extended up to 10000 cycles of cycling stability, reaching an extreme energy density of 55.6 Whkg^{-1} at a power density of 1061 Wkg^{-1}.

Li et al. also prepared porous nanotubes of NiCo₂O₄ (NCO-NTs), which are fabricated by a single-spinneret electrospinning method and then finally sintered in air [66]. The synthesized nanotubes exhibited 1D architecture with a permeable structure and hollow cores. The precursor solution was injected into a syringe with a flow rate of 1 mL/h applying a voltage of 12 kV and the distance between the aluminum collector and the needle tip was 14 cm, finally the electrospun fibers were sintered at a temperature of 450 °C for 2 h in air. The SEM and TEM analyses as presented in Fig. 9 showed nanofibers were uniform smooth surfaces and even diameters of around 250 nm. The 1D nanostructure is clearly visible after the calcination process. Owing to their enhanced specific surface area and hollow one-dimensional nanostructures, the synthesized NCO-NT electrodes display 1647 Fg^{-1} specific capacitance at 1 Ag^{-1}, excellent retention of capacitance of 77.3% at 25 Ag^{-1} with only 6.4% stability loss after 3000 cycles, thereby indicating its impending for superior electrochemical capacitors (Fig. 9).

Carbon nanofibers/transition metal oxide spinels via electrospinning and proper annealing process can also offer good conductivity and efficiency in supercapacitor application [67]. Nie et al. reported vanadium/cobalt oxide on electrospun carbon nanofibers for the electrochemical application [68]. Cobalt monoxide is a p-type semiconductor, but its poor conductivity and low electrochemical reaction kinetics have made the material solely inappropriate for the energy storage application [69]. The group addressed this disadvantage by incorporating vanadium, which has low electronegativity and thus can enhance the electronic conductivity of CoO. Furthermore, integrating the vanadium/cobalt oxide with carbon nanofiber can produce more reaction sites, enhance the transport of electrons and also increase the strength of the system [70]. The VCO/CNF-based composite was made up by facile electrospinning technique, which presented that the morphology and the dimensions are dependent on the carbonization temperature. Figure 10 depicts the SEM and TEM morphology of the composites. The SEM image in 10B shows a rough surface with multiple mesoscopic voids that can accelerate the accessibility of the electrolytic ions. Figure 10C also reveals that the metal/metal oxide particles with a wide range of diameter are uniformly distributed in the interior and exterior of the CNFs. The VCO/CNFs displayed 1.83 Fcm^{-2} of areal capacitance at 8 mA cm^{-2} of current density, 44.2 $\mu Whcm^{-2}$ areal energy density at 2.8 $mWcm^{-2}$ power density, an extended cycling stability with a retention of 95.2% for 10,000 cycles and excellent flexibility. Therefore, the group reported a binder-free non-polarity membrane electrode via electrospun engineering.

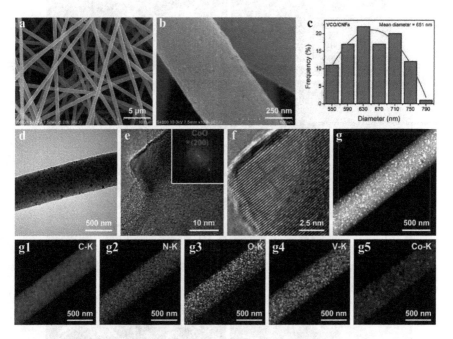

Fig. 10 Morphological depictions of the VCO/CNFs: **A** and **B** SEM images corresponding to low and high magnifications, respectively. **B** Wide range of distribution of diameter diagram. **D** TEM and **E** and **F** HRTEM images (corresponding FFT pattern, Inset in **E**). **G** HAADF-STEM version and (G1-G5) elemental mappings of C, N, O, V and Co elements. Reproduced with permission from [68]

Suktha et al. studied the mixed oxidation states of manganese oxides. They synthesized Mn_3O_4 nanofibers by electrospinning technique and showed the charge storage mechanism, which involved the phase transformation of Mn_3O_4 to $Na_8MnO_x \cdot nH_2O$ [71]. The Mn_3O_4 nanofibers were produced by electrospinning process using a solution of $(Mn(OAc)_2)$ in polyacrylonitrile (PAN). The structure of the electrospun nanofibers of $PAN/Mn(OAc)_2$ is presented in Fig. 11. They reported that the diameters of the as-spun fibers ranged from 400 nm to 1700 nm for different wt% of $Mn(OAc)_2$. The MnO_x nanofibers were then calcined at 500 °C for 2 h and after the removal of the PAN, the nanofibers of MnO_x were reduced but on increasing the concentration of the $Mn(OAc)_2$, the MnO_x nanofibers formed larger nanofibers. The charge storage performance of the Mn_3O_4 nanofibers was tested with 10 wt% of $Mn(OAc)_2$, the specific capacitance of Mn_3O_4 symmetric supercapacitor calculated was 289 F/g at 1.25 mAcm^{-2}, and 37.4 Whkg^{-1} specific energy with a maximum

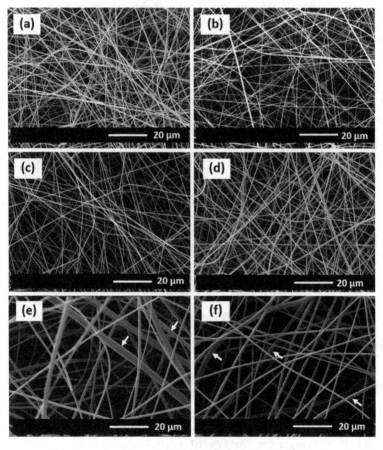

Fig. 11 Diverse $Mn(OAc)_2$ loading contents of the as-electrospun PAN nanofibers: **a** 0, **b** 5, **c** 10, **d** 20, **e** 30 and **f** 36 wt%. Reproduced with permission from [71]

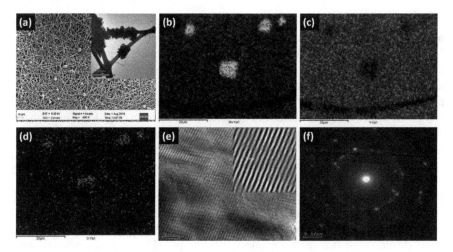

Fig. 12 **a** FESEM image of MnO/Mn−V−O@C composite (inset depicting a TEM image of the composite), energy dispersive X-ray (EDX) mapping images of **b** Mn, **c** V, and **d** O, and **e** and **f** HRTEM and SAED pattern of synthesized MnO/Mn−V−O@C composite. Reproduced with permission from [72]

specific power of 11.1 kWkg^{-1}. The group concluded that the amount of $Mn(OAc_2)$ precursor decides the structure of the crystals of the final product obtained, i.e. Mn_3O_4 nanofiber. Also, based on XAS, a phase transformation from Mn_3O_4 to $Na_\delta MnO_x \cdot nH_2O$ is the basic surface redox reaction that creates adsorption/desorption of solvated ions and such charge storage mechanism of the as-calcined Mn_3O_4 is beneficial for the electrochemical performance.

Samir et al. studied the assembly of binder-free electrospun, mesoporous fibrous electrodes of $Mn_{0.56}V_{0.42}O@C$ [72]. In the electrospun technique, the graphite sheets were used to collect the electrospun Mn − V@C mesoporous nanofibers serving as a material for electrode in supercapacitor devices. The graphite sheets acted as the conductive carbonaceous matrix that can enhance the electronic/ionic-transfer pathways along with supportive fibrous network. Again, the binder-free electrodes also exhibited very low charge-transfer resistance. A conventional electrospinning setup was again used in which the viscous solution was emitted through a needle tip under a very high working voltage of 22 kV, distance between the needle tip and the collector was 15 cm, and a flow rate of 1 mL/h. The FESEM and TEM study in Fig. 12 revealed the formation of nanofibers with a high aspect ratio of an average diameter of 350−550 nm and high surface roughness. The carbon-based matrix holds the Mn−V−O nanofibers with condensed MnO octahedrons making it better for ionic and electronic conductivity. The fabricated $MnO/Mn_{0.56}V_{0.42}O@C$ electrodes presented a specific capacitance of 756.9 F g^{-1} at 5 mV s^{-1}. The electrospun fibers also showed pseudocapacitive behavior with reduced resistance to charge transfer and thereby boosted specific capacitance of 668.5 F g^{-1} at 1 A g^{-1}.

Co$_3$O$_4$ is another transition metal oxide, which is also found as an electroactive material due to its non-toxicity, regular abundance and good redox properties [73]. Various morphologies have been observed such as nanowires/nanofibers, nanorods, nanotubes, nanoparticles, nanoflakes and nanosheets. Kumar et al. have synthesized Co$_3$O$_4$ nanofibers by electrospinning technique and studied their electrochemical properties [74]. The SEM morphology of the electrospun Co-Ac/PVP precursor composite showed smooth fibers with diameters 250–350 nm (Fig. 13). These fibers obtained have very small grains with an average size of <35 nm of cobalt oxides, indicating the nanostructure morphology of the fiber. The BET specific surface area of Co$_3$O$_4$ nanofibers is 67.0 mg^{-2} consisting of interparticle spacing and multiple voids, which is favorable for the unhindered diffusion of electrolyte at the electrode/electrolyte interface, refining the capacitive performance of the electrode material for energy storage application. The specific capacitance of Co$_3$O$_4$ nanofibers is found to be 407 Fg^{-1} with very low Rs and R$_{ct}$ values. Therefore, the electrospun fibers display high rate electrochemical stability, pseudocapacitance behavior and admirable cyclic performance.

Another work on Co$_3$O$_4$ was reported by Barik et al., where they have reported the electrospun 1D nanofibers of Co$_3$O$_4$@C for flexible supercapacitors, which are binder-free and also with no external carbon source [75]. In the electrospinning process, the precursor solution was prepared of cobalt acetate with PVP polymer

Fig. 13 SEM images of Co$_3$O$_4$ nanofibers depending on calcination **a** Before **b** After calcination at 475 °C for 5 h. Reproduced with permission from [74]

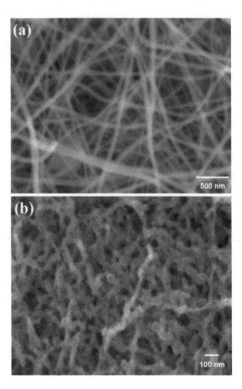

solution in the weight percentage ratio of 1:1. The nanofibers were obtained after annealing them at 300 and 500°C temperatures, between which the 500 C annealed nanofibers showed a better response. Figure 14 displays the structural morphology of the PVP, composite of Co(acac)2−PVP and nanofibers of Co_3O_4 @C. The pristine uniform and bead-free morphology of Co_3O_4 @C electrospun polymers possess an average diameter of 150 nm that are intersected with grain-like structures.

Close investigation of electrospun nanofibers by the HRTEM (Fig. 15) showed that Co_3O_4 @C nanofibers looked as extended rod-shaped structures comprising of a large number of interconnected grains that are composed of small sphere-shaped particles of Co_3O_4 in the diameter range of 50–70 nm. Such 1D nanofibers with active sites or pores will be beneficial for capacitance behavior.

The nanofibers of Co_3O_4 @C electrode material attained a greater specific capacitance of 731.2 Fg^{-1} at a current density of 8 Ag^{-1}, in a three-electrode configuration delivering 100% retention after 3000 cycles at a current density of 9 Ag^{-1}. The supercapacitor device fabricated was flexible and can be bent with consistent electrochemical performance as shown in Fig. 16. It can be twisted up to an angle of 120° without any change in cyclic voltammogram leading to supercapacitive behavior and stability.

Sol-gel electrospinning technique is another multifunctional and feasible strategy for spawning unbroken fibers with an extensive choice of diameters, from tens of nanometers to a number of micrometers [76]. Applying this method, Ghaziani et al. studied the morphology by appropriately choosing the polymer, solvent, different experimental parameters, needle tip–collector distance, the flow rate of the sol and the voltage applied [77]. Their results showed that the morphology of the nanofibers obtained via electrospinning is dependent on the calcination temperature and the temperature plays a substantial role, thus affecting their properties including the optical band gap, specific capacitance and cycle stability. An applying voltage of 17.5 kV and a distance between the needle tip and collector (Al foil) was kept 10 cm with a flow rate of 0.4 mLh^{-1}. The electrospinning process was further followed by annealing the nanofibers in order to remove the polymer matrix at various temperatures viz., 500, 600 and 700 C for 2 h at a heating rate of 2 C min^{-1}, generating the $MgCo_2O_4$ nanostructure. The morphological features of the electrospun $MgCo_2O_4$ nanofibers were investigated by FESEM analysis as in Fig. 17. Before calcination,

Fig. 14 Morphological demonstration of **a** PVP polymer, **b** Co(acac)$_3$–PVP composite and **c** electrospun annealed nanofibers of Co_3O_4 @C. Reproduced with permission from [75]

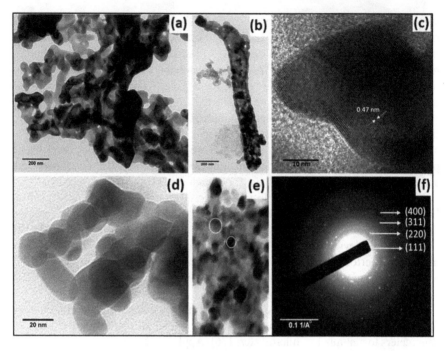

Fig. 15 Structural study of annealed nanofibers of Co_3O_4@C at 500 C through HRTEM and TEM: **a** connected grains, **b** assemblage of spheres to generate grains, **c** fringes and **d, e** grain and discrete grains. **f** SAED pattern. Reproduced with permission from [75]

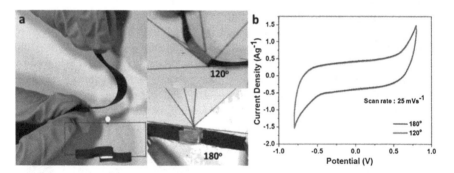

Fig. 16 a Digital photograph of the flexible bendable substrate (i.e. carbon cloth) utilized for assembling the all-solid-state supercapacitor device. **b** CVs at a scan rate of 25 mVs^{-1} carried out at diverse winding angles. Reproduced with permission from [75]

the electrospun nanofibers were smooth and uniform (Fig. 17a), whereas after calcination, the nanofibers showed contraction, which suggests the exclusion of PVP and decay of the nitrates of cobalt and magnesium precursors (Fig. 17b–d). At elevated temperature 700 °C, ling nanofibers tended to breakdown into small nanorods. The

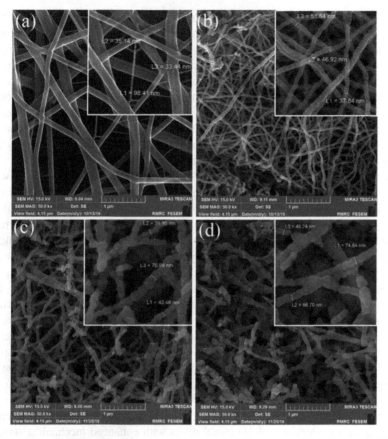

Fig. 17 FESEM images of $MgCo_2O_4$ nanofibers: **a** as-spun, **b–d** after calcination at 500, 600 and 700 °C, respectively. Reproduced with permission from [77]

electrochemical study showed that the MC-500 electrode had a very good supercapacitive performance with a potential specific capacitance of 210 Fg^{-1} at 0.5 Ag^{-1} current densities and also retention of capacitance of 88% capacity after 3000 cycles at 3 Ag^{-1}. Henceforth, it has been found that structural morphology is directed by the calcination temperature, which plays a crucial part in enhancing the electrochemical presentation of the electrode materials, with a trend that increasing the calcination temperature decreases the specific capacitance contribution.

2 Conclusion and Future Outlook

The vast surface area of the electrospun nanofibers and the spongy, permeable nanowoven mats with big surface-to-volume ratios are the smart properties, which make the 1D electrospun spinel structures properly oriented for supercapacitor applications. The extremely permeable structure initiating from the unsystematic entanglement of 1D nanofibers significantly eases mass transportation, a prerequisite for effective energy storage applications. Penetrability can be induced by the incorporation of salts, porogens and phase separation [78]. The development of the pores on the electrospun nanofibers is accredited to the phase separation during the process of solvent evaporation, which resulted in the formation of matrix from the solidified polymer-rich phase and pores from the solvent-rich phase [79]. Furthermore, supercapacitor devices made of electrospun nanofibers have good flexibility and lightweight, making them eye-catching for energy applications [80].

Thus, from the above discussions, it is clear that electrospinning is a very well-organized and proficient procedure for the synthesis of 1D nanostructured materials. The chapter highlighted the rational design and controllable porosity of 1D nanostructure, which are the prerequisites for an efficient supercapacitor electrode. There are certain parameters that must be met, which include temperature, humidity, atmosphere, needle tip, needle tip-collector distance, viscoelastic solution flow rate and applied voltage. Finally, the calcination step or multi-step post-spinning hat treatment is necessary to obtain the fibers with the required nanostructures. This chapter summarized different well-constructed 1D spinels, which can act as intrinsic and extrinsic pseudocapacitive materials by achieving small flow paths, greater adsorption of ions and rapid electrolyte admittance at redox-active reaction sites.

Even though outstanding results have been obtained in fabricating 1D spinel nanostructures, there are still several challenges for promoting the electrospun 1D spinels from laboratory to industry where they can be employed for supercapacitor application. Initial studies also suggested that coalescing electrospinning with other synthesis methods can offer a stimulating forthcoming novel porous 1D structure. For instance, development of 1D spinel fibers with enhanced mechanical stability for the construction of stretchy, bendable and free-standing electrodes is still a challenge. More methodical and organized studies, refining the hypothetical and investigational parameters, are required to augment the strategy of these nanostructures for precise solicitations. In the electrospun nanomaterial electrodes, the synergistic effect between the various components is the essential criteria. For example, nanocomposite comprising of shell component and core component with high electrochemical activity and high conductivity, respectively, can benefit the entire system with desired properties. Also, doping of other materials like nitrogen, fluorine, phosphorous, sulfur also improves the charge mobility, which can also be done with the electrospinning technique. In addition, doping is an alternative adaptable approach to expand the electrochemical performance of the electrodes for SCs. For example, N-doping of carbonaceous materials advances the charge movement in CNFs, leading to an augmentation of the specific capacitance. Flexible electrospun nanofibers containing metal-active chemical components are still a trial due to their reduced mechanical properties. The key point to commercialize the supercapacitor electrode is to increase the production rate of the materials. Such a production rate can be achieved by increasing multiple nozzles to generate manifold jets during the electrospinning process, which is also regarded as a promising approach to achieve the objective [81].

The use of needleless electrospinning using cylinder to generate jets is advantageous as it prevents unstable jet-motion and generates uniform nanofibers on the collectors. One such electrospinning equipment (Nanospider) has been marketed for the hefty manufacture of various kinds of polymeric and ceramic nanofibers. For large-scale and cost-effective production, the synthesis methodologies must be simplified and engineered. Henceforth, it can be believed that attaining high-performance supercapacitor devices based on 1D nanostructure spinel will be an imperative track for the forthcoming energy storage technologies.

References

1. Service RF (2006) New' Supercapacitor' Promises to pack more electrical punch. 313: 902–902
2. Simon P, Gogotsi Y (2010) Materials for electrochemical capacitors. Nanosci Technol Collect Rev Nat J 7:320–329
3. Wang G, Zhang L, Zhang J (2012) Chem Soc Rev 41:797–828
4. Yin X, Li H, Wang H, Zhang Z, Yuan R, Lu J, Song Q, Wang J G, Zhang L, Fu Q (2018) ACS applied materials & interfaces,10:29496–29504.
5. Guo X, Zhang G, Li Q, Xue H, Pang H (2018) Energy Storage Mate 15:171–201
6. Yuan R, Li H, Yin X, Wang P, Lu J, Zhang L (2021) J Mater Sci Technol 65:182–189
7. Mai L, Tian X, Xu X, Chang L, Xu L (2014) Chem Rev 114:11828–11862
8. Wang Y, Zeng J, Li J, Cui X, Al-Enizi AM, Zhang L, Zheng G (2015) J Mater Chem A 3:16382–16392
9. Xiao Y, Huang J, Xu Y, Zhu H, Yuan K, Chen Y (2018) J Mater Chem A 6:9161–9171
10. Liu B, Zhang J, Wang X, Chen G, Chen D, C, Zhou, Shen G (2012) Nano Lett 12:3005–3011
11. Xia Y, Yang P, Sun Y, Wu Y, Mayers B, Gates B, Yin Y, Kim F, Yan H (2003) Adv Mater 15:353–389
12. Li D, Xia Y (2004) Adv Mater 16:1151–1170
13. Greiner A, Wendorff JH (2007) Angew Chem Int Ed 46:5670–5703
14. Yan W, Kim JY, Xing W, Donavan KC, Ayvazian T, Penner RM (2021) Chem Mater 24:2382–2390
15. Lu X, Wang C, Wei Y (2009) Small 5:2349–2370
16. Wang H-G, Yuan S, Ma D-L, Zhang X-B, Yan J-M (2015) Energy Environ Sci 8:1660–1681
17. Dong Z, Kennedy SJ, Wu Y (2011) J Power Sources 196:4886–4904
18. Kumar PS, Sundaramurthy J, Sundarrajan S, Babu VJ, Singh G, Allakhverdiev SI, Ramakrishna S (2014) Energy Environ Sci 7:3192–3222
19. Cavaliere S, Subianto S, Savych I, Jones DJ, Rozière J (2011) Energy Environ Sci 4:4761–4785
20. Guan C, Liu J, Cheng C, Li H, Li X, Zhou W, Zhang H, Fan HJ (2011) Energy Environ Sci 4:4496–4499
21. Wang Q, Chen D, Zhang D (2015) RSC Adv 5:96448–96454
22. Wu Y, Chen C, Jia Y, Wu J, Huang Y, Wang L (2018) Appl Energy 210:167–181
23. Pampal ES, Stojanovska E, Simon B, Kilic A (2015) J Power Sources 300:199–215
24. Zhang X, Ji L, Toprakci O, Liang Y, Alcoutlabi M (2011) Polym Rev 51:239–264
25. Aravindan V, Sundaramurthy J, Suresh Kumar P, Lee Y-S, Ramakrishna S (2015) Madhavi S. Chem Commun 51:2225–2234
26. Liu Q, Zhu J, Zhang L, Qiu Y (2018) Renew Sustain Energy Rev 81:1825–1858
27. Sarbatly R, Krishnaiah D, Kamin Z (2016) Mar Pollut Bull 106:8–16
28. Ahmed FE, Lalia BS (2015) Hashaikeh. Desalination 356:15–30
29. Vol'fkovich YM, Serdyuk TM (2002) Russ J Electrochem 38:935–959
30. Hyun T-S, Kang J-E, Kim H-G, Hong J-M, Kim I-D (2009) Electrochem Solid-State Lett 12:A225

31. Wei Q, Xiong F, Tan S, Huang L, Lan EH, Dunn B, Mai L (2017) Adv Mater 29:1602300
32. Li L, Peng S, Cheah Y, Ko Y, Teh P, Wee G, Wong C, Srinivasan M (2013) Chem A Eur J 19:14823–14830.
33. Yu Y, Gu L, Zhu C, Van Aken PA, Maier J (2009) J Am Chem Soc 131:15984–15985
34. Futaba DN, Hata K, Yamada T, Hiraoka T, Hayamizu Y, Kakudate Y, Tanaike O, Hatori H, Yumura M, Iijima S (2006) Nat Mater 5:987–994
35. Liu J, Chen M, Zhang L, Jiang J, Yan J, Huang Y, Lin J, Fan HJ, Shen ZX (2014) Nano Lett 14:7180–7187
36. Wang K, Meng Q, Zhang Y, Wei Z, Miao M (2013) Adv Mater 25:1494–1498
37. Chen L-F, Huang Z-H, Liang H-W, Guan Q-F, Yu S-H (2013) Adv Mater 25:4746–4752
38. Yu M, Zeng Y, Zhang C, Lu X, Zeng C, Yao C, Yang Y, Tong Y (2013) Nanoscale 5:10806–10810
39. Lu X, Zhai T, Zhang X, Shen Y, Yuan L, Hu B, Gong L, Chen J, Gao Y, Zhou J, Tong Y, Wang ZL (2012) Adv Mater 24:938–944
40. Wei D, Scherer MRJ, Bower C, Andrew P, Ryhänen T, Steiner U (2012) Nano Lett 12:1857–1862
41. Shen L, Che Q, Li H, Zhang X (2014) Adv Func Mater 24:2630–2637
42. Dong X, Guo Z, Song Y, Hou M, Wang J, Wang Y, Xia Y (2014) Adv Func Mater 24:3405–3412
43. Lu X, Zeng Y, Yu M, Zhai T, Liang C, Xie S, Balogun M-S, Tong Y (2014) Adv Mater 26:3148–3155
44. Xiao X, Ding T, Yuan L, Shen Y, Zhong Q, Zhang X, Cao Y, Hu B, Zhai T, Gong L, Chen J, Tong Y, Zhou J, Wang ZL (2012) Adv Energy Mater 2:1328–1332
45. Chen H, Jiang J, Zhang L, Wan H, Qi T, Xia D (2013) Nanoscale 5:8879–8883
46. Zhu T, Wu HB, Wang Y, Xu R, Lou XW (2012) Adv Energy Mater 2:1497–1502
47. Xu J, Wang Q, Wang X, Xiang Q, Liang B, Chen D (2013) Shen G. ACS Nano 7:5453–5462
48. Yu N, Zhu M-Q, Chen D (2015) J Mater Chem A 3:7910–7918
49. Banerjee A, Bhatnagar S, Upadhyay KK, Yadav P, Ogale S (2014) ACS Appl Mater Interfaces 6:18844–18852
50. Brezesinski T, Wang J, Tolbert SH, Dunn B (2010) Nat Mater 9:146–151
51. Ko SH, Lee D, Kang HW, Nam KH, Yeo JY, Hong SJ, Grigoropoulos CP, Sung HJ (2011) Nano Lett 11:666–671
52. Hosono E, Saito T, Hoshino J, Okubo M, Saito Y, Nishio-Hamane D, Kudo T, Zhou H (2012) J Power Sources 217:43–46
53. Yu Z, Tetard L, Zhai L, Thomas J (2015) Energy Environ Sci 8:702–730
54. Chen HM, Chen CK, Liu R-S, Zhang L, Zhang J, Wilkinson DP (2012) Chem Soc Rev 41:5654–5671
55. Zhang G, Xiao X, Li B, Gu P, Xue H, Pang H (2017) J Mater Chem A 5:8155–8186
56. Hong YJ, Yoon J-W, Lee J-H, Kang YC (2015) Chem A Europ J 21:371–376
57. Li L, Peng S, Wang J, Cheah YL, Teh P, Ko Y, Wong C, Srinivasan M (2012) ACS Appl Mater Interfaces 4:6005–6012
58. Lee J-S, Lee Y-I, Song H, Jang D-H, Choa Y-H (2011) Curr Appl Phys 11:S210–S214
59. Yun H, Zhou X, Zhu H, Zhang M (2021) J Colloid Interface Sci 585:138–147
60. Yu H, Zhao H, Wu Y, Chen B, Sun J (2020) J Phys Chem Solids, 140:109385.
61. Bhagwan J, Sivasankaran V, Yadav KL, Sharma Y (2016) J Power Sources 327:29–37
62. Bhagwan J, Sahoo A, Yadav KL, Sharma Y (2017) J Alloy Compd 703:86–95
63. Bhagwan J, Sahoo A, Yadav KL, Sharma Y (2015) Electrochim Acta 174:992–1001
64. Peng S, Li L, Hu Y, Srinivasan M, Cheng F, Chen J, Ramakrishna S (2015) ACS Nano 9:1945–1954
65. Chen W, Rakhi RB, Hu L, Xie X, Cui Y, HN, Alshareef (2011) Nano Lett 11:5165–5172
66. Li L, Peng S, Cheah Y, Teh P, Wang J, Wee G, Ko Y, Wong C, Srinivasan M (2013) Chem A Europ J 19:5892–5898
67. Xia L, Li X, Wu Y, Hu S, Liao Y, Huang L, Qing Y, Lu X (2020) Chem Eng J 379:122325
68. Nie G, Zhao X, Jiang J, Luan Y, Shi J, Liu J, Kou Z, Wang J, Long Y-Z (2020) Chem Eng J 402:126294

69. Lin J, Liu Y, Wang Y, Jia H, Chen S, Qi J, Qu C, Cao J, Fei W, Feng J (2017) J Power Sources 362:64–72
70. Zhang S, Gao G, Hao J, Wang M, Zhu H, Lu S, Duan F, Dong W, Du M, Zhao Y (2019) ACS Appl Mater Interfaces 11:43261–43269
71. Suktha P, Phattharasupakun N, Dittanet P, Sawangphruk M (2017) RSC Adv 7:9958–9963
72. Samir M, Ahmed N, Ramadan M, Allam NK (2019) ACS Sustain Chem Eng 7:13471–13480
73. Wei TY, Chen CH, Chang KH, Lu SY, Hu CC (2009) Chem Mater 21:3228–3233
74. Kumar M, Subramania A, Balakrishnan K (2014) Electrochim Acta 149:152–158
75. Barik R, Raulo A, Jha S, Nandan B, Ingole PP (2020) ACS Appl Energy Mater 3:11002–11014
76. Xue J, Wu T, Dai Y, Xia Y (2019) Chem Rev 119:5298–5415
77. Ghaziani MM, Mazloom J, Ghodsi FE (2021) J Phys Chem Solids 152:109981
78. Bognitzki M, Czado W, Frese T, Schaper A, Hellwig M, Steinhart M, Greiner A, Wendorff JH (2001) Adv Mater 13:70–72
79. Chen H, Di J, Wang N, Dong H, Wu J, Zhao Y, Yu J, Jiang L (2011) small,7:1779–1783.
80. Freitag M, Teuscher J, Saygili Y, Zhang X, Giordano F, Liska P, Hua J, Zakeeruddin SM, Moser JE, Grätzel M, Hagfeldt A (2017) Nat Photonics 11:372–378
81. Theron SA, Yarin AL, Zussman E, Kroll E (2005) Polymer 46:2889–2899

Polymer and Ceramic-Based Hollow Nanofibers via Electrospinning

Priyanka Mankotia, Kashma Sharma, Vishal Sharma, Rakesh Sehgal, and Vijay Kumar

Abstract Polymer and ceramic-based hollow nanofibers (NFs) are gaining a lot of attention in the scientific community for a variety of applications in the areas of sensing, environment, energy, and health. They possess various properties such as large surface area, high porosity, and sensitivity. The ceramic and polymer-based hollow NFs exhibit higher space charge modulation depth, larger electronic transport properties, and shorter ion diffusion length. A lot of work has been done on using ceramic-based hollow NFs while in the case of polymer-based hollow NFs, no significant research has been conducted to date. This chapter discusses the different types of hollow NFs produced by electrospinning and their application in various fields. The area of NF fabrication is still advancing and a lot more research can be done efficiently to explore more of their applications in various sectors.

Keywords Nanofibers · Hollow · Ceramics · Polymers · Electrospinning

P. Mankotia · V. Sharma
Institute of Forensic Science and Criminology, Panjab University, Chandigarh 160014, India

K. Sharma
Department of Chemistry, DAV College, Sector-10, Chandigarh, India
e-mail: kashma@davchd.ac.in

R. Sehgal (✉)
Department of Mechanical Engineering, National Institute of Technology Srinagar, Srinagar 190006, Jammu and Kashmir, India

V. Kumar (✉)
Department of Physics, National Institute of Technology Srinagar, Hazratbal 190006, Jammu and Kashmir, India

Department of Physics, University of the Free State, P.O. Box 339, Bloemfontein Z9300, South Africa

1 Introduction

Nowadays, the use of 1D nanostructures such as nanofibers (NFs), nanowires, nanorods, nanotubes, and nanobelts is drawing great attention from diverse scientific communities. They possess unique properties such as size effects, surface effects, and superparamagnetism, owing to their enthralling applications in various sectors [1]. The utilization of the hollow NFs having considerable length, larger surface area per unit mass, and smaller diameter (10–100 nm) has been widely highlighted as all these structural features mentioned above provide robustness in their utilities. To date, various methods have been developed for the synthesis of NFs from polymers, ceramics, metals, and glass viz. drawing, phase separation, template synthesis, self-assembly, and electrospinning (ES). The method of producing NFs through the drawing process is very much similar to the dry spinning method, which can make very long single NFs via step-by-step synthesis. However, it is to be noted that only a viscoelastic material that can go through intense deformations along with being cohesive enough to deal with the stresses developed during pulling and NFs can be developed through drawing. The method of template synthesis utilizes a nonporous membrane as a template to generate NFs of hollow (a tubule) shape. Fabrication of fibrils prepared from raw materials like electronically conducting polymers, carbons, semiconductors, and metals can be efficiently done through the template synthesis method. However, one shortcoming of the above method is that it is unable to prepare one-by-one continuous NFs. The phase separation method comprises steps like dissolution, gelation, extraction employing a different solvent, freezing, and drying resulting in the formation of nanoscale porous foam. This process is time-consuming and takes a relatively long time to transfer the solid polymer into the nanoporous foam. In self-assembly, individual pre-existing components organize themselves into required patterns and functions. However, it is also time-consuming in the processing of continuous polymer NFs [2]. The methods are useful for the assembling as well as the production of NFs in 2D and 3D structures, which can then be utilized for many applications [1]. In the past 20 years, ES technique has been widely used for the synthesis of hollow NFs due to its extraordinary capabilities [3–5]. ES is a top-down nano-manufacturing technique that is used most often for the preparation of polymeric or ceramic NFs [6, 7]. It is considered to be a straightforward method for the production of ultrafine fibers having a desirable morphological surface. Fibers ultrafine in nature are developed through the application of a very strong electrical field on ceramics or polymer solution. Various studies have been conducted [8, 9] and a wide range of electrospun NFs synthesized from various polymers as well as ceramics have been successfully manufactured including neat, nanohybrid, and biohybrid [10–17]. It is also considered to be a noble method for the manufacture of continuous ultrafine fibers having a diameter in a range of 10–100 μm solely based upon the forcing of the polymer solution through a spinneret by the application of electrical force [18, 19]. A fine mat of small fibers is then produced with small pore sizes, high specific surface areas (10–100 m^2/g), and higher porosity. Therefore, they can serve as superb materials for a variety of applications like tissue engineering, textiles, drug delivery, filtration, reinforcement materials, etc. [18, 20]. The utilization of electrostatic forces for the formation of ultrafine fibers started approximately 100 years ago and got first patented by Formhals in 1934 [20]. It was in 1990 this technique was named as electrostatic spinning. The reason behind the gained escalated interest in this technique is because of its simplicity, versatility, the potentiality for various applications, and the production of fibers with highly desirable properties like an extremely large surface area to volume ratio, flexible surface functions, and superior mechanical performances. There are a variety of parameters and processing variables affecting the ES process and those are represented as follow [21–23]:

i. The system parameters include molecular weight, the architecture of the polymer, distribution of molecular weight, and various properties of polymer a solution like conductivity, dielectric constant, surface tension, viscosity and the charge that is carried by the spinning jet.

ii. The process parameters like rate of flow, electric potential, and concentration, the distance between the collection screen and capillary, optimum parameters such as temperature, air velocity and humidity inside the chamber, and motion of the target screen. For example, the polymer solution should be in high concentration to create polymer entanglements still not that high to prevent the viscosity of polymer motion prompted via electric field. The solution should also possess a low surface tension, high charge density, and high viscosity in order to prevent the jet from disrupting into droplets before the evaporation of the solvent. Certain morphological changes can take place when the distance between the syringe needle and substrate is decreased. If the distance is increased or the electric field is decreased, the density of beads also decreases regardless of the polymer concentration in the solution. The morphology of the surface of NFs can be influenced in periodic ways through the applied field thereby making a wide variety of new shapes over the surface. ES, a polymer solution, can also manufacture thin fibers along with creating round NFs. Moreover, branched fibers, flat ribbons, and fibers that are usually split longitudinally thereby forming larger fibers have also been observed via the ES process [24, 25].

2 Electrospinning-Based Methods for the Production of Hollow Nanofibers

As described above, ES is the most common and widely used process for the production of hollow NFs. It is a relatively simple and straightforward method for producing NFs from organic or inorganic components in comparison to the template synthesis and self-assembly method, which rely upon the complex molecules [26]. It is considered to be an efficient and versatile technique that forms NFs of numerous structures like beads on a string, cylinder, ribbon, porous, core–shell, helical, hollow, side by side, and grooves [27–29]. These days' electrospun NFs are in high demand because of their flexibility, porosity, surface area, and morphology [30, 31]. A variety of ES techniques and procedures have been developed for the synthesis of ceramic hollow

NFs since 2010 and are still an active area of research to date. A few of the methods are discussed as follows.

2.1 Electrospinning with a Single Spinneret

It is also known as the single nozzle ES procedure that functions via forming hollow NFs from single or multiple components [32–36]. However, in multi-components ES, polymers comprising of high and low viscosity move toward the internal and external layers [37]. The homogeneity of the solutions is influenced by two factors, i.e., the solubility of components and phase separation of the utilized blends [38]. The output of this technique is comparatively lower by changing between 1 and 5 mL/h with a flow rate of 0.1–1 g/h. This makes it quite challenging to work with depending upon the properties of the solution as well as operating factors [39, 40].

2.2 Microfluidic Electrospinning

Hollow NFs can be easily produced when the needle of the coaxial spinning apparatus comprises of two or more than two channels made [41, 42]. The apparatus in multifluid or microfluidic ES comprises numerous inner capillaries provided with an outer nozzle. The external and internal solutions are taken care of independently into the vessels, which at that point build up a compound Taylor cone, which is extended under an applied electric field and gets solidified into multichannel NFs [43]. In contrast with the ordinary coaxial ES, this method decreases the interaction between the sheath and center fluids that are profoundly miscible or can even go through a quick detachment of stage through the presentation of an additional middle fluid that acts as a separator. In this way, a wide scope of fluid parts can be considered [41].

2.3 Coaxial Electrospinning with a two-Capillary Spinneret

The most widely applied technique of ES for the manufacture of hollow NFs made up of polymers [44–47], ceramics [47, 48], metals [48, 49], and carbon [50–52] is coaxial ES. The technique involves the addition of solutions of two different compositions in a spinneret, which consisted of two coaxial capillaries thereby forming a core of the inner layer and sheath of the outer layer NFs structure. Furthermore, the core is removed via calcinations, solvent extraction, or washing, which leads to the development of hollow NFs [52, 53]. For an efficient ES process, two critical concerns ought to be thought about. The first one is that the core and sheath solutions ought to be immiscible [53, 55]. The second factor is that the sheath solution must be sufficiently spinnable to be able to enforce shear stress over the core solution while pulling the mixed droplet [56].

2.4 Triaxial Electrospinning

It uses a spinneret having three concentric needles. The three different solutions are added and pumped together which then enters the tip of the spinneret. This leads to the deformation of the solution into a Taylor cone under the influence of electrostatic field followed by the emergence of a Triaxial jet, which experiences bending instability, solvent evaporation, and whipping motion, which gets deposited over the collectors as dry fibers [57, 58]. The NFs produced via this method are three-layered comprising the inner core, intermediate, and outer sheath layers. The intermediated layer creates a barrier between the sheath and core regions [59]. This technique has been utilized to prepare hollow NFs from various materials [60, 61].

2.5 Emulsion Electrospinning

This technique is somewhat similar to the solution ES in having a mixture of two immiscible liquid phases but is different in chemistry [62, 63]. The formation of discontinuous core–sheath NFs takes place through this method [64]. Two different types of polymer are dissolved inside a suitable solvent, mixed evenly, and are settled forming an emulsion. Inside the emulsion, the core and sheath fragments tend to form dispersed droplets and a continuous phase, respectively [65, 66]. Emulsifier is generally used for maintaining the stability of emulsion during the process of jet formation. For the deformation to accomplish, the viscosity of the drop phase must be optimum [65, 66]. This method is eco-friendly in nature as it uses water instead of the organic solvents and due to its large dielectric constant, very small NFs are produced quickly [67].

3 Hollow Nanofibers Based on Different Materials

Metal oxide-based hollow NFs have gained a lot of interest in the category of ceramic-based hollow NFs due to their unusual physical and chemical properties like absorptivity, conductivity, etc. [68]. They also possess unique catalytic, electrochemical, and electrical properties associated with high surface area to volume ratio. Some of the ceramic-based metal oxide hollow NFs have been described below.

3.1 TiO$_2$ Hollow NFs

Titanium dioxide is a non-toxic, chemically stable, and photocatalytic semiconductor used in gas sensing [69] and solar cell applications [70]. Therefore, various methods like hydrothermal [71] and sol–gel [72] have been developed for the synthesis of TiO$_2$-based hollow NFs with varying morphologies having applications in various fields. Many researchers till now have done the synthesis of porous and hollow titanium dioxide nanostructures that are capable of showing enhanced photocatalytic activity. The photocatalysis is usually affected by the crystallinity and pore characteristics of the titanium dioxide materials. The porous and hollow TiO$_2$ nanostructure consists of a high surface, which eventually causes an increase in photocatalytic degradation [73–75]. Some more of the interesting features of titanium dioxide include its optimum dielectric properties, large refractive index (2.52-anatase and 2.49-rutile), electrical conductivity, and a large electrical bandgap (3–3.5 eV) [76]. Due to a wide bandgap, titanium dioxide is able to absorb ultraviolet (UV) light thereby showing an optimum photocatalytic activity. This property gets effectively amplified when structures of large surface area are produced (e.g., hollow NFs), which leads to the better degradation of inorganic as well as organic molecules [77]. Yongliang et al. [78] demonstrated an easy method for the synthesis of titanium dioxide hollow NFs through direct annealing of electrospun composite fibers. The method did not require any special equipment or successive coating process. They prepared titanium dioxide NFs through ES combined with heat treatment removing the PVP and converting the solid composite NFs into hollow titanium dioxide NFs. They also concluded that the important parameters for the production of hollow NFs are the decomposition rate of PVP and gel rigidity [78]. Zhang et al. [79] reported the synthesis of micro and nano-scale hollow titanium dioxide fibers via coaxial ES method and applied it in the gas sensing applications. The diameter of the hollow NFs was controlled by altering the content of PPV in solutions. They observed clear sensing signals for carbon monoxide at room temperature because of the enhancement in surface-to-volume ratio generated by the inner surfaces [79]. Choi et al. [80] reported a versatile microemulsion ES method for the tailoring of hollow NFs of titanium dioxide. They utilized alkoxide as a precursor to synthesize the hollow NFs. The developed fibers showed an increased working efficiency for the photodecomposition of methylene blue dye in comparison to the conventionally synthesized fibers. The hollow structure of the fibers helped in the up-regulation of material performance making it suitable for a variety of applications including catalysis, absorption, sensing, etc. [80]. Guangfei et al. [81] prepared electrospun anatase-phase titanium dioxide NFs with different morphologies. They synthesized the NFs via pyrolysis at 500°C. Two types of techniques were employed for making two different varieties of titanium dioxide-based hollow NFs, i.e., coaxial ES for making hollow or tubular NFs and etching treatment in aqueous NaOH for making porous NFs. The results demonstrated that the so-developed titanium dioxide hollow NFs having a diameter in the range of 300–500 nm and a wall thickness of approximately 200 nm comprised of BET specific surface area of ~27.3 m^2/g, which was reported to be two times as that of solid NFs (~15.2 m^2/g) with diameters ranging from approximately 200 to 300 nm along with the lengths of at least tens of microns. It is also to be noted that the electrospun anatase-phase titanium dioxide NFs having a good morphology and comparatively high BET specific surface area will definitely exceed other nanostructures like films and powder of titanium dioxide for a variety of applications especially in dye-sensitized solar cell and photocatalysis [81]. Homaeigohar et al. [82] prepared hollow titanium dioxide NFs loaded with platinum nanoparticles. A mixed-phase of anatase–rutile in a ratio of 70:30 was utilized for the preparation of the fibers. Through the incorporation of Pt inside the NFs, the bandgap gets declined from 3.09 to 2.77 eV. A photocatalytic process occurred under the visible light as a result of the above modification. This system was studied for the degradation of azo dye orange II. The degradation of dye molecules was possible with the synthesized doped hollow nanofiber via a pseudo-first-rate constant of 0.0069 min^{-1}, which was reported to be 11.5 and 3.63 times larger than that of the unloaded hollow NFs and Pt/P25 (TiO$_2$ nanoparticles), respectively. The major factors that affect the performance included the quantity of platinum loaded, pH, calcination temperature, and the light source. It can be concluded from the results that the hollow NFs can be calcinated at 350 °C with the addition of 2 wt. % platinum and a successful photocatalytic activity can be achieved in acidic conditions under solar light, therefore, producing excellent hollow nanofiber [82].

3.2   Cobalt Ferrite (CoFe$_2$O$_4$) and Strontium Ferrite (SrFe$_{12}$O$_{19}$)-Based Hollow NFs

Cobalt ferrite is a kind of magnetic material having large magnetocrystalline anisotropy, mechanical and chemical stability, and unique non-linear spinware properties [83–87]. It is known that one-dimensional magnetic hollow nanostructure is found to possess a unique magnetism property because of their longest axial ratio. CoFe$_2$O$_4$ NFs have been prepared by Cheng et al. [88] via direct annealing using ES. Figure 1 shows the SEM and TEM images of precursor NFs as well as the annealed CoFe$_2$O$_4$ NFs after calcination. A simple procedure requiring only calcination of PVP/nitrate salts composite fibers without any coating procedure and spinner was employed. The structure of hollow NFs was found to be well kept after annealing at different temperatures. The gases flowing from PVP decomposition caused the nanoparticles to flow from inside to the outside of composite materials thus forming a hollow structure. The increase in calcination temperature increased the volume of CoFe$_2$O$_4$ NFs thereby increasing the magnetocrystalline anisotropy of CoFe$_2$O$_4$ nanoparticles, as shown in Fig. 1. The CoFe$_2$O$_4$ hollow NFs synthesized through ES can be effectively employed in electromagnetic, drug delayed-release and spintronic devices [88].

Strontium ferrite is a type of hard hexagonal ferrite useful for the synthesis of various absorbing materials because of its low cost, large magnetization, high

Fig. 1 **a** SEM and **b** TEM images of CoFe$_2$O$_4$ hollow fibers calcined at 500 °C; **c** SEM and **d** TEM images of CoFe$_2$O$_4$ hollow fibers calcined at 600 °C; **e** SEM and **f** TEM images of CoFe$_2$O$_4$ hollow fibers calcined at 700 °C. Reused with permission from Ref. [88]

stability, and high magnetic anisotropy [88–90]. To date, one-dimensional NFs having a large aspect ratio are utilized for a wide variety of applications like filtration media, chemical sensors, and microwave absorbing materials [91–94]. In recent years, ES is used as an effective method for the fabrication of hollow NFs. Wang et al. [95] fabricated strontium ferrite-based hollow NFs using the coaxial ES method as shown in Fig. 2. The material density of the synthesized hollow NFs was calculated

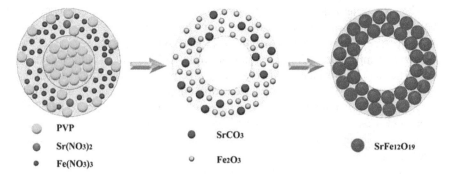

PVP

Sr(NO$_3$)$_2$

Fe(NO$_3$)$_3$

SrCO$_3$

Fe$_2$O$_3$

SrFe$_{12}$O$_{19}$

Fig. 2 Formation mechanism of the hollow SrM NFs. Reused with permission from [95]

to be around 48.5% lower than that of solid strontium ferrite NFs having a similar diameter. The increase in calcination temperature also increased the magnetism properties of strontium ferrite NFs. The loss of electromagnetic radiation was also less, which made these NFs more desirable in properties. They concluded that the lower density of strontium ferrite NFs enhanced their ability to perform absorption, which is considered to be one of the most significant factors while designing hollow NFs [95].

3.3 V_2O_5 and Au/V_2O_5 NFs

Among all the previously reported metal oxides, vanadium pentoxide (V_2O_5) has drawn a lot of attention because of its inherent multiple valencies, broad optical band gap, excellent chemical treatment, and thermoelectric performance [96–99]. Gold is considered to be a well-founded spill over inducer and catalyst, which helps in providing active sites when introduced inside V_2O_5 thereby facilitating the electron transfer [100]. Zeng et al. [101] prepared hollow V_2O_5 and Au/V_2O_5 NFs via combining the combing emulsion ES (EE) and post calcination treatment. The occurrence of stretching and de-emulsifying forces during ES leads to the turning of polystyrene drops into rods, therefore, acting as templates for the synthesis of hollow NFs through calcination. Formation of a metal salt poor zone took place while the polystyrene drops got stretched enough to form an elliptical shape within the fiber during the process of ES. Subsequently, the formation of PVP or the metal salt layer over the surface of the whole fiber took place, which formed the NFs after calcination. They utilized these NFs for designing ethanol-based sensors as shown in Fig. 3. It was observed that the samples showed a very fast response and recovery rate with

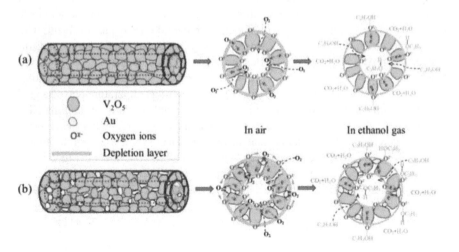

Fig. 3 The morphologies and sensing mechanism of **a** bare V_2O_5 nanotubes and **b** Au/V_2O_5 nanotubes. Reused with permission from [101]

a great sensing response against ethanol. They concluded that hollow metal oxide NFs prepared through fabricated emulsion ES are very promising in manufacturing highly effective gas sensors [101].

3.4 LiFePO4 and Porous Alumina Hollow NFs

$LiFePO_4$ is considered to be a very powerful cathode material being low in cost, non-toxic, highly safe, long cycle life, having high thermal stability, and causing less pollution [102, 103]. However, there lies a disadvantage to using $LiFePO_4$ as a conductor as it offers low conductivity and slow-release kinetics of lithium-ion diffusion between different phases [104]. To circumvent this, various conductive additives like carbon are being used for improving the electrical conductivity of $LiFePO_4$ [105]. Carbon-coated $LiFePO_4$ not only enhances the electronic conductivity but also successfully forbids $LiFePO_4$ crystal grain growth during the process of heat treatment, is effective in the reduction of particle size, helps in the compensation for the charge balance during the Li-ion intercalation or deintercalation process, and induces a high rate performance of $LiFePO_4$ [106, 107]. Various authors reported the synthesis and electrochemical properties of hollow NFs [108, 109].

Alumina is considered to be one of the most outstanding ceramic oxides being widely employed as adsorbents, catalysts, and reinforcing agents for composite materials [110–112]. It is also beneficial for the active adsorption of toxic metals like arsenic depending upon the surface area available. However, there still lies a question to find a desirable approach for the development of aluminum oxides having hierarchically porous and hollow nanostructures, which further provides high surface areas being specific toward the uptake of pollutants. Liu et al. [113] fabricated hollow Al_2O_3 NFs utilizing single capillary ES of aluminum nitrate $[Al(NO_3)_3]$/polyacrylonitrile (PAN) precursor solution along with dimethylformamide (DMF) as a solvent, with sintering at various temperatures. This method involved an unbalanced counter diffusion between Al^{3+} and Al_2O_3 particles through a reaction interface. They observed that both the $Al(NO_3)_3$/PAN composite NFs and alumina NFs had continuous configuration and were randomly oriented (Fig. 4). The sintering temperature played a vital role in optimizing the diameter, morphology, and crystal structure of alumina NFs, and an observation was made that a decrease in diameter occurred when the temperature was increased. They explained that the mechanism of synthesis of porous hollow alumina NFs was most likely due to the "Kirkendall effect" as shown in Fig. 5. They concluded that in order to make hollow ceramic oxide NFs, it is not a specific requirement that the polymer must acquire a higher initial decomposition temperature than that of the final decomposition temperature of the metal salt, which in turn augments this method for the synthesis of more porous ceramic oxide hollow NFs [113].

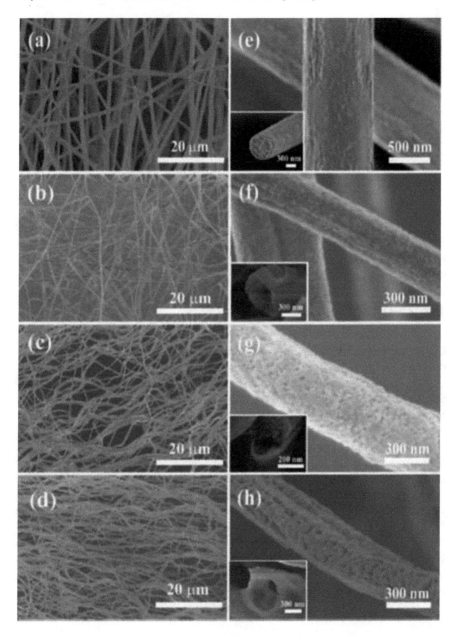

Fig. 4 SEM images of Al(NO$_3$)$_3$/PAN composite NFs (**a, e**) and alumina NFs sintered at different temperatures (**b, f**): 500 °C, (**c, g**): 1000 °C, (**d, h**): 1300 °C. Insets show the SEM images of nanofiber cross-section. Reused with permission from [113]

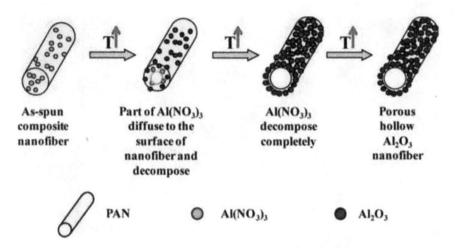

Fig. 5 Possible formation mechanism of porous hollow alumina NFs. Reused with permission from [113]

3.5 CuO and Cu-Based Hollow NFs

Copper oxide NFs are considered to be important one-dimensional metal oxide materials being widely used in electrical transport, field emission emitters, bacterial inactivation, hydrogen detectors, and photo electrochemical detectors [114–121]. Additionally, reduction of copper oxide to copper NFs can be done easily, which can then be used for certain applications in transparent electrodes. Xiang et al. [122] presented a noble method for the synthesis of porous hollow copper oxide and CuO NFs through single-spinneret ES of polyvinyl pyrrolidone(PVP)/copper acetate ($Cu(CH_3COO)_2$) precursor solution accompanied by thermal treatment. One of the main reasons for using this method was that this method did not require any post coating treatment or special designing of the spinneret [120, 121]. Apart from this, they also demonstrated that porous hollow copper NFs can be prepared by reducing copper oxide NFs. Moreover, the diameter and shell thickness of the NFs could be easily modified by particularly varying the PVP concentration along with the ratio of PVP to $Cu(CH_3COO)_2$ in the solution. The so-formed $PVP/Cu(CH_3COO)_2$ composite NFs had diameters ranging from 600 to 800 nm. The possible mechanism for the formation of porous CuO-based hollow NFs is represented in Fig. 6 [122].

3.6 SnO_2–ZnO and γ-Al_2O_3 Hollow NFs

The tin dioxide (SnO_2) and zinc oxide (ZnO) are the semiconductor metal oxides contributing to the sensing performances of various compounds [123, 124]. To date,

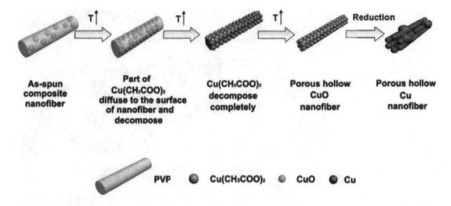

Fig. 6 Possible formation mechanism of porous hollow CuO and Cu NFs. Reused with permission from [122]

many successful gas sensors have been fabricated from one-dimensional SnO_2–ZnO heterostructures [125–128]. Wei et al. [129] developed SnO_2–ZnO hollow NFs through a single capillary ES technique and then used them for the detection of toluene. The process of calcination was carried out at 600 °C for about 3 h in which metal chlorides were successfully converted into metal oxides and PVP was removed from the as-spun NFs. The FE-SEM images of SnO_2–ZnO hollow NFs before and after calcination are shown in Fig. 7. The possible mechanism of the formation of hollow NFs was because of the evaporation effect produced from the induced phase during the process of ES [127, 128]. The schematic representation of the formation of SnO_2–ZnO is displayed in Fig. 8 [129]. They have concluded that the formation of SnO_2–ZnO hollow NFs provides excellent sensitivity and good stability to the sensing of toluene at 190 °C [129].

γ-Al_2O_3 is also named as active alumina. It is extensively being used as an adsorbent because of its high mechanical strength, larger surface area, and tremendous thermal and mechanical stability [130, 131]. γ-Al_2O_3 has a spinel structure consisting of a cubic close-packed oxygen anion layer, having Al cations distributed randomly over the octahedral and tetrahedral sites [132, 133]. Peng et al. [134] worked on the development of porous hollow γ-Al_2O_3 NFs through single capillary ES of $Al(NO_3)_3$/polyacrylonitrile (PAN) mixture solution, accompanied by sintering treatment. They observed two ES phenomena during the process of ES. First, the composite NFs got accumulated over the rotating drum, which was stashed with aluminum foil because of the dragging forces occurring between the rotating drum and the NFs. Second, the freshly prepared solution was easily electrospun further being able to form a gel phase if kept for long. The diameter of the g-Al_2O_3 NFs and the pores on its surface could be tailored by altering the ratio of weight of $Al(NO_3)_3$.$9H_2O$ to PAN (Fig. 9). The performed adsorption experiments proved that the developed hollow NFs exhibited high adsorption efficiency [134].

Fig. 7 FE-SEM images of SnO$_2$–ZnO hollow NFs: **a** before calcination, **b** after calcination at low magnification and **c** at high magnification, **d** the ruptured sections of hollow fibers (the inset shows the corresponding TEM pattern). Reused with permission from [129]

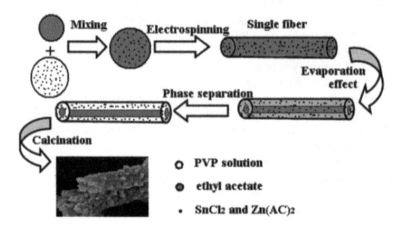

Fig. 8 The illustrative scheme for the formation mechanism of SnO$_2$–ZnO hollow. Reused with permission from [129]

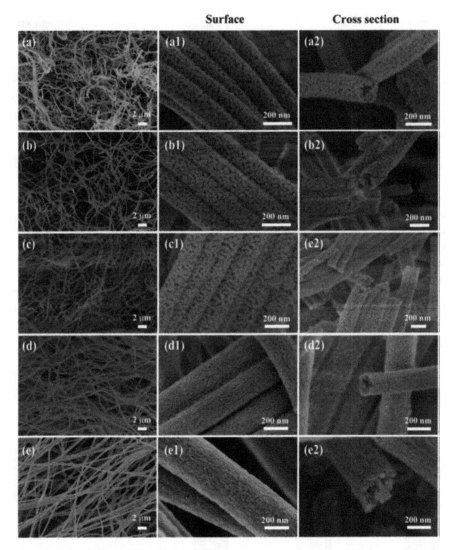

Fig. 9 SEM low-resolution images and high-resolution images of surface and cross-section of the porous hollow g-Al_2O_3 NFs sintered at 800 °C with different weight ratios of $Al(NO_3)_3 \cdot 9H_2O$ to PAN. **a** 1:10, **b** 2:10, **c** 3:10, **d** 5:10, and **e** 10:10. Reused with permission from [134]

3.7 VN and MnO₂ Nanosheets/Cobalt Doped Carbon-Based Hollow NFs

VN is considered to be a very tough refractory material having metallic properties, great metal conductivity, and stability towards heat [135]. It also exhibits high stability in aqueous electrolytes and can be developed with a high surface area [136].

It has been found that VN nanomaterials can be tailored using various methods like temperature-programmed ammonia reduction of V_2O_5, two-step ammonolysis reaction of VCl_4 in chloroform (anhydrous), magnesium sputtering, and microwave plasma torch [135, 137–139]. Zhao et al. [140] described the synthesis of porous VN hollow NFs via ES followed by annealing and nitride processing. Furthermore, the effect of various factors on the morphological structure and electrochemical performance of the NFs was also studied in detail. They carried electrochemical characterization of the porous textured nitrides, results of which showed admissible rate potentiality in 2M KOH aqueous electrolytes. The performance of VN is likely to be connected to its crystalline state along with the composition of its surface layer. The mechanism of formation of hollow VN and V_2O_5 NFs is shown in Fig. 10. The results indicated that the phase structure of the synthesized hollow NFs belonged to a cubic structure and the wall of hollow NFs comprised of nanoparticles that exhibited a porous structure. The value of maximum specific capacitance was calculated to be around 115F/g at a current density of 1A/g in 2M KOH electrolyte solution. Additionally, the capacitance fade of vanadium hollow NFs was linked to the presence of V_2O_5 having high resistance during cycling [140].

MnO_2 is considered to be an excellent material for applications in supercapacitors [141, 142] and electrocatalysts for the oxygen reduction reactions [143]. However, the

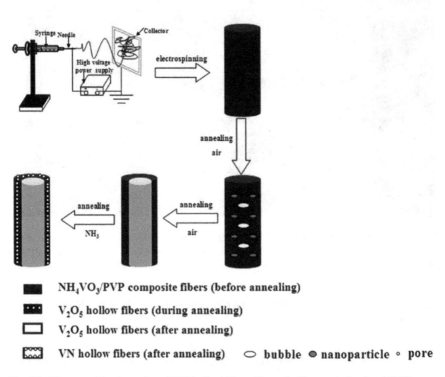

Fig. 10 Diagram of the formation of VN hollow fibers. Reused with permission from [140]

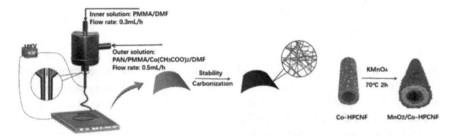

Fig. 11 Schematic illustration of the synthesis procedure of MnO2/Co-HPCNFs. Reused with permission from [144]

electrical resistance of manganese oxide restricts its use for a lot of potential applications. To overcome this problem, MnO_2 is sometimes doped with heteroatoms. It is also noted that the addition of cobalt metal to MnO_2 solution can be beneficial in enhancing the conductivity and catalytic property of the so-formed MnO_2 architecture. The tremendous properties of MnO_2 were exploited by Yang et al. [144] to create a hybrid material of MnO_2/cobalt doped with hollow carbon NFs through a redox reaction between the former two components. In the first step of the synthesis, they prepared hollow carbon NFs using coaxial ES. In the second step, honeycomb double-layered manganese oxide NFs over the surface of hollow carbon NFs by dipping the NFs mats in the solution of potassium magnate after which they were washed and dried were further investigated by XRD and Raman spectroscopy. Figure 11 demonstrates the entire procedure for the synthesis of hollow NFs [144]. The even distribution of cobalt, manganese, carbon, nitrogen, and oxygen over the surface of hollow NFs was demonstrated using the above-mentioned techniques. The formation of cobalt-doped hollow porous NFs coated with honeycomb double-layered MnO_2 nanosheets was observed on both inner and outer surfaces. Since the structure of NFs formed was unique in itself; it added various advantages to the hollow NFs including highly specific surface area contributing to an increase in electrochemical property in zinc-air battery and supercapacitors [145]. It was observed that the developed hollow NFs showed that a large amount of manganese oxide loaded on the surface of cobalt-based hollow NFs exhibited a high capacitance as it was accessible to the ions thus contributing to its excellent rate capability. Due to the excellent properties of the synthesized hollow NFs, a self-made primary zinc-air battery was made, using the NFs as ORR electro catalyst. A desirable oxygen reduction reaction (ORR) catalytic activity equivalent to 0.842 V was obtained, which provided long-term stability to the NFs [146]. It can be concluded that the zinc-air battery developed using the manganese oxide/cobalt doped with hollow carbon NFs excels the conventional platinum/carbon-based zinc-air battery in regard to the performance rates. The battery also exhibits a superior discharge voltage plateau, more robustness, stability, and high power density. Therefore, MnO_2/cobalt-doped hollow carbon NFs can be successfully used in developing more efficient zinc-air battery [144].

3.8 Indium Oxide-Based Hollow NFs for the Detection of Acetone

Acetone has a high evaporation rate and when inhaled directly causes many health problems such as headache, narcosis, and fatigue thus posing a risk to human health and safety. The requirement of a sensor that can work efficiently in sensing the presence of acetone is the need of the era [147, 148]. To date, numerous sensors for the detection of acetone have been manufactured including metal oxide semiconductor sensors [149], nanorods [150], hollow spheres [150], nanosheets [152], and NFs [32]. ES is considered to be the best method for the synthesis of various oxide NFs [153]. Attempts have been made for many years for the improvement of gas sensing properties of the previously developed oxide semiconductors NFs but not much success has been reported. Liang et al. [154] worked on the development of hollow indium oxide-based hollow NFs through ES. In the first step, solid indium oxide NFs were prepared by ES. A homogenous solution comprising of PVP, In(NO$_3$)$_3$ xH$_2$O, C$_2$H$_5$OH, and N, N dimethylformamide were loaded in a syringe and electrospun followed by drying. In this way, solid NFs were prepared by heat treatment. The NFs consisted of interconnected In$_2$O$_3$ nanocrystals having an average diameter of 20 nm as shown in Fig. 12. The SEM images of as prepared and annealed In$_2$O$_3$ at various times are shown in Fig. 13. Once the hollow NFs were developed, they were dispersed with distilled water through ultrasonication [155]. The developed sensor was then further given a heat treatment at 550 °C for 2 hours in order to increase its sensitivity. The response of the sensor toward the sensing of acetone was recorded. A 5 ppm solution of acetone was prepared, and the response of the sensor toward its sensing was measured at varying temperatures, and it was understood that the response increased with an increase in temperature up to 300 °C and then decreased. Indium oxide hollow NFs as sensors for acetone detection reported a maximum response of 151 at 300 °C. This indicated that the hollow NFs are efficient in the detection

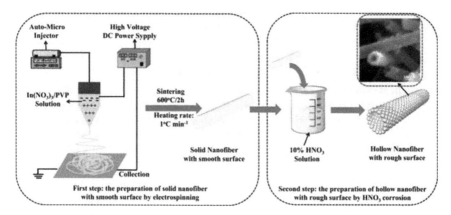

Fig. 12 Schematic representation of the steps involved in the synthesis of Indium oxide based hollow NFs. Reused with permission from [154]

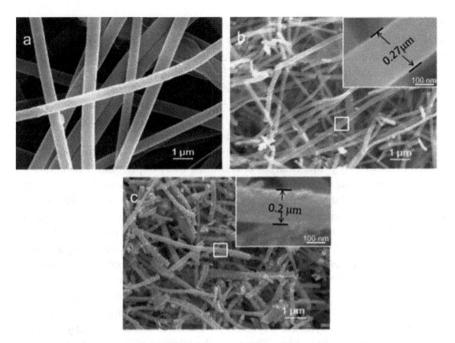

Fig. 13 SEM images of **a** as-spun In-precursor NFs, **b** and **c** In_2O_3 after heat treatment at 600 °C for 2 h; **b** In_2O_3-0 h; **c** In_2O_3-2 h. Reused with permission from [154]

of acetone. The limit of detection of the sensor toward the detection of acetone was calculated to be 20 ppb. The other sensors developed to date like SnO_2 [156], WO_3 [157], ZnO [158], and In_2O_3 [159]. Therefore, the hollow nanofiber-based sensor displayed high potential applicability in the field of safety and health [154].

3.9 Praseodymium-Doped $BiFeO_3$ Hollow NFs for Formaldehyde Sensor

Formaldehyde is one of the most commonly used chemicals in textiles, timber, and various chemical industries [159, 160]. Despite its various uses, formaldehyde is labeled to be a fatal gas that is capable of killing living organisms and is therefore considered to be a carcinogenic agent [160]. The use of perovskite-based composite oxide as sensors for the detection of various types of gases including the toxic one is now an active area of research. Among these, $BiFeO_3$ has become an attractive material for sensing various gases [161, 162]. One limitation, however, limited the use of $BiFeO_3$ as a sensor, which is the presence of smaller surface area and extremely fast electron pair recombination. Tie et al. [163] doped praseodymium into $BiFeO_3$ hollow NFs for gas sensing. The synthesis of Pr-doped $BiFeO_3$ hollow NFs took place via ES and calcination. The morphological examination of Pr-doped $BiFeO_3$ hollow

NFs depicted numerous hollow pores thereby confirming the formation of hollow nanostructures. The so formed mesoporous structure and wide-area thus resulted in the easy transportation and diffusion of targeted gas through the NFs structure ameliorating the gas sensing performance [164]. The sensing performance was recorded to be the highest at an optimum temperature of 190 °C. The gas-sensing performance of the developed hollow NFs toward formaldehyde, ethanol, and acetone was recorded. The reason behind sensitivity toward two other compounds is because of the different reducibility between the testing gases and their adsorption capacity on the sensing material [49]. The addition of praseodymium definitely improved the sensing performance of hollow NFs. The changes in resistance of Pr-doped BiFeO$_3$ hollow NFs toward formaldehyde were noted for a period of 28 days, and it was observed that the sensor was highly stable and sensitive showing a constant value of 17. The sensing mechanism of the current nanofiber is based upon the p-type semiconducting mechanism. In this, adsorption of oxygen molecules over the surface of hollow NFs takes place, which, in turn, creates a high potential barrier. When formaldehyde comes in contact with the sensor, it immediately reacts with the oxygen ions present over the surface, which helps in the releasing back of entrapped electrons to BiFeO$_3$ thereby enlarging the electron depletion layer and increasing the resistance of the sensor. When the NFs return to the air, the aforementioned chemical reaction occurring between formaldehyde and oxygen anions tends to reappear thereby returning the resistance to the initial value. Therefore, the addition of Pr contributed to more oxygen vacancies leading to an enhanced sensing mechanism. The small grain size and unique hollow structure of the NFs also ameliorated the sensing properties [163]. Various hollow NFs and their applications in different areas are summarized and tabulated in Table 1.

4 Potential Applications of Polymer and Ceramic-Based Hollow NFs

A wide variety of applications related to electrospun hollow NFs have been reported to date. Some of them are in the field of drug delivery, wound dressing, tissue engineering, enzyme immobilization, desalination, filtration, antibacterial dressing, etc. A few of these are explained in detail as follows.

The drugs are coated over the NFs and then delivered to the target sites on the basis of a fact that the rate of dissolution of the drug increases with the increase in surface area of the drug as well as the carrier molecules [176–178]. The ceramic and polymer-based hollow NFs have been used for delivering various drugs like anticancer drugs, DNA, RNA, proteins, and antibiotics [179]. Serum bovine albumin protein (SBA) was added into polysulfone (PSU) NFs via ES. The mechanism of coating the surface of NFs with serum bovine albumin has been explained in detail by Chung et al. [180], Sokolsky-Papkov et al. [181] Additionally, electrospun NFs are useful for the enhancement of therapeutic efficiency and reduction in the toxic properties of

Table 1 Hollow NFs and their applications in different areas

Type of hollow NFs	Precursors	Application	References
SnO_2/TiO_2	Titanium isopropoxide/PVP/acetic acid/ethanol	Photocatalytic activity	[163]
TiO_2	PVP/tetrabutyltitanate/ethanol/acetic acid/polyacrylonitrile/ethylene glycol/dimethylformamide/paraffin oil/methylene chloride	Photocatalytic activity, gas sensing, fast lithium storage	[49, 164–166]
Carbon nanotube TiO_2	Polyacrylonitrile/dimethyl formamide/tetrabutyltitanate	Adsorption and photocatalytic activity	[167]
$CoFe_2O_4$	$PVP/Co(NO_3)_2\cdot6H_2O$/water/ethanol/ferric nitrate	Electromagnetic and spintronic devices	[87]
$SrFe_{12}O_{14}$	Ferric nitrate/strontium nitrate/PVP/dimethylformamide	Improvement of microwave absorption properties	[168]
$BaTiO_3$	Acetic acid/ethanol/PVP/titanium isopropoxide	Electrodes for lithium ion batteries	[169]
MnO_2-doped Fe_2O_3	Ferric citrate/deionized water/manganese acetate/citric acid	Catalytic activity	[168]
Fe_3O_4/Eu $(BA)_3$phen/PVP	Dimethylformamide/chloroform/ferric oxide nanoparticles	Magnetic luminescent flexible fiber production	[101]
Chromium-doped spinel	Ethanol/deionized water/PVP/$Zn(NO_3)_2\cdot6H_2O$/$Mg(NO_3)_2\cdot6H_2O$/$Al(NO_3)_3\cdot9H_2O$	Photoluminescence property of fibers	[170]
YF^3:Eu^{3+}	Europium oxide/DMF/ammonium hydrogen fluoride/nitric acid/ethanol/PVP/deionized water	Luminescent fibers	[171]
Te	PVP/nickel acetate	Nitrogen dioxide sensor	[172]
Au/V_2O_5	Gold chloride trihydrate/PVP/polystyrene/vanadylacetylacetonate	Ethanol sensor	[174]
Mn-doped SnO_2	Ethanol/PVP/DMF/$SnCl_2\cdot2H_2O$/$Mn(CH_3COO)_2\cdot4H_2O$	Magnetic semiconductor devices	[175]

drugs in which these NFs become a carrier for drug delivery devices [182, 183]. Similarly, Kenawy et al. utilized polycaprolactone (PCL) as a biodegradable polymer and polyurethane as a non-biodegradable polymer and then used their mixtures as a carrier for various drugs, and not much difference was observed in the rate of drug release, however, an increase in the mechanical properties of the polymer blends was observed [184]. Haider et al. synthesized PLGA NFs as a carrier growth factor calcium apatite [185, 186].

It is used for the regeneration of damaged tissues and organs utilizing the body as a host. Currently, scaffold-based tissue engineering is being used for various applications particularly for bone regeneration, neuroscience, and reconstruction of skin. Electrospun NFs are mostly employed as matrices for tissue engineering applications [187]. These so-formed NFs are effective in providing cell to cell and cell to matrix adhesion causing excellent growth. Therefore, researchers are using natural polymers like alginate, silk, collagen, fibrinogen, protein, and starch for tissue engineering because of their excellent biocompatibility and degradability [188–192]. Polylactic acid and polyglycolic acid are excellent biomaterials for tissue regeneration due to their rapid rate of degradation and smooth spinnability. Other materials include the blended silk fibroin protein polycaprolactone, which is also efficiently used for bone regeneration [193].

These are essential for the protection of injury for infection by elimination of any kind of microbial infection. When applied to the site of injury, these provide a moist environment for the enhancement of the healing process. It is found that the wound dressings synthesized via ES possess more advantages than prepared via conventional methods [20, 194]. It includes larger surface area, fiber pores, and rapid stimulation response to fibroblast cells making them efficient to be used for skin healing and cleansing [195]. Numerous factors for skin healing are added to NFs during the ES process [196]. Usually, the antibacterial substance is encapsulated inside the nanomatrix and is then dispersed freely in the ES solution accompanied by ES. Powell et al. synthesized collagen NFs scaffold extracted from bovine collagen [197], and the results indicated that the scaffold was biocompatible in comparison to those prepared via conventional methods [197]. Gelatin-based NFs produced through ES showed antibacterial activity against the common wound bacteria [198].

The immobilization of enzymes into any soluble matrix is important in order to improvise its workability, maintaining important properties like bioprocessing and long-duration calls [199]. The methods of immobilization preparation include porous membranes, particles, and gel matrix. Jia et al. synthesized polystyrene-based NFs for the immobilization of chymotrypsin, and they observed an enhancement in its working efficiency by 65% increment [200]. Ibrahim et al. [201] synthesized poly (acrylonitrile-co-maleic acid) electrospun NFs for the immobilization of lipase, and the results indicated that the NFs retained their activity several times in comparison with the hollow nanofiber membrane [201].

Numerous antibacterial hybrid electrospun NF scaffolds have been synthesized by researchers like polyacrylonitrile/silver PAN/Ag NFs acquire antibacterial activity against both the classes of gram-positive and gram-negative bacteria. These are immobilized for an extensible range of antimicrobial amidoxime. The antibacterial action of amidoxime relies upon its binding to metal ions such as Mg^{2+} and Ca^{2+}, which disturbs the bacterial balance causing bacterial death [202, 203]. There are various research papers that discuss the antibacterial action mechanism along with the unique properties of natural biopolymer [203].

The highly porous structure of ceramic hollow NFs makes them an ideal material for filtration applications [204]. Wen et al. [205] prepared a filtration device consisting of ceramic membranes with a meshwork of metal oxide NFs. They prepared a hierarchical structure comprising of a porous substrate coated with a separation layer utilizing titanate NFs and smaller boehmite NFs [206, 207]. These ceramic-based hollow NFs produced via ES are able to overcome a lot of limitations faced by conventional ceramic separation membranes like the sudden loss of flux leading to the formation of dead-end pores making zero contribution in filtration process. On the contrary, ceramic-based hollow NFs produced via ES are able to distinguish between larger and smaller voids without the formation of any dead-end pores thus making them more efficient. The porosity of the separation layer in the case of the conventional ceramic membrane was reported to be 36% while in the case of ceramic hollow NFs, it was calculated to be 70%. Therefore, the change in membrane texture significantly increases the performance of the membrane, and it is able to successfully filter out 95% of particles which is quite higher than the previously developed filters. It can, therefore, be concluded that an assembly of ceramic-based hollow NFs provides excellent sensitivity and selectivity toward filtration and separation [208].

5   Conclusion

Nanomaterials have gained significant importance due to the variety of applications they are capable of being employed in. Researchers are actively utilizing various hollow NFs for their potential applications in environmental remediation, energy, and biomedicine. The hollow NFs possess various attractive properties such as their size effects, surface effects, larger length to diameter ratios, superparamagnetism, larger surface area per unit mass, and small diameter. It is because of these enthralling properties that nanomaterials have gained so much importance and utilization for various applications by researchers all over the world. The chapter discusses various types of polymer and ceramic-based hollow NFs produced by the ES method. There are various types of ES methods developed for the synthesis of numerous types of hollow NFs as discussed in detail in the review. Out of all the ES techniques, co-axial ES was indeed found to be the best method for the synthesis of NFs with higher efficacy for practical applications. In addition, ES is found to be a cheap method for NFs production thereby making it a more reliable and utilizable technique. The applications of the so-formed hollow NFs developed so far have been discussed in detail. A lot of research still needs to done and a lot of applications are yet to be discovered using the tremendous properties of hollow NFs and their functionalization for the betterment of the society.

References

1. Sun B, Long YZ, Zhang HD, Li MM, Duvail JL, Jiang XY, Yin HL (2014) Prog Polym Sci 39:862–889
2. Huang ZM, Zhang YZ, Kotaki M, Ramakrishna S (2003) Comp Sci Tech 63:2223–2253
3. Doshi J, Reneker DH (1995) J Electrost 35:151–160
4. Reneker DH, Chun I (1996) Nanotech 7:216
5. Fong H, Chun I, Reneker D (1999) Polym 40:4585–4592
6. Dzenis Y (2004) Sci 304:1917–1919
7. Greiner A, Wendorff JH (2007) Angew Chem Int Ed 46:5670–703
8. Kim JS, Reneker DH (1999) Polym Compos 20:124–131
9. Bognitzki M, Czado W, Frese T, Schaper A, Hellwig M, Steinhart M, Greiner A, Wendorff JH (2001) Adv Mater 13:70–72
10. Homaeigohar S, Dai T, Elbahri M (2013) J Colloid Interf Sci 406:86–93
11. Homaeigohar S, Disci-Zayed D, Dai T, Elbahri M (2013) Bioinspired Biomim Nanobiomater 2:186–193
12. Homaeigohar S, Elbahri M (2014) Mater 7:1017–1045
13. Homaeigohar S, Zillohu AU, Abdelaziz R, Hedayati MK, Elbahri M (2016) Mater 9:848
14. Homaeigohar SS, Buhr K, Ebert K (2010) J Membr Sci 365:68–77
15. Homaeigohar SS, Elbahri M (2012) J Colloid Interf Sci 372:6–15
16. Fang X, Reneker DJ (1997) Macromol Sci B 36:169–173
17. Fong H, Reneker DHJ (1999) Polym Sci Polym Phys 37:3488–3493
18. Beachley V, Wen X (2010) Prog Polym Sci 35:868–892
19. He JH, Wan YQ, Yu JY (2004) Int. J. Nonlinear Sci Numer Simul 5:253–262
20. Huang Z-M, Zhang Y-Z, Kotaki M, Ramakrishna S (2003) Compos Sci Technol 63:2223–2253
21. Frenot A, Chronakis IS (2003) Curr Opin Colloid Interf Sci 8:64–75
22. Li D, Xia Y (2004) Adv Mater 16:1151–1170
23. Deitzel JM, Kleinmeyer J, Harris D, Tan NCB (2004) Polymer 42:261–272
24. Koombhongse S, Liu W, Reneker DH (2001) J Polym Sci Polym Phys Ed 39:2598–2606
25. Xinhua Z, Kim K, Shaofeng R, Hsiao BS, Chu B (2002) Polym 43:4403–4412
26. Xia X, Dong X, Wei Q, Cai Y, Lu K (2012) Express Polym Lett 6:169–176
27. Niu H, Lin T (2012) J Nanomater 2012:12
28. Shadi L, Karimi M, Ramazani S, Entezami AA (2014) J Mater Sci 49:4844–4854
29. Ding Y, Hou H, Zhao Y, Zhu Z, Fong H (2016) Prog Polym Sci 61:67–103
30. Rodríguez K, Gatenholm P, Renneckar S (2012) Cellulose 19:1583–1598
31. Mokhena T, Jacobs V, Luyt A (2015) Express Polym Lett 9:839–880
32. Wei S, Zhou M, Du W (2011) Sens Actuat B Chem 160:753–759
33. Qi R, Guo R, Shen M, Cao X, Zhang L, Xu J, Yu J, Shi X (2010) J Mater Chem 20:10622–10629
34. Zander NE, Strawhecker KE, Orlicki JA, Rawlett AM, Beebe TP (2011) J Phys Chem B 115:12441–12447
35. Liu P, Zhu Y, Ma J, Yang S, Gong J, Xu J (2013) Colloids Surf A 436:489–494
36. Hou H, Shang M, Wang L, Li W, Tang B, Yang W (2015) Sci Rep 5:15228
37. Zhang H, Zhao C, Zhao Y, Tang G, Yuan X (2010) Sci China Chem 53:1246–1254
38. Huang J, You T (2013) Advances in NFs. InTech, Rijeka, Croatia
39. Zhou FL, Gong RH, Porat I (2009) J Mater Sci 44:5501–5508
40. Nurwaha D, Han W, Wang XJ (2013) Eng Fibers Fabr 8:42–49
41. Chen H, Wang N, Di J, Zhao Y, Song Y, Jiang L (2010) Langmuir 26:11291–11296
42. Zhang X, Gao X, Jiang L, Qin J (2012) Langmuir 28:10026–10032
43. Zhao T, Liu Z, Nakata K, Nishimoto S, Murakami T, Zhao Y, Jiang Fujishima L (2010) J Mater Chem 20:5095–5099
44. Liu Y, Ma Q, Yang M, Dong X, Yang Y, Wang J, Yu W, Liu G (2016) Chem Eng J 284:831–840
45. Lee GH, Song J-C, Yoon K-B (2010) Macromol Res 18:571–576
46. Pakravan M, Heuzey M-C, Ajji A (2012) Biomacromol 13:412–421

47. Choi SJ, Chattopadhyay S, Kim JJ, Kim S-J, Tuller HL, Rutledge GC, Kim I-D (2016) Nanoscale 8:9159–9166
48. Chang W, Xu F, Mu X, Ji L, Ma G, Nie J (2013) Mater Res Bull 48:2661–2668
49. Zhang J, Choi SW, Kim SS (2011) J Solid State Chem 184:3008–3013
50. Lee BS, Son SB, Park KM, Lee G, Oh KH, Lee SH, Yu WR (2012) ACS Appl Mater Interf 4:6702–6710
51. Lee BS, Yang HS, Jung H, Mah SK, Kwon S, Park JH, Lee KH, Yu WR, Doo SG (2015) Eur Polym J 70:392–399
52. Kaerkitcha N, Chuangchote S, Sagawa T (2016) Nanoscale Res Lett 11:186
53. Ning J, Yang M, Yang H, Xu Z (2016) Mater Des 109:264–269
54. Lee BS, Jeon SY, Park H, Lee G, Yang HS, Yu WR (2013) Sci Rep 4:6758
55. Loscertales IG, Barrero A, Márquez M, Spretz R, Velarde-Ortiz R, Larsen G (2004) J Am Chem Soc 126:5376–5377
56. Qian W, Yu DG, Li Y, Liao YZ, Wang X, Wang L (2014) Int J Mol Sci 15:774–786
57. Liu W, Ni C, Chase DB, Rabolt JF (2013) ACS Macro Lett 2:466–468
58. Yu DG, Li XY, Wang X, Yang JH, Bligh S, Williams GR (2015) ACS Appl Mater Interf 7:18891–18897
59. Khalf A, Singarapu K, Madihally SV (2015) React Funct Polym 90:36–46
60. Zanjani JSM, Okan BS, Letofsky-Papst I, Yildiz M, Menceloglu YZ (2015) Eur Polym J 62:66–76
61. Zanjani JSM, Saner Okan B, Menceloglu YZ, Yildiz MJ (2015) Reinf Plast Compos 34:1273–1286
62. Vonch J, Yarin A, Megaridis CJ (2007) Undergrad Res 1
63. Samanta A, Nandan B, Srivastava RK (2016) J Colloid Interf Sci 471:29–36
64. Wei J, Shi H, Zhou M, Song D, Zhang Y, Pan X, Zhou J, Wang T (2015) Appl Catal 499:101–108
65. Elahi F, Lu W, Guoping G, Khan F (2013) Bioeng Biomed Sci J 3:1–14
66. Samanta A, Takkar S, Kulshreshtha R, Nandan B, Srivastava RK (2016) Mater Sci Eng 69:685–691
67. Pal J, Singh S, Sharma S, Kulshreshtha R, Nandan B, Srivastava RK (2016) Mater Lett 167:288–296
68. Wang W, Zhang L, Tong S, Li X, Song W (2009) Biosens Bioelectron 25:708–714
69. Buso D, Post M, Cantalini C, Mulvaney P, Martucci A (2008) Adv Funct Mater 18:3843
70. Kim YJ, Lee MH, Kim HJ, Lim G, Choi YS, Park NG, Kim K, Lee WI (2009) Adv Mater 21:1
71. Liu SW, Yu JG, Mann S (2009) Nanotechnology 20:325606
72. Song XF, Gao L (2007) J Phys Chem C 111:8180
73. Liu B, Nakata K, Sakaie M, Saito H, Ochiai T, Murakmi T, Takagi K, Fujishima A (2011) Langmuir 27:8500–8508
74. Iskandar F, Nandyanto ABD, Yun KM, Hogan CJ, Okuyama K, Biswas P (2007) Adv Mater 19:1408–1412
75. Ming H, Ma Z, Huang H, Lian S, Li H, He X, Yu H, Pan H, Liu Y, Kang Z (2011) Chem Commun 28:8025–8027
76. Batool SS, Imran Z, Rafiq MA, Hasan MM, Willander M (2013) Ceram Int 39:1775–1783
77. Tian J, Zhao Z, Kumar A, Boughton RI, Liu H (2014) Chem Soc Rev 43:6920–6937
78. Yongliang C, Wenzhi H, Zhang Y, Ling Zhu L, Liu Y, Fanc X, Cao X (2010) Cryst Eng Comm 12:2256–2260
79. Zang J, Choi SW, Kim SS (2011) J Solid State Chem 184:3008–3013
80. Choi K, Lee SH, Parka JU, Choi DY, Hwang CH, Lee IH, Chang M (2013) Mater Lett 112:113–116
81. Guangfei H, Yibing Cai G, Zhao Y, Wang X, Lai C, Xi M, Zhu Z, Fong HJ (2013) Coll Interf Sci 398:103–111
82. Homaeigohar S, Davoudpour Y, Habibi Y, Elbahri M (2017) The electrospun ceramic hollow NFs. Nanomater 7:1–32

83. Yang Z, Lu J, Ye W, Yu C, Chang Y (2017) Preparation of Pt/TiO$_2$ hollow nanofibers with highly visible light photocatalytic activity. Appl Surf Sci 392:472–480
84. Hu G, Choi JH, Eom CB, Harris VG (2000) Suzuki. Phys Rev B: Condens Matter 62:779
85. Huang W, Zhu J, Zeng HZ, Wei XH, Zhang Y, Li YR (2006) Appl Phys Lett 89:262506
86. Lisfi A, Williams CM, Nguyen LT, Lodder JC, Coleman A, Corcoran H, Johnson A, Chang P, Kumar A, Morgan M (2007) Phys Rev B: Condens Matter Mater Phys 76:054405
87. Carey MJ, Maat S, Rice P, Farrow RCF, Marks RF, Kellock A, Nguyen P, Gurney BA (2002) Appl Phys Lett 81:1044
88. Cheng Y, Zou B, Yang J, Wang C, Liu Y, Fan X, Zhu L, Wang Y, Ma H, Cao X (2011) Cryst Eng Comm 13:2268
89. Hosseini SH, Sadeghi M (2014) Curr Appl Phys 14:928–931
90. Li CJ, Mater GR (2011) Res Bull 46:119–123
91. Pullar RC, Bdikin IK, Bhattacharya AK (2012) J Eur Ceram Soc 32:905–913
92. Wang GZ, Gao Z, Tang SW, Chen CQ, Duan FF, Zhao SC, Lin SW, Feng YH, Zhou L, Qin Y (2012) ACS Nano 6:11009–11017
93. Dong CS, Wang X, Zhou PH, Liu T, Xie JL, Deng LJ (2014) J Magn Magn Mater 354:340–344
94. Yang J, Zhang J, Liang CY, Wang M, Zhao PF, Liu MM, Liu JW, Che RC (2013) ACS Appl Mater Inter 5:7146–7151
95. Wang Z, Zhao L, Wang P, Guo L, Yu J (2016) J Alloys Comp 687:541–547
96. Wollenstein J, Scheulin M, Herres N, Becker WJ, Bottner H (2003) Sens Mater 15:239
97. Takahashi K, Limmer SJ, Wang Y, Cao GZ (2004) J Phys Chem B 108:9795
98. Cao AM, Hu JS, Liang HP, Wan LJ (2005) Angew Chem Int Ed 44:4391
99. Raible I, Burghard M, Schlecht U, Yasuda A, Vossmeyer T (2005) Sens Actuat B 106:730
100. Raj AD, Pazhanivel T, Kumar PS, Mangalaraj D, Nataraj D, Ponpandian N (2010) Curr Appl Phys 10:531
101. Zeng W, Chen W, Li Z, Zhang H, Li T (2015) Mater Res Bull 65:157–162
102. Liu ZL, Tay SW, Hong L, Lee JY (2011) J Solid State Electrochem 15:205
103. Zhang WJ (2011) Struct J Power Sourc 196:2962
104. Lee J, Teja AS (2006) Mater Lett 60:17–18
105. Prosini PP, Zane D, Pasquali M (2001) Electrochim Acta 46:3517
106. Sanchez MAE, Brito GES, Fantini MCA, Goya GF, Matos JR (2006) Solid State Ionics 177:497
107. Lin Y, Gao MX, Zhu D, Liu YF, Pan HG (2008) J Power Sourc 184:444
108. Wang YZ, Zhang LX, Wang YB (2013) New Chem Mater 41:92
109. Wei BB, Wu YB, Yu FY, Zhou Y (2016) Int J Minerals Metallurgy Mater 23:474
110. Mahapatra A, Mishra BG, Hota G (2011) Ceram Int 37:2329–2333
111. DeRosa F, Kibbe MR, Najjar SF, Citro ML, Keefer LK, Hrabie JA (2007) J Am Chem Soc 129:3786–3787
112. Colle RD, Longo E, Fontes SR (2007) J Membrane Sci 289:58–66
113. Liu P, Zhu Y, Ma J, Yang S, Gong J, Xu J (2013) Colloids and Surf A: Physicochem Eng Aspects
114. Wu H, Lin DD, Pan W (2006) Appl Phys Lett 89:133125
115. Chen J, Huang NY, Deng SZ, She JC, Xu NS, Zhang WX, Wen XG, Yang SH (2005) Appl Phys Lett 86:151107
116. Torres A, Ruales C, Pulgarin C, Aimable A, Bowen P, Sarria V (2010) ACS Appl Mater Interf 2:2547
117. Chen L, Shet S, Tang HW, Wang HL, Deutsch T, Yan YF (2010) HH. J Mater Chem 20:6962
118. Turner J, Al-Jassim M (2010) J Mater Chem 20:6962
119. Hoa ND, Quy NV, Jung H, Kim D, Kim H, Hong SK (2010) Sens Actuat B 146: 266
120. Sander MS, Gao H (2005) J Am Chem Soc 127:12158
121. Li D, Xia YN (2003) Nano Lett 3:555
122. Xiang H, Long Y, Yu X, Zhang X, Zhao N, Xu J (2011) Cryst Eng Comm 13:4856
123. Lee DS, Jung JK, Lim JW, Huh JS, Lee DD (2001) Sens Actuat B 77:228–236
124. Song XF, Liu L (2009) Sens Actuat A 154:175–179

125. Choi SW, Park JY, Kim SS (2009) Nanotechnology 20:465603
126. Park JA, Moon J, Lee SJ, Kim SH, Chu HY, Zyung T (2010) Sens Actuat B 145:592–595
127. Dayal P, Kyua T (2006) J Appl Phys 100: 043512-1–043512-6
128. Li XH, Shao CL, Liu YC, Chu XY, Wang CH, Zhang BX (2008) J Chem Phys 129:114708-1–114708-5
129. Wei S, Zhang Y, Zhou M (2011) Solid State Commun 151:895–899
130. Lee HJ, Yamauchi H, Suda H, Haraya K (2006) Sep Purif Technol 49:49–55
131. Mitra A, Jana D, De G (2012) Micropor Mesopor Mat 158:187–194
132. Wang YG, Bronsveld PM, DeHosson JTM, Djuricic B, McGarry D, Pickering S (1998) J Am Ceram Soc 81:1655–1660
133. Engelhart W, Dreher W, Eibl O, Schier V (2011) Acta Mater 59:7757–7767
134. Peng C, Zhang J, Xiong Z, Zhao B, Liu P (2015) Microporous Mesoporous Mater 215:133–142
135. Hong YC, Shin DH, Uhm HS (2007) Mater Chem Phys 101:35–40
136. Pande P, Rasmussen PG, Thompson LT (2012) J Power Sourc 207:212–215
137. Newport A, Carmalt CJ, Parkin IP, Neill SA (2004) Eur J Inorg Chem 21:4286–4290
138. Suszko T, Gulbinski W, Urbanowicz A, Gulbinski W (2011) Mater Lett 65:2146–2148
139. Choi BD, Blomgren GE, Kumta PN (2006) Adv Mater 18:1178–1182
140. Zhao J, Liu B, Xu S, Yang J, Lu Y (2015) J Alloys Comp 651:785–792
141. Subramanian V, Zhu HW, Vajtai R, Ajayan PM, Wei BQ (2005) J Phys Chem B 109:20207–20214
142. Wei W, Cui X, Chen W, Ivey DG (2011) Chem Soc Rev 40:1697–1721
143. Benbow EM, Kelly SP, Zhao L, Reutenauer JW, Suib SL (2011) J Phys Chem C 115:22009–22017
144. Yang X, Peng W, Fu K, Mao L, Jin J, Yang S, Li G (2020) Electrochimica Acta 340:135989
145. Zhang L, Yang X, Cai R, Chen C, Xia Y, Zhang H, Yang D, Yao X (2019) Nanoscale 11:826–832
146. Wang Z, Peng S, Hu Y, Li L, Yan T, Yang G, Ji D, Srinivasan M, Pan Z, Ramakrishna S (2017) J Mater Chem 5:4949–4961
147. Upadhyay S, Mishra R, Sahay P (2015) Actuat B Chem 209:368–376
148. Teleki A, Pratsinis S, Gouma P (2008) Chem Mater 20:4794–4796
149. Liu F, Chu X, Dong Y, Zhang W, Sun W, Shen L (2013) Sens Actuat B: Chem 188:469–474
150. Kim S, Park S, Park S, Lee C (2015) Sens Actuat B: Chem 209:180–185
151. Wang L, Lou Z, Fei T, Zhang T (2012) Sens Actuat B: Chem 161:178–183
152. Fan H, Jia X (2011) Solid State Ion 192:688–692
153. Park J, Choi S, Kim S (2000) Nanotechnology 21
154. Lianga X, Jin G, Liu F, Zhang X, An S, Ma J, Lu G (2015) Cer Inter 41:13780–13787
155. Kim Y, Hwang I, Kim S, Lee C, Lee JH (2008) Sens Actuat B: Chem 135:298–303
156. Xu X, Zhao P, Wang D, Sun P, You L, Sun Y, Liang X, Liu F, Chen H, Lu G (2013) Sens Actuat B: Chem 176:405–412
157. Qi Q, Zhang T, Liu L, Zheng X, Yu Q, Zeng Y, Yang H (2008) Sens Actuat B: Chem 134:166–170
158. Pramod NG, Pandey SN (2014) Ceram Int 40:3461–3468
159. Zhang G, Han X, Bian W, Zhan J, Ma X (2016) RSC Adv 6:3919–3926
160. Fu X, Yang P, Xiao X, Zhou D, Huang R, Zhang (2019) J Alloys Compd 797:666–675
161. Hernández-Rodríguez MA, Lozano-Gorrín AD, Martín IR, Mendoza UR, Lavin V (2018) Sens Actuat B Chem 255:970–976
162. Chakraborty S, Pal MJ (2019) Alloys Compd 787:1204–1211
163. Tie Y, Ma SY, Pei ST, Zhang QX, Zhu KM, Zhang R, Xu XH, Han T, Liu, WW (2020) Sens Actuat B: Chem 308:127689
164. Tang K, Yu Y, Mu X, van Aken PA, Maier J (2013) Electrochem Commun 28:54–57
165. Choi KI, Lee SH, Park JY, Choi DY, Hwang CH, Lee IH, Chang MH (2013) Mater Lett 112:113–116
166. Jung JY, Lee D, Lee YS (2015) J Alloys Compd 622:651–656
167. Lee S, Ha J, Choi J, Song T, Lee JW, Paik U (2013) ACS Appl Mater Interf 5:11525–11529

168. Zhan S, Qiu M, Yang S, Zhu D, Yu H, Li YJ (2014) Mater Chem A 2:20486–20493
169. Yu W, Ma Q, Li X, Dong X, Wang J, Liu G (2014) Mater Lett 120:126–129
170. Dong G, Xiao X, Peng M, Ma Z, Ye S, Chen D, Qin H, Deng G, Liang Q, Qiu J (2012) RSC Adv 2:2773–2782
171. Li D, Wang J, Dong X, Yu W, Liu G (2013) J Mater Sci 48:5930–5937
172. Park H, Jung H, Zhang M, Chang CH, Ndifor-Angwafor NG, Choa Y, Myung NV (2013) Nanoscale 5:3058–3062
173. Wang Z, Zhao L, Wang P, Guo L, Yu J (2016) J Alloys Compd 687:541–547
174. Mohanapriya P, Sathish CI, Pradeepkumar R, Segawa H, Yamaura K, Watanabe K, Natarajan T, Jaya NV (2013) J Nanosci Nanotechnol 13:5391–5400
175. Kenawy ER, Bowlin GL, Mansfield K, Layman J, Simpson DG, Sanders EH, Wnek GEJ (2002) Contr Release 81:57–64
176. Liu W, Thomopoulos S, Xia Y (2012) Adv Healthcare Mater 1:10–25
177. Hu X, Liu S, Zhou G, Huang Y, Xie Z, Jing XJ (2014) Contr Release 185:12–21
178. Ma Z, Kotaki M, Ramakrishna S (2006) J Membr Sci 272:179–187
179. Nasreen SAAN, Sundarrajan S, Nizar SAS, Balamurugan R, Ramakrishna S (2013) Membranes 3:266–284
180. Chung HJ, Park TG (2007) Adv Drug Deliv Rev 59:249–262
181. Sokolsky-Papkov M, Agashi K, Olaye A, Shakesheff K, Domb AJ (2007) Adv Drug Deliv Rev 59:187–206
182. Kenawy ER, Abdel-Hay FI, El-Newehy MH, Wnek GE (2009) Mater Chem Phys 113:296–302
183. Haider A, Gupta KC, Kang IK (2014) BioMed Res Int
184. Haider A, Gupta KC, Kang IK (2014) Nanoscale Res Lett 9:1
185. Vasita R, Katti DS (2006) Int J Nanomed 1:15
186. Almany L, Seliktar D (2005) Biomaterials 26:2467–2477
187. Pavlov MP, Mano JF, Neves NM, Reis RL (2004) Macromol Biosci 4:776–784
188. Prabhakaran MP, Vatankhah E, Ramakrishna S (2013) Biotechnol Bioeng 110:2775–2784
189. Yoo CR, Yeo IS, Park KE, Park JH, Lee SJ, Park WH, Min BM (2008) Int J Biol Macromol 42:324–334
190. Liu W, Lipner J, Moran CH, Feng L, Li X, Thomopoulos S, Xia Y (2015) Adv Mater 27:2583–2588
191. Wong HM, Chu PK, Leung FK, Cheung KM, Luk KD, Yeung KW (2014) Prog Nat Sci: Mater Int 24:561–567
192. Rim NG, Shin CS, Shin H (2013) Biomed Mater 8:014102
193. Gao Y, Bach Truong Y, Zhu Y, Louis Kyratzis I (2014) J Appl Polym Sci 131
194. Si Y, Tang X, Yu J, Ding B (2014) Electrospun NFs for energy and environmental applications. Springer, pp 3–38
195. Powell HM, Supp DM, Boyce ST (2009) Biomater 29:834–843
196. Rujitanaroj P, Pimpha N, Supaphol P (2008) Polym 49:4723–4732
197. Xie J, Hsieh YL (2003) J Mater Sci 38:2125–2133
198. Jia H, Zhu G, Vugrinovich B, Kataphinan W, Reneker DH, Wang P (2002) Biotechnol Prog 18:1027–1032
199. Li SF, Chen JP, Wu WTJ (2007) Mol Catal B Enzym 47:117–124
200. Zhang L, Luo L, Menkhaus TJ, Varadaraju H, Sun Y, Fong H (2011) J Membr Sci 369:499–505
201. Ibrahim HM, Mostafa M, Kandile NG (2020) Int J Biol Macromol 149:664–671
202. Hassabo AG, Nada AA, Ibrahim HM, Abou-Zeid NY (2015) Carbohydr Polym 122:343–350
203. El-Bisi MK, Ibrahim HM, Rabie AM, Elnagar K, Taha GM, El-Alfy EAD (2016) Pharma Chem 8:57–69
204. Zhang R, He Y, Xu LJ (2014) Mater Chem A 2:17979–17985
205. Wen Z, Tian-Mo L (2010) Physica B Conden Matter 405:1345–1348
206. Zhang XW, Ji LW, Toprakci O (2011) Polym Rev 51:239–264
207. Ke XB, Zheng ZF, Liu HW (2008) J Phys Chem B 112:5000–5006
208. Hui WU, Wei PAN, Dandan LIN, Heping LIJ (2012) Adv Cer 1:2–23

Surface Engineering of Nanofiber Membranes via Electrospinning-Embedded Nanoparticles for Wastewater Treatment

Jagdeep Singh, Sourbh Thakur, Rakesh Sehgal, A. S. Dhaliwal, and Vijay Kumar

Abstract Nanofibers (NFs) are fibers with diameters in the nanometer range and have found numerous applications due to their unique properties. Researchers are still trying to improve the properties of electrospun-based fibers by using unique nano-materials for solving environmental problems especially the treatment of wastewater. The modification of NFs has been carried out by decorating and embedding the various types of nanoparticles, such as noble metals, carbon nanomaterials, and metal oxide nanoparticles onto the surface of the membrane. The decorated surface of the NFs membrane possesses high surface area, surface energy, additional functionality, and anti-fouling properties that make them a suitable candidate for wastewater treatment application. This chapter highlights the modern trends in the surface engineering of NFs via electrospinning embedded nanoparticles (NPs) for wastewater treatment. The shape and size of Ag and Au NPs prepared under different reducing and stabilizing agents are also reviewed. The electrospun polymer NFs embedded with different NPs and surface modifications of NF membranes are discussed. The critical issues related to the use of electrospun polymer NFs embedded with different NPs for wastewater treatment along with a concluding note on possible future directions on this have also been included.

J. Singh · A. S. Dhaliwal
Department of Physics, Sant Longowal Institute of Engineering and Technology, Longowal, Punjab 148106, India

S. Thakur (✉)
Center for Computational Materials Science, Institute of Physics, Slovak Academy of Sciences, Bratislava, Slovakia

R. Sehgal (✉)
Department of Mechanical Engineering, National Institute of Technology Srinagar, Srinagar, Jammu and Kashmir 190006, India

V. Kumar
Department of Physics, National Institute of Technology Srinagar, Hazratbal, Jammu and Kashmir 190006, India

Department of Physics, University of the Free State, P.O. Box 339, Bloemfontein Z9300, South Africa

Keywords Surface engineering · Nanofibers · Electrospinning · Wastewater treatment

1 Introduction

Water is an elixir of life and it is important to mankind for drinking as well as for many other daily activities and we cannot sustain our life without it. Due to the rapid growth of the global population, modernization, and industrialization, there is tremendous depletion in the quality and quantity of water. Out of total water available on the earth, only 2% is fresh and potable, another 98% is saline. Moreover, out of the small portion of freshwater, only 0.77% of water is easily accessible because a major part of the world's population has limited access to pure drinking water. Wastewater or environmental pollution is mostly initiated from household products, industrial wastage and medical waste, municipals, which contaminate the water. The various pollutants from different origins, such as heavy metal ions, synthetic dyes, and other organic and inorganic materials caused a serious deterioration in the water quality [1–3]. This contaminated/unhealthy water causes various perilous diseases among people. Hence, to lessen its drawbacks for the protection of the environment, the technology to retrieve water from wastewater is one of the major concerns for the scientific community. A number of techniques (chemical, physical, and biological) such as chemical precipitation [4], adsorption [5], sonochemical [6], and electrochemical degradation [7] are available to tackle such types of pollutants and retrieve water from wastewater efficiently. The various advantages and disadvantages of these technologies are presented in Table 1. These methods provide a slow rate of purification, require accessories and expensive equipment, and therefore encourage the scientific community to go for an alternative method. Due to simplicity, high efficiency, and cost-effectiveness, membrane filtration technology has been demonstrated to be an effective and viable technique for the exclusion of several pollutants [8–10].

In the modern era, nanotechnology, an emerging science; has fascinated scientists due to its novel size-dependent properties [14, 15]. Researchers are trying to improve the quality of current technology by inventing unique materials. Nanofiber is one of the great inventions for its unique characteristics. Nanofibers can be considered as two separate words "nano" and "fibers". Historically, "nano" is used to describe the reference unit, nanometer, and the word fiber is defined as a slender, elongated, thread-like object or structure. Polymer nanofiber is the 1D nanomaterials having a highly porous structure and large surface-to-volume ratio. Polymeric nanofibers are better to use these days because of their certain properties like enormous specific surface area, high tensile strength, huge stiffness, comprehensive flexibility, sustainability, and unique dynamic mechanical, electrical and thermal properties [16, 17]. Polymeric nanofibers have found applications in the field of decontamination, catalysis, aerospace, fuel cell, solar energy, filtration, superabsorbents, energy storage, or as scaffolds for tissue engineering and wound dressings [18–22]. A lot of methods

Table 1 Various techniques for the wastewater treatment with advantages and disadvantages

Techniques	Methods	Advantages	Disadvantages	References
Chemical techniques	Chemical precipitation Neutralization Redox method Electrolytic Coagulation	Simple technology Economic Very efficient for metal and fluoride elimination	Chemical consumption Physiochemical monitoring of the effluent (pH) Unproductive in exclusion of the metal ions at low concentration High cost Cause secondary pollutants	[11]
Physical techniques	Ion exchange Air flotation Membrane separation Adsorption	Stable treatment methods, Simple operation and management, Wide variety of target contaminants Highly effective process (adsorption) with fast kinetics	Expensive Non-destructive processes Non-elective methods, Performance depends upon the type of material Rapid saturation and clogging of the reactors Not effective with certain types of dyestuffs and some metals	[12]
Biological techniques	Bioreactors Biological activated sludge Biofilm methods Anaerobic biological treatment Natural biological treatment Enzymatic decomposition	Efficiently eliminates Biodegradable organic matter, Microorganism degrade organic pollutants into harmless substances Low cost and stable effect	Essential to create a feasible environment Needs organization and care of the microorganism and physicochemical parameters Slow process, little biodegradability of certain molecules, Poor decolorization	[13]

are available in the literature for the production of nanofibers such as melt fibrillation [23], template synthesis [24], island-in-sea [25], magneto spinning [26], and gas jet techniques [27]. Electrospinning is the most embraced technique to produce polymer nanofibers. It has a low cost, a large selection of materials, strong versatility, and simple modification methods [28]. The concept of nanotechnology has led to a new membrane at the nanoscale with enhanced performance standards and with new functionality, such as large surface area and surface energy, high permeability, catalytic reactivity, and fouling resistance. Nanostructure membranes or mats based on nanofibers are amazing materials that could be synthesized via in-situ and ex-situ incorporation of nanoparticles/nanofiller into the polymer. The embedding and decoration of these nanoparticles/nanofiller add new functionalities to the resulting nanocomposite as well as change the physical, chemical, optical and electronic properties of the solution. This chapter highlights all the strategies investigated by the various researchers in the surface engineering of nanofiber membranes via electrospinning embedded nanoparticles as nanofiller such as noble metal (Ag and Au), TiO_2, SiO_2, carbon nanotubes, graphene, and graphene oxide. The change in the inherent properties of the polymer by the incorporation of embedded nanoparticles as nanofiller is also explored. Furthermore, the utilization of the nanofibers membrane for wastewater treatment is also explored. The various techniques and methods to minimize the problem of membrane fouling are also discussed in detail.

2 Basics of Electrospinning

The electrospinning (ES) phenomenon was reported for the first time by an astronomer, William Gilbert. In the 1960s, Sir Geoffrey Taylor analyzed the phenomenon comprehensively, studied mathematical aspects, and afterward, he put forward the modeled cone of fluid formed by an electric field. This cone has a certain specific shape now known as "Taylor cone" [29, 30]. He detected that when brought near an electric field, the drop of water turned into conical in shape and the leftover water of this drop get ejected in the form of smaller droplets. ES is a multipurpose technique to produce nanofibers via an electrically charged jet of the polymer solution. This process has four important aspects; a direct current power supply with high voltage (10–50 kV), a metallic needle having a blunt tip, a grounded conductive collector, and a syringe that holds polymer solution, as demonstrated in Fig. 1. First, the polymeric material whose nanofiber one wants to develop is dissolved and taken in a syringe as displayed in Fig. 1.

The polymeric solution contained in the syringe will be impelled out through the metallic tip at a controlled feeding rate, and there is the formation of polymer droplet at the tip of the pipette. The applied high potential creates a high electric field, so, there is the production of charges on the surface of polymer droplet and the Taylor cone is also formed. When the repulsive force is enough to chock the surface tension of polymer droplets then the polymer droplets get extended and its conical shape is obtained. Hence, the nanofiber or mat is formed on the grounded conductive collector.

The advantages of this technique are that by changing the experimental parameters such as polymer concentration, needle diameter, molecular weight, solution conductivity, applied voltage, solution viscosity, air humidity, flow rate, and the working distance between the tip of the needle and the collector, the surface morphology of the electrospun nanofibers can be easily controlled [31–34]. The different distances

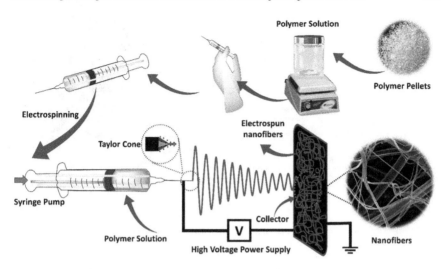

Fig. 1 The graphical representation of electrospinning process

among the needle and collector corresponding to the varying morphology of fibers are produced by different deposition times and evaporation rates. It was shown that the fiber diameter reduced as the distance increased [35]. The implemented voltage is another vital factor in the electrospinning techniques, which ends in the production of nanofibers. Sufficient voltage is required to overcome the surface tension of the polymer solution. Usually, 10–50 kV DC voltage is applied in this method. The decrease in the Taylor cone and an increase in jet speed at the same feed rate are obtained by increasing the applied voltage. These effects result in a fall in the fiber diameter and a rise in evaporation of the solvent [36]. The concentration of polymer solution additionally determines the morphology of nanofibers, breaking of polymer chains into pieces brought about by low weight concentration of the polymeric solution, which, in turn, was due to applied high voltage and surface tension of the solution. It prompts the development of beads or chain-type fibers. Alternatively, an increasing weight concentration of the polymeric solution tends to rise in the viscosity and resulted in an increase in the chain structure of fibers [37]. Flow and feed rate also regulate the surface morphology and pore size of nanofibers. Various studies represent that with the increase in the flow rate, the fiber diameter and pore size also rise up [38]. However, a flow rate above the critical value leads to the formation of fibers that lack the ability to dry. Moreover, the conductivity of polymer solution also affects the quality of nanofiber, an increment in the conductivity of the solution, charge on the surface of droplet additionally expands, which produces Taylor cone yet the fiber breadth diminishes [39]. Also, the selection of polymer prompts diverse webs of nanofibers. The determination of solvents for the electrospinning process is another significant reason for acquiring smooth and beadles electrospun NF [40]. The polymer ought to be totally soluble in a solution that has a reasonable boiling point for simpler evaporation like distilled water. The molecular weight

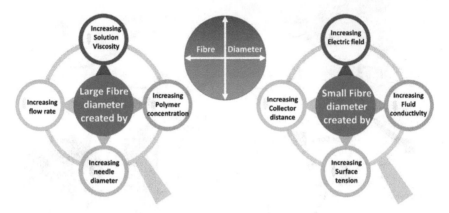

Fig. 2 Effect of various synthesis parameters in the electrospinning process

of the selected polymer has a significant influence on properties like conductivity, viscosity, or surface tension [41]. Low molecular weight polymer will, in general, form globules rather than fibers, and the expanding value of the molecular weight of polymer yields fibers with bigger diameters. The rise in temperature leads to falling in fiber diameter and high humidity, which causes pores on the surface of the fibers. Figure 2 sums up the impact of different ES limits on the NF formation, structure, and morphology.

3 Synthesis of Au or Ag Nanoparticles

During the last decades, the synthesis of Ag and Au NPs has been carried out with different methods for getting different shapes and sizes. Surface plasmon resonance (SPR) is the most important optical property of metallic NPs, which comprises collective oscillations of conduction electrons excited by the EM wave. This property is directly related to shape, size, and the adjacent medium. The nanostructure having different shapes and morphologies possesses different electronic, optical, chemical, and magnetic properties, which may be appropriate for various applications. The reductive procedure can be categorized into physical and chemical methods. Initially, silver/gold salts and capping compounds need to be dissolved in solvents. Then, the solution needs to be reduced with reducing agents (sodium borohydride, sodium citrate, alcohols, and poly (vinyl pyrrolidone)) to get stable colloidal NPs with different sizes and shapes. After the completion of the reaction, these precipitates can be collected through centrifugation. Several materials have been employed for the synthesis and stabilization of the silver and gold nanoparticles (Table 2).

Table 2 Shape and size of Ag and Au nanoparticles prepared under different reducing and stabilizing agents

Precursor	Reducing agent	Stabilizing agent	Shape/size (nm)	References
$AgNO_3$	Trisodium citrate	Trisodium citrate	Spherical/30–60	[42]
$AgNO_3$	NaBH$_4$	Dodecanoic acid (DDA)	Spherical/7	[43]
$AgNO_3$	Ethylene glycol	–	Nanowires/diameters of 30–40 nm	[44]
$AgNO_3$	Ascorbic acid	–	Nanowires/diameters of 30–40 nm	[45]
$AgNO_3$	Ethylene glycol monoalkyl ethers	PVP	Nanoprisms	[46]
HAuCl$_4$	NaBH$_4$	Citrate	Spherical/13 nm	[47]
HAuCl$_4$	CTAB	Ascorbic acid	Nanoflower/45 nm	[48]
HAuCl$_4$	Chitosan	–	Spherical/10–15 nm	[49]
HAuCl$_4$	Tannic acid	–	9 nm	[50]

Ag and Au NPs have different properties such as higher surface area, surface energy, SPR, and stability, which make them versatile materials for many applications in plasmonics, biomedical research, sensing, and catalysis, etc. NFs show distinct properties for example extremely high surface-to-weight ratio, low density, highly porous structure, and tight pore size for various advanced applications. The combination of Au or Ag NPs with polymer NFs has abundant potential to improve its properties. The addition of Au and Ag NPs with nanofiber can change its surface morphology and increase the surface area and energy. This addition and decoration of Ag and Au NPs into the NF can be done in two different methods, ex-situ and in-situ. The systematic representation of in-situ and ex-situ methods for the synthesis of NFs is displayed in Fig. 3. In the ex-situ, i.e., the first method, Ag and Au NPs with unlike morphologies are synthesized separately and the prepared NPs can be added into polymer precursor solution. In the second method, stabilized NPs are added to the polymeric solution and there is no need to add any precursor and stabilizing agent. Moreover, the in-situ method comprises the presence of precursors in the polymeric solution. For this, the silver/gold salt precursor is dissolved into the polymeric solution with a reducing and stabilizing agent. The reduction process transfers the Ag/Au ions into Ag and Au NPs. Normally, to complete the reduction of Ag/Au ions into Ag/Au NPs, self-reducing polymers such as chitosan and PVP are used here. The biomolecules present in these polymers are the cause of the reduction of ions. Moreover, other methods such as heating, UV irradiation, microwave irradiation, gamma irradiation are also employed for the reduction of ions into NPs [51–53].

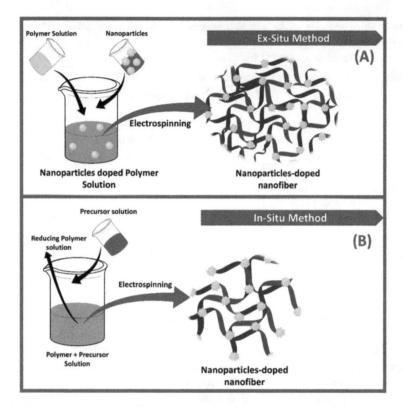

Fig. 3 The systematic representation of in-situ and ex-situ methods

4 Electrospun Polymer Nanofibers Embedded with Different Materials

4.1 Electrospun Polymer Nanofibers Embedded with Ag and Au Nanoparticles

The electrospun polymer embedded in noble metal nanoparticles or decorated with it could be synthesized using the above-stated in-situ and ex-situ methods. In the ex-situ method, the metal NPs are synthesized using a reducing and stabilizing agent and then dispersed into the polymer precursor solution [54]. Jin et al. [55] reported a one-step method to synthesize Ag NPs in a poly(vinylpyrrolidone) matrix. Wang et al. [56] carried out the synthesis of polyvinyl pyrrolidone/ silver nanocomposite by PVP ethanol solution without any reducing agent. Ethanol is used as a solvent for electrospinning and it reduces the silver ions. The characterization performed in this study approves that the average diameter of the nanofiber has been about 80 nm with the average diameter of the silver nanoparticles of about 8 nm. Aadil et al. [57]

synthesized poly (vinyl alcohol)-lignin nanofiber mats loaded with Ag NPs. The in-situ synthesis of Ag NPs has been carried out using alkali lignin extracted from Acacia wood as a reducing agent and then the synthesis of ultrafine nanofiber mats of Acacia lignin combined with Ag NPs is done by using the electrospinning technique. The size of nanofiber synthesized using electrospinning is in the range of 100–1000 nm and this size does not fall in the nano range. Hence, different designs in electrospinning are formed through phase separation splitting by exposure to the high electric field. In this method, nanonets are formed, which is also called a spider web or spider net form. Pant et al. [58] synthesized Ag NP-embedded polyurethane nanofiber/nanonet structured membrane. The uniform distribution of Ag nanoparticles on the surface of PU nanofiber is achieved by the in-situ method. Figure 4 shows the FE-SEM and TEM micrographs of synthesized nanofiber and from it is clear that the presence of spiderweb-like nanonet structure is accompanied by the main fiber, which offers a large surface area.

Nguyen et al. [59] fabricated Ag NPs loaded nanowire mat based on PVA using the electrospinning method. The synthesis has been optimized by changing the microwave irradiation time and electro-spinning parameters. The shape and size are manipulated by altering the microwave energy. By increasing the irradiation time, the size of particles is optimized for suitable applications. The mechanical strength

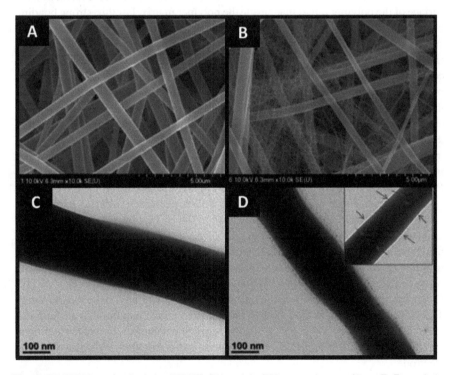

Fig. 4 FE-SEM images of pristine PU NFs (**A**) and Ag/PU composite nanofibers. **B–D** are their respective Bio-TEM images. The inset in **D** shows the Ag NPs embedded in the nanofiber [58]

is also examined by the stress–strain curve of the PVA nanofibrous mat and after the introduction of Ag NPs. The result shows that the strength of the pure PVA fibrous mat is improved by the loading of Ag NPs. Hence, the nanocomposite fibrous mats were more brittle and stronger than the pure PVA fibrous mat. Dong and co-worker [60] fabricated Ag porous films by the heat treatment of $AgNO_3$ doped poly(vinyl alcohol) (PVA) electrospun nanofibers with smooth and super-long nanofibers with an average diameter of about 500 nm. Moreover, after 2 h of heat treatment at 600 °C in air, the morphologies are completely degraded and the porous film is formed having a pore size of about microns to 10 mm. Tijing et al. [61] conducted the in-situ synthesis of Ag NPs inside the polymer. For this, $AgNO_3$ precursor is dissolved in the polymer solution and afterward electrospinning, the synthesized mat is exposed to UV light to synthesize the Ag NPs. Various other authors have also reported the production of Ag NP-embedded nanofiber membrane by electrospinning method [62–67].

Similar to the above-mentioned approaches to embedding the NPs into a polymer matrix, various researchers reported several nanofibers embedded with Au NPs. Au NPs are synthesized using various methods and then dispersed into the polymer precursor solution [68–76]. In order to functionalize or decorate the nanoparticles on the surface of nanofiber, different techniques such as dipping, spraying, and layer-by-layer deposition are used [77–79]. By these methods, different NPs are located on the surface of the polymer rather than inside the polymer matrix. For this, strong bonding between NPs and polymer surfaces is required. Otherwise, the NPs released from the surface of the polymer and can cause secondary pollution. Another method to functionalize nanofibers is the coaxial electrospinning technique. In this technique, two concentric nozzles are used instead of a single nozzle. The inner nozzle provides the polymeric solution while the outer nozzle functionalizes the outer surface by spraying the NPs on it. Hence using this method, nanoparticles decorated nanofiber can be easily fabricated simultaneously. Various researchers have developed a nanofiber decorated with NPs [80–85]. Yu et al. [86] studied the development of polyacrylonitrile (PAN) nanofibers coated with Ag NPs with a modified coaxial electrospinning process. The result confirms the uniform distribution of Ag NPs on the surface of PAN nanofibers. Wang et al. [87] decorated Au NPs on the surface of PVA electrospun nanofiber. The synthesized nanofiber of PVA possesses sulfur as a functional group through it and the Au NPs are attached to the surface by gold–sulfur bonding interactions. Figure 5, shows the steps involved in the procedure to decorate the nanoparticles on the surface of PVA. First, PVA electrospun fibrous is prepared by in situ cross-linking and dipped into the acidic solution to stimulate the cross-linking between hydroxyl groups of the PVA and the aldehyde groups of GA. The MPTES (3-mercaptopropyltrimethoxysilane) introduces the thiol groups on the PVA/GA mat. Therefore, Au NPs are homogenously immobilized on the mat through an Au NPs–sulfur bonding interaction. Figure 6 displays a graphic demonstration of the hydrogen peroxide sensor based on the horseradish peroxidase/Au NPs–PVA/glassy carbon electrode.

Dong et al. [88] reported the decoration of Ag, Au, and Pt NPs on the surface of nylon 6 fiber. Ag, Au, and Pt NPs are synthesized by means of sodium citrate and

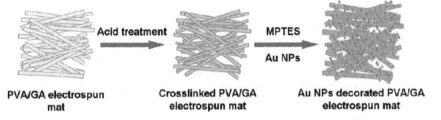

Fig. 5 Schematic fabrication of water-stable functional Au NPs–PVA/GA electrospinning nanofibrous mats. Reprinted with permission from [87]. Copyright (2012) American Chemical Society

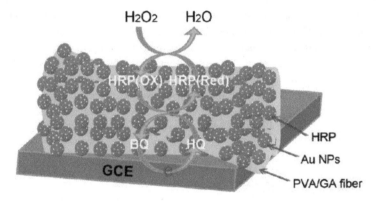

Fig. 6 Schematic presentation of the hydrogen peroxide biosensor. Reprinted with permission from [87]. Copyright (2012) American Chemical Society

nylon 6 nanofiber mats, formed by ES. The decoration of metal nanoparticles is determined by the hydrogen bond interactions between the amide groups in the nylon 6 backbone and the –COOH groups capped on the surface of the metal NPs. Son et al. [89] fabricated assemble noble metal nanostructures by means of electrospun catechol-rafted PVA nanofibers. The –OH groups of PVA can be used to functionalize its surfaces. The synthesized PVA-g-catechol NFs have been employed as a chemically reactive template for the reduction of metal ions into solid metal nanostructures. The SEM and elemental mapping confirm the uniform distribution of Ag, Au, and Pt NPs on the surface of PVA-g-catechol nanofibers.

4.2 Electrospun Polymer Nanofibers Embedded with Carbon Nanoparticles

The synthesis and characterization of electrospun nanofiber embedded or decorated with carbon nanoparticles have received significant attention from many researchers. Carbon nanotubes (CNTs) possess remarkable thermal, mechanical, and electrical properties, and combining them inside the polymer NF has shown improved properties with relatively low percolation thresholds. The most difficult challenge in the path to developing nanofiber embedded with carbon nanomaterial is to prevent bundling and agglomeration. This problem is explored by synthesis via electrospinning method, by uniform dispersion of CNT onto the surface of the nanofiber. Chen et al. [90] synthesized high strength polyimide nanofiber membrane containing multiwall carbon nanotubes (MWNTs) by using the electrospinning method. The thermal and mechanical properties of the polyimide matrix have been expressively enhanced with the addition of MWNTs. Xiao et al. [91] also synthesized mechanical strong nanofiber by mixing multiwalled carbon nanotubes (MWCNTs) with PAA/PVA polymer solution. The MW-CNTs amalgamated PAA/PVA nanofibers are first cross-linked and then employed as a reducing reactor to reduce Fe(III) ions into zero-valent iron nanoparticles. Weng et al. [92] functionalized MW-CNTs and strengthened the poly(methyl methacrylate) (PMMA) nanofiber mats with average diameters range of 370–800 nm. Moreover, the electrical and mechanical properties of the synthesized NF mat are better as compared with native PMMA nanofibers.

4.3 Electrospun Polymer Nanofibers Embedded with Metal Oxide Nanoparticles

Metal oxide NPs played a vital role in the formation of new materials for a range of applications. The metal oxide NPs have a large surface area, which is required in various advanced applications. A large number of synthesis methods are reported for the synthesis of metal oxide NPs such as co-precipitation [93], thermal decomposition [94], hydrothermal [95], sol–gel method [96], ultrasound irradiation [97], and biological method [98]. Co-precipitation and thermal decomposition methods are the most important methods for the synthesis of metal oxide NPs [99]. Various metal oxide NPs embedded/decorated in NFs such as ZrOlasma/wet chemistry, coating, nanofibers/activated carbon composite [100], polyacrylonitrile nanofiber webs comprising titanium dioxide (TiO_2) [101], cellulose acetate/silica composite nanofibrous membrane [102], TiO_2 NFs doped-activated carbon [103], polyethylene terephthalate (PET) with CuO nanoparticles [104], metal oxide/silica NFs [105], metal oxide-coated polymer NFs [106], Fe_2O_3 doped on ZnO NFs [107], ultrafine metal oxide-decorated hybrid carbon NFs [108] are also available in the literature. Therefore, various nanofibers embedded with noble metal (Ag, Au, and Pt), carbon nanomaterial (CNTs and MW-CNTs), and metal oxide nanoparticles (TiO_2, Fe_2O_3, CuO, etc.) are available in the literature and have been used as a promising material for various advanced applications including wastewater treatment. This review highlights the various electrospun nanofibers. Various researchers fabricated the electrospun nanofibers embedded with nanoparticles. But still, more research work is needed to develop modified nanofibers. Hence, the various modification techniques to develop nanofibrous membranes with better properties are also explored in the next section. Furthermore, the applications of various electrospun nanofibers membrane in the field of wastewater treatment are described in the next sections.

5 Surface Modification of Nanofiber Membranes

The surface modification is employed to obtain enough performance for a particular application. The surface modification has been carried out either chemically or physically by varying the atoms/molecules in the existing surface, changing the surface morphology, or coating over the surface with materials. The physical modification includes methods such as coating, adsorption, and blending. Moreover, the chemical techniques, for example, the introduction of functional groups by plasma treatment or wet-chemistry reactions, or yet grafted polymers covalently bonded to the surface and layer by layer electrostatic interaction. The surface modification techniques are divided into two categories: (i) post-treatment technique and (ii) one-step treatment during ES. In post-treatment methods, the synthesized membrane has been treated after ES using plasma/wet chemistry, coating, and grafting method. Moreover, in one-step treatment, the NPs are added to the precursor before the synthesis. The embedding of NPs to the polymer solution before ES gives them significant properties. The embedded NPs offer a large surface area, surface energy and help to increase the surface roughness of the fibers and thus amplify the wettability of electrospun nanocomposites. The improved properties of electrospun membranes with various types of NPs are studied by various authors [109–115]. There are many other studies related to the improvement of the properties of nanofibrous membranes, reported by various authors, available in the literature [116–119]. In these studies, researchers have tried to increase the mechanical properties as well as to increase inter-fiber adhesion. Membrane fouling is an exceptionally complex phenomenon that has linked with the accumulation of unwanted materials/particles such as colloids, macromolecules, and salts on the surface of the membrane or inside the membrane pores. Various types of fouling such as organic fouling, inorganic fouling or scaling, colloidal fouling, and biofouling have been accommodated on the membrane surface as shown in Fig. 7.

With the purpose to cut the tissue of membrane fouling, several techniques are invented in the last few decades. The surface modification by the decoration of NPs onto the membrane surface and by the coating of the hydrophilic polymer layer is one of the suitable methods to prevent the problem of fouling. But the coating method involves high cost, complexity, and pollutant production. Hence, the embedding and decoration of nanoparticles onto the surface of the membrane is an emerging technique, and several investigations on the preparation of mixed matrix membranes (MMMs) with nanofiller are available in the literature. The nanoparticles based on

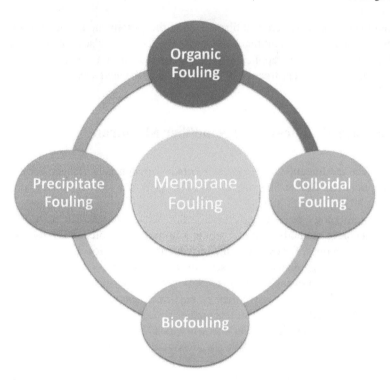

Fig. 7 Various types of fouling accommodated on the membrane surface

noble metals, carbon, and metal oxides have also overcome the fouling problems. The decorated and embedded nanoparticles into the nanofiber membrane can increase the hydrophilicity, contact angle, surface roughness, or imposing bactericidal agents and charge on the surface of the membrane, which may cut the membrane biofouling.

6 Applications of Nanofiber Membranes for Wastewater Treatment

Membrane technology has become the most significant method for the purification of wastewater. This technique is utilized for the elimination of suspended solids, microorganisms, organic, inorganic pollutants, and heavy metal ions dissolved in the aqueous solution [120]. Several membranes processes based on the different types of membranes such as microfiltration (MF), ultrafiltration (UF), nanofiltration (NF), and reverse osmosis (RO) are available in the literature for the treatment of water and desalination and Table 3 shows the properties and purposes of these membrane types and the capability of the membranes for the elimination of certain pollutants from the wastewater as described in Fig. 8 [121, 122].

Table 3 Various membranes techniques with all specifications

Membrane type	Pore sizes (nm)	Material passed	Material retained	References
Nanofiltration (NF)	0.1–1	Water, sugar, and monovalent ion	Solutes, MW > 500, di- and multivalent ion	[123, 124]
Reverse osmosis (RO)	Not relevant	Water	All dissolved and suspended solute	[125]
Microfiltration (MF)	>50	Water, salt, and macromolecules	Particles, bacterial, yeast, etc	[126]
Ultrafiltration (UF)	1–10	Water, salt, and sugar	Macromolecules, colloids, lattices, and solute >10,000	[127]
Forward osmosis (FO)	Not relevant	Water, salt, and sugar	Organics, minerals, and other solids	[128]

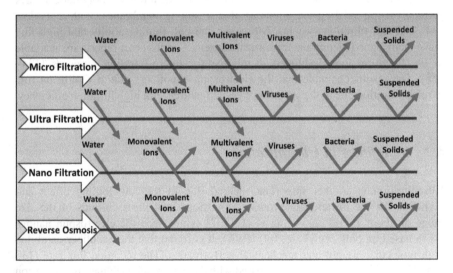

Fig. 8 Capability of the various membranes for the removal of certain pollutant from wastewater

6.1 Nanofibers in Microfiltration

The utilization of electrospun nanofibrous membrane has been taken place for wastewater treatment at the microfiltration level. For this, cellulose nanofiber membranes with a diameter ranging between 200 nm and 1 μm have been synthesized and treated in sodium hydroxide solution in H_2O/C_2H_5OH to get regenerated cellulose NF mesh. The filtration result confirms the reduction in pressure drop and an increment in flux as compared with the viable microfiltration membranes and after regeneration using elution '*buffer*, it presented reusability [129]. Gopal et al. [130] designed an electro-spun nanofibrous membrane, i.e., polysulfone to eliminate microparticles from the solution. The synthesized membrane was capable of removing 99% of 10, 8, and 7 μm particles. In another study, the formation of an electrospun polyvinyli-dene fluoride membrane having a fiber diameter of about 380 nm has been done. This membrane is used to isolate 1, 5, and 10 μm polystyrene particles. The results point out the rejection of about a maximum of 90% particles from feed solutions to the membranes to decrease the possibility of fouling [131]. Kaur et al. [132] fabri-cated an electrospun membrane of polysulfone and polyvinylidene fluoride with a high angle of contact by using diverse surface transforming macromolecules. The angle of contact of the blended EM gets reduced from 140° to 54°. More-over, in comparison to non-blended EM, the water flux is increased to 20% in the case of blended EM. Moradi and Zinadini [133] reported a large value of flux and anti-fouling graphene oxide (GO) NP set in PAN nanofiber microfiltra-tion membranes. The incorporation of GO NPs into the PAN polymeric matrix pointedly amplified the permeation flux of the resulting membrane in both cross-flow and dead-end filtration systems. Moreover, the anti-fouling ability and high flux make them a good membrane for water treatment. Various other reports are available in the literature that developed membrane for micro-filtration purposes [134–136]. The results conclude that the electrospun nanofibers play a significant role in microfiltration due to the superior physical features with high filtration efficiency.

6.2 Nanofibers in Ultrafiltration

Ultrafiltration is another important method due to the low fouling effect and high flux for the filtration of proteins, emulsions, and virus colloids of the size range of about 1–100 nm. Wang et al. [137] synthesized the electrospun nanofi-brous based on poly (vinyl alcohol) (PVA). It is found that the nanofiber exhibited high molecular weight and mechanical performance. The recorded flux rate (130 L/m^2) is significantly higher as compared with the commercial UF membranes. Yoon and Lee [138] reported membrane based on polyacrylonitrile coated with the thin top layer of chitosan. The average diameter of the synthesized membrane ranges from 124 to 720 nm. It also exhibits a higher flux rate and filtration efficiency as compared with the commercial filtration system. Dobosz et al. [127] reported a high-porous nanofiber based on cellulose and polysulfone with improved flux and fouling resistance. The fouling resistance is improved by about 90% as compared with the control membranes. This is possibly due to the decrease in the contact time. Polyaniline/polysulfone membranes have been fabricated and the performance of filtration has been tasted using cross-flow equipment by taking PEG aqueous solution, water, and bovine serum albium as feed solution. The results revealed that the synthesized membrane has higher hydrophilicity and permeability. Moreover, the antifouling characteristics of PANI/PS membranes displayed a better performance than PS substrate during the filtration [139].

Reverse osmosis (RO) is an important technique for the production of fresh-water by distillation of seawater using a membrane having pore sizes of 0.1–1 nm [140]. Tian et al. [141] and Bui et al. [142] reported electrospun nanofiber as forward osmosis membrane based on polyethersulfone (PES) and polysulfone (PSf). The MPD and TMC monomers have been utilized to formulate an active barrier layer onto the fibrous supporting layer. High water flux and low salt flux are the major possessions of the membrane.

6.3 Nanofiltration

Nanofiltration techniques have been extensively applied to effectively remove heavy metals, dyes, and other impurities present in wastewater due to its high efficiency, low cost, small pore sizes, and ease [143, 144]. Ritcharoen et al. [145] reported membrane for nanofiltration based on cellulose acetate fiber mat coated with chitosan/sodium alginate and poly(styrene sulfonate). Yoon et al. [137] reported polyacrylonitrile support with high flux thin film nanofibrous composite (TFNC) membranes in the nanofiltration process. The active layer onto the membrane has been produced by IP of trimesoylchloride, piperazine, and some additives and utilized in the nanofiltration of divalent salts ($MgSO_4$). The TFNC membrane has shown higher rejection rates (2–22%) as well as higher permeate fluxes (21–42%) than those of TFC membranes. It is also directed that the concentration of piperazine has played an important part in controlled flux and rejection capacity. The effect of nanofibers structure and film composition in the nanofiltration process is examined by Kaur et al. [132]. In this study, they explored the impact of fiber diameters on filtration and study the separation performance with the help of dead-end filtration cells of 2000 ppm concentration of $MgSO_4$ as feed solution. The consequences revealed that with a reduction in the diameter of the fiber, the pore size also decreased, which amplified flux with high salt rejection. The results revealed that nanofiltration with an increment in the applied hot pressure membrane showed lower flux and higher rejection than those of TFNC membranes [146]. Another study related to the improvement in the performance of membrane based on PET nonwoven by spun used the DMF and DMF/NMP mixture onto the membrane as a support layer. The supporting layer upsurges the adhesion between the forming PET nonwoven scaffold and PES nanofibrous layer. The produced membrane showed almost double flux compared with that of NF-90 with an equal salt rejection ratio [147]. Nanostructure membrane can also be modified using the nanomaterials such as noble metal oxide NPs, carbon nanomaterials, and metal NPs. Nanotechnology provides advanced performance nanofibers material with amazing properties such as enhanced catalytic reactivity, fouling resistance, high permeability, and water treatment [114]. The high efficiency, low capital need, quality of treated water, and effective disinfection are other features of this technique [123]. These modified membranes have been used for removing various particles, bacteria, and viruses, natural organic matter, water hardness metals, organic and inorganic substances. The detailed discussion of the nanofiber membrane decorated or embedded with different nanomaterials for wastewater treatment is discussed in the following sections.

6.4 Electrospun Noble Metal (Ag, Au, and Pt) Nanofiber Composites for Wastewater Treatment

Nanoparticles based on noble metals, i.e., Ag, Au, Pt, and Pd, are essential for modern nanotechnology because of their outstanding optical, physical, and chemical properties [148, 149]. The modification of nanofibers by decorating onto the surface or by embedding inside it is important for improving its properties. Taurozzi et al. [150] synthesized a membrane based on polysulfone with incorporated Ag particles in order to decrease biofouling. The availability of the Ag NPs implanted in the membranes has been quantitatively evaluated by the extent of growth inhibition of biofilm because of ionic Ag released by the nanocomposites. Gunawan et al. [151] fabricated a novel membrane using Ag/MWNTs coated on PAN hollow fiber membrane for biofouling control and water disinfection. Ag NPs of optimized sizes have been deposited on the surface of polyethylene glycol-grafted MWNTs. The Ag/MWNTs layer coated on the outside surface acts as a disinfection barrier. The nonstop filtration test against E. coli confirms the much-boosted antifouling properties and antimicrobial activities of the membrane against E. coli. Most of the E. coli cells were observed on the composite membrane with dented cell walls. This is caused by the direct contact of the Ag NPs and cells. Furthermore, Ag NPs amalgamated into PSf ultrafiltration membranes possess antimicrobial properties toward a large number of bacteria, comprising Escherichia coli, Pseudomonas mendocina, and bacteriophage. Interestingly, the utilization of nanoparticles with various polymeric phases yields similar properties [152]. Dasari et al. [153] also revealed that the existence of Ag NPs in NFs leads in lesser fouling of the NFs. Sui et al. [154] deposited the Ag NPs onto the PAA/PVA/PW12 fibrous by layer-by-layer (LBL) self-assembly technique for the degradation of MB solution. Also, the effects of charge, the number, and components of the LBL film on the photodegradation of MB dye have been optimized. It is found that the negatively charged surface has higher activity as compared with the positively charged surface. The resultant NF has excellent photocatalytic activity, stability, and recyclability. Pant et al. [155] fabricated Ag-impregnated TiO_2/nylon-6 nanocomposite mats with good photocatalytic and antibacterial properties. TiO_2 NPs present in the nylon-6 solution are able to form a spider-wave-like structure while ES enabled the UV photoreduction of $AgNO_3$ to Ag. The antibacterial and photocatalytic activity of the resultant NF mat revealed that the mats containing NPs are more effective as compared with the mats without Ag NPs. Ag NP decorated cellulose nanofibrous membrane has decent antibacterial functions and separation efficiency is a promising candidate for drinking water as reported by Chen and

Peng [156]. In another study, a nano-enabled membrane was fabricated from two components, i.e., a silver nanoparticle, poly (4-vinyl-N-hexylpyridinium bromide) and silver bromide embedded nanoparticles. The cationic polymer/silver bromide nanoparticle composite membranes showed potent long-lasting antibacterial activity toward gram-positive and gram-negative bacteria [157]. Lv et al. [158] synthesized Ag NP-decorated porous ceramic complex for water treatment. The synthesized composite can be kept for long durations and it is enduring under washing, even with ultrasonic irradiation, without losing NPs from the surface of the membrane with strong sterilization and antibacterial property. Moreover, the complete reduction of 4-nitrophenol has been carried out by using nanomembrane made up of alumina and polymers by LBL adsorption of citrate-stabilized Au NPs and polyelectrolytes [159].

6.5 Electrospun Carbon Nanofiber (CNF) Composites for Wastewater Treatment

The first carbon nanofiber (CNF) reported by Thomas Edison was contrived by bamboo and carbonizing cotton and used as a filament of an electric bulb in 1879 [160, 161]. The fabrication of carbon nanofibers is done by electrospinning method by the use of the precursors of the CNFs. By changing the kind of polymer solution and processing parameters, the properties of synthesized CNFs are regulated. PAN is often employed in the synthesis of electrospun CNF materials [162]. Moreover, polybenzimidazole (PBI), PVA, poly (vinylidene fluoride) (PVDF) polyimides (PIs), lignin, and phenolic resin have also been used. CNFs membranes have been used for the elimination of many organic pollutants such as organic dyes [163, 164], organic solvents [164, 165], and volatile organic compounds [166, 167]. The potential material based on carbon is electrospun CNTs, which have been utilized for the treatment of wastewater due to their properties [168, 169]. Singh et al. [170] reported unmodified electrospun CNF for the exclusion of the disinfection by-products from water by regulating carbonization temperatures during synthesis to <500 °C. The extent of graphitization of the polymeric precursor may be prevented by low carbonization temperatures. Thus, sorption capacity is decreased. Moreover, unmodified CNTs are relatively weak (brittle material). Hence modification of CNFs for this application is required. In order to improve the material flexibility, silica (SiO_2) nanoparticles have been embedded in the CNFs matrix or by the introduction of macrospores via insertion of detachable parts in the precursor solution [171–173]. It is seen that the composite membrane is harder than the pristine CNF membrane when the embedded SiO_2 concentration is retained at 2.7wt.%, beyond which the membrane toughness decreases [171]. The modified nanofiber embedded with Fe_2O_3, SiO_2, $CaCO_3$, $ZnCl$, tetra ethoxy orthosilicate [173, 174] nanoparticles based on the polymer such as poly(methyl methacrylate) [175], terephthalic acid, and poly(styrene-co-acrylonitrile) [176] are used for the removal of organic dyes, oil [151, 177, 178].

CNTs are widely used 1D nanomaterials because of the unique mechanical, magnetic, and electrical properties of CNTs [179]. Moreover, the high surface area and high adsorption ability make CNTs very good candidates [180]. By considering the properties of CNTs, several researchers have studied the influence of embedding CNTs within the nanofibers [90, 92]. The embedded CNTs inside nanofibers can improve their mechanical strength, electrical and thermal properties [181, 182]. Ji et al. [183] fabricated PAN NFs with enhanced mechanical properties by embedding the MWCNTs inside them. Srivastava et al. [184] reported nanocomposites incorporating aligned CNT walls. The study showed applications for the membrane to eliminate heavy hydrocarbons from petroleum and to filter the bacterial contaminants such as Escherichia coli or poliovirus of nanometer-size (25 nm) from water. In vitro testing has shown substantial antimicrobial activity of carbon nanotube composite films against Staphylococcus warneri, Staphylococcus aureus, and Staphylococcus warneri colonization. Xiao et al. [91] fabricated MWCNT-reinforced electrospun polymer NFs comprising zero-valent iron NPs for environmental remediation. The synthesized nanofibrous mat shows the outstanding proficiency for decolorization of model dyes such as methyl blue, acid fuchsine, and acridine orange with a more than 90% decolorization percentage. Moreover, the synthesized nanofibrous mat is found to be effective for the degradation of trichloroethylene with a degradation efficiency is about 93%. Singh et al. [170] testified the confiscation of carcinogens disinfection by-products (DBPs) present in water by using MWCNTs assimilated in the carbonized nanofibrous membranes (CNMs). The adsorption capacity for chloroform and monochloroacetic acid is about 554 mg/g and between 287 and 504 mg/g, respectively. Moreover, Salipira et al. [185] have also reported that cyclodextrin polyurethanes comprising of CNTs is effective for the removal of trichloroethylene as compared with the granulated activated carbon because of its higher surface area. Wang et al. [186] reported a novel GO-based nanofiltration membrane on a highly porous GO@PAN mat for water treatment application. A high rejection performance is shown by the membrane (nearly 100% rejection of Congo red, and 56.7% for Na_2SO_4).

6.6 Electrospun Metal Oxide Polymer Composites for Wastewater Treatment

Various types of metal oxide nanoparticles such as ferric oxide, titanium oxides, manganese oxides, magnesium oxides, cerium oxides, and aluminum oxide-embedded nanofibers are used as a promising material for the removal of various pollutants from wastewater because of its nontoxic nature, high photosensitivity, and large bandgap [187–194]. Mu et al. [195] fabricated one-dimensional ZnO-carbon nanofiber (CNF) heteroarchitectures via ES for the degradation of Rhodamine B. They reported that the synthesized nanofiber has higher photocatalytic activities as compared with the pure ZnO. Furthermore, it could be easily recycled without the reduction of photocatalytic activity because of its one-dimensional nanostructure property. PA6@FexOy nanofibrous membrane has been synthesized via electrospinning technique combined with a hydrothermal strategy for the removal of chromium Cr (VI) ions from wastewater [196] with an adsorption capacity of 150 mg Cr/g nanofibrous membrane. Furthermore, the Freundlich adsorption isotherm indicates the multi-layer adsorption onto the surface of the nanofiber membrane. Wang et al. [197] prepared a novel PAN/TiO$_2$ electrospun system for the degradation of dyes. The removal of the dye occurs by the interaction of methylene blue with SiO$_2$ in montmorillonite. The rate of adsorption upsurges with an increase in montmorillonite, and the rate of degradation increases with an increment in spin-coating layers of TiO$_2$. Liu et al. [198] prepared polyaniline (PANI)-coated TiO$_2$/SiO$_2$ nanofiber membranes by a combination of ES, calcination, and in situ polymerization with enhanced visible light photocatalytic degradation activity. The figure depicts the time-dependent UV–vis spectra of the MO (methyl orange) solution. It is clearly indicating that the efficient degradation of MO dye has taken place. Furthermore, the digital photos (inset figure) also visually confirm the degradation of MO dye. Joo Kim et al. [199] also testified multifunctional TiO$_2$-fly ash/polyurethane nanocomposite membrane for the effective removal of heavy metals (Hg, Pb) as well as the removal of MB dyes with good antibacterial activity and enhanced water flux.

Moreover, various researchers modified the silica with amino or thiol groups earlier to ES via hydrolysis poly-condensation. Then the functionalized nanofibers have been used for the removal of Cr(III) or Cu, respectively [200, 201]. Similarly, Taha et al. reported the amino-functionalized cellulose acetate/silica composite for the effective removal of Cr(VI). Here, the silica component served both to support the surface functional groups and to improve material stability [102]. Dastbaz and Keshtkar [202] verified an alternate approach to the co-electrospinning of silica by integrating SiO$_2$ nanoparticles surface-functionalized with aminopropyltriethoxysilane (APTES) to merge amine functional groups into electrospun PAN. The loading and functionalization of nanoparticles inside the nanofiber have been optimized for improving the adsorption capacity for Cd^{2+} , U^{6+} , Ni^{2+} , and Th^{2+} ions from water. Also, the regeneration study confirms that the nanofiber could be used for industrial applications repetitively without any notable decline in its adsorption capacity. Teng et al. [203] reported mesoporous polyvinyl alcohol (PVA)/SiO$_2$ composite nanofiber membrane functionalized with cyclodextrin groups for water purification purposes. The produced nanofiber membranes have a decent performance in adsorption of indigo carmine dye with the most adsorption capacity is about 495 mg/g. Moreover, the membranes have worthy recycling properties for practical usage. Hu et al. [204] reported thermally stable and flexible CuO nanocrystal-decorated SiO$_2$ nanofibers for the removal of Rhodamine B. The nitrogen adsorption–desorption isotherm has been carried out for surface properties and results confirm that both the samples, SiO$_2$ and SC fibers, exhibited a type-IV isotherm, and total pore volume and the specific surface area of the SiO$_2$ fibers are recorded to be 0.012 cc g^{-1} and 12.02 m^2 g^{-1}, respectively. While the SC (0.25) nanofiber is extra porous and its total pore volume and specific surface area increased to 0.073 cc g^{-1} and 55.59 m^2 g^{-1}, respectively. Taha et al. [102]

reported novel NH_2-functionalized cellulose acetate (CA)/silica composite nanofibrous membranes for the removal of Cr (VI) ions with maximum adsorption capacity is about 19.46 mg/g. Afkhami et al. [205] reported sodium dodecyl sulfate (SDS)-coated nano-alumina with polyvinylidene fluoride membrane for effective removal of metal cations Cd(II), Pb(II), Co(II), Cr(III), Mn(II), and Ni(II) from wastewater samples. The consequences showed that the composite membrane had high adsorption capacity for Cr(III), Cd(II), and Pb(II) and in mixed ion systems. Desorption experimentations by elution of the adsorbent with a mixture of menthol and nitric acid showed that the modified alumina NPs could be used again without significant property losses even after three adsorption–desorption cycles. Thus, modified nano-alumina with SDS is favorable and useful for the exclusion of these metal ions. The high adsorption capacity makes it a promising candidate material for Cr(III), Pb(II), and Cd(II) removal. Furthermore, the ES NFs with active Fe(II) and Fe(III) as nanofillers inside the PAN and PVA have been synthesized for the effective removal of Cr(VI) and Ar ions from wastewater [206, 207]. The main disadvantage of these NFs is the leaching of iron salts from the nanofibers over time. Thus, the NPs are a promising method to make the NF with an active site due to its durability. In this view, Patel et al. modified PAN composites with surface-deposited iron oxide NPs. The PAN nanofiber-embedded iron oxide nanoparticles possessed about half the capacity for the removal of Congo Red dye [208]. Xiao et al. [91] fabricated electrospun NFs embedded with zerovalent iron NPs for potential environmental applications. The NF mat containing iron NPs has a porous structure and capable of quick decolorization of an organic dye (acid fuchsine) with percentage degradation is about 95.8% within 40 min. Moreover, a resultant NF mat has been employed for the remediation of many other contaminants such as PCB, TCE, and toxic metal ions (e.g., arsenic). The resultant NF possesses high efficiency for the removal of Rhodamine B dye and Cu^{2+} ions from the aqueous solution [209]. Horzum et al. [210] synthesized chitosan fiber-supported zero-valent iron nanoparticles for the removal of inorganic arsenic from aqueous solutions. Zhu et al. [211] synthesized membrane based on chitosan with magnetic NPs (γ-Fe_2O_3). The SEM and TEM studies revealed that synthesized membranes have numerous pores and folds on the surface, which give active sites for dye entrapment. This membrane exhibited good adsorption ability and adsorbed up to 70% methyl orange at pH6.

7   Conclusions and Outlook

Electrospun nanofibrous membranes have many interesting and controllable properties that provide good separation efficiencies when it is used as filtration and adsorption membrane. Adsorption membranes have applications in the removal of various organic/inorganic and heavy metal ions from wastewater, while filtration membranes can be used for the separation of various bacteria, viruses, and oil–water emulsion. Due to the advancement of technology, the nanofibrous membranes can be easily modified by researchers in terms of precision for using the electrospinning method and efficiency for the removal of various pollutants. Amazing nanomaterials and surface modification techniques have been realized for developing novel nanofibrous membranes with high surface area, porous structure, high tensile strength, huge stiffness, comprehensive flexibility, and sustainability. The surface modification also reduces its fouling nature and overcomes the various limitations while used as a device for wastewater treatment application. In this review, an attempt has been made to review different polymeric electrospun nanofibrous membranes utilized in water purification applications. However, the surface modification of electrospun nanofibrous membranes by using pre and posttreatment techniques by embedding and decorating various nanostructures based on noble metals (Ag, Au, and Pt), carbon nanostructures (CNT, GO, MWCNT), and metal oxide nanoparticles onto the native membranes are also explored. The utilization of various modified nanofibrous membranes for the removal of different pollutants from wastewater is also highlighted in this chapter. It may be concluded that some details about various nanofibrous membranes with good removal efficiencies are available in the literature. However, systematic research work is required for the development of nanofibrous membranes with improved properties for wastewater treatment on an industrial scale with recyclability and reusability.

References

1. Singh J, Dhaliwal AS (2020) Plasmon-induced photocatalytic degradation of methylene blue dye using biosynthesized silver nanoparticles as photocatalyst. Environ Technol (UK) 41:1520
2. Barakat MA (2011) New trends in removing heavy metals from industrial wastewater. Arab J Chem 4:361
3. Sharma P, Kaur R, Baskar C, Chung WJ (2010) Removal of methylene blue from aqueous waste using rice husk and rice husk ash. Desalination 259:249
4. Zhu MX, Lee L, Wang HH, Wang Z (2007) Removal of an anionic dye by adsorption/precipitation processes using alkaline white mud. J Hazard Mater 149:735
5. Singh J, Dhaliwal AS (2021) Effective Removal of Methylene Blue Dye Using Silver Nanoparticles Containing Grafted Polymer of Guar Gum/Acrylic Acid as Novel Adsorbent. J Polym Environ 29:71
6. Pooresmaeil M, Mansoori Y, Mirzaeinejad M, Khodayari A (2018) Efficient removal of methylene blue by novel magnetic hydrogel nanocomposites of poly(acrylic acid). Adv Polym Technol 37:262
7. Liu YX, Liao ZY, Wu XY, Zhao CJ, Lei YX, Bin Ji D (2015) Electrochemical degradation of methylene blue using electrodes of stainless steel net coated with single-walled carbon nanotubes, desalin. Water Treat 54:2757
8. Shannon MA, Bohn PW, Elimelech M, Georgiadis JG, Mariñas BJ, Mayes AM (2008) Science and technology for water purification in the coming decades. Nature 452:301
9. Tijing LD, Woo YC, Yao M, Ren J, Shon HK (2017) 1.16 Electrospinning for membrane 1329 fabrication: strategies and applications. In: Drioli E, Giorno L, Fontananova (eds) Compr Membr Sci Eng, 2nd edn. Elsevier, Oxford, 418
10. Wang K, Ma Q, Wang SD, Liu H, Zhang SZ, Bao W, Zhang KQ, Ling LZ (2016) Electrospinning of silver nanoparticles loaded highly porous cellulose acetate nanofibrous membrane for treatment of dye wastewater. Appl Phys A Mater Sci Process 122:1
11. Crini G, Lichtfouse E (2019) Advantages and disadvantages of techniques used for wastewater treatment. Environ Chem Lett 17:145

12. Chakraborty S, Rusli H, Nath A, Sikder J, Bhattacharjee C, Curcio S, Drioli E (2016) Immobilized biocatalytic process development and potential application in membrane separation: a review. Crit Rev Biotechnol 36:43
13. Nawaz MS, Ahsan M (2014) Comparison of physico-chemical, advanced oxidation and biological techniques for the textile wastewater treatment. Alexandria Eng J 53:717
14. Singh J, Dhaliwal AS (2019) Novel green synthesis and characterization of the antioxidant activity of silver nanoparticles prepared from nepeta leucophylla root extract. Anal Lett 52:213
15. Daraee H, Eatemadi A, Abbasi E, Aval SF, Kouhi M, Akbarzadeh A (2016) Application of gold nanoparticles in biomedical and drug delivery. Artif Cells Nanomed Biotechnol 44:410
16. Huang ZM, Zhang YZ, Kotaki M, Ramakrishna S (2003) A review on polymer nanofibers by electrospinning and their applications in nanocomposites. Compos Sci Technol 63:2223
17. Jung JW, Lee CL, Yu S, Kim ID (2016) Electrospun nanofibers as a platform for advanced secondary batteries: a comprehensive review. J Mater Chem A 4:703
18. Zhu J, Bahramian Q, Gibson P, Schreuder-Gibson H, Sun G (2012) Chemical and biological decontamination functions of nanofibrous membranes. J Mater Chem 22:8532
19. Nagata S, Atkinson GM, Pestov D, Tepper GC, McLeskey JT (2013) Electrospun polymer-fiber solar cell. Adv Mater Sci Eng 2013
20. Nemati S, Jeong Kim S, Shin YM, Shin H (2019) Current progress in application of polymeric nanofibers to tissue engineering. Nano Converg 6
21. Ambekar RS, Kandasubramanian B (2019) Advancements in nanofibers for wound dressing: a review. Eur Polym J 117:304
22. Singh J, Dhaliwal AS (2018) Synthesis, characterization and swelling behavior of silver nanoparticles containing superabsorbent based on grafted copolymer of polyacrylic acid/guar gum. Vacuum 157:51
23. Ibrahim YS, Hussein EA, Zagho MM, Abdo GG, Elzatahry AA (2019) Melt electrospinning designs for nanofiber fabrication for different applications. Int J Mol Sci 20
24. Soltani S, Khanian N, Choong TSY, Rashid U (2020) Recent progress in the design and synthesis of nanofibers with diverse synthetic methodologies: characterization and potential applications. New J Chem 44:9581
25. Nakata K, Fujii K, Ohkoshi Y, Gotoh Y, Nagura M, Numata M, Kamiyama M (2007) Poly(Ethylene Terephthalate) nanofibers made by sea-island-type conjugated melt spinning and laser-heated flow drawing. Macromol Rapid Commun 28:792
26. Kenry, Lim CT (2017) Nanofiber technology: current status and emerging developments. Prog Polym Sci 70(1)
27. Benavides RE, Jana SC, Reneker DH (2012) Nanofibers from scalable gas jet process. ACS Macro Lett 1:1032
28. Meng J, Liu X, Niu C, Pang Q, Li J, Liu F, Liu Z, Mai L (2020) Advances in metal-organic framework coatings: versatile synthesis and broad applications. Chem Soc Rev 49:3142
29. Taylor G (1964) Disintegration of water drops in an electric field. Proc R Soc London Ser Math Phys Sci 280:383
30. Peijs T (2017) Electrospun polymer nanofibers and their composites. Compr Compos Mater II 162
31. Ahmed FE, Lalia BS, Hashaikeh R (2015) A review on electrospinning for membrane fabrication: challenges and applications. Desalination 356:15
32. Tijing LD, Choi JS, Lee S, Kim SH, Shon HK (2014) Recent progress of membrane distillation using electrospun nanofibrous membrane. J Memb Sci 453:435
33. Casper CL, Stephens JS, Tassi NG, Chase DB, Rabolt JF (2004) Controlling surface morphology of electrospun polystyrene fibers: effect of humidity and molecular weight in the electrospinning process. Macromolecules 37:573
34. Vasireddi R, Kruse J, Vakili M, Kulkarni S, Keller TF, Monteiro DCF, Trebbin M (2019) Solution blow spinning of polymer/nanocomposite micro-/nanofibers with tunable diameters and morphologies using a gas dynamic virtual nozzle. Sci Rep 9
35. Mazoochi T, Hamadanian M, Ahmadi M, Jabbari V (2012) Investigation on the morphological characteristics of nanofiberous membrane as electrospun in the different processing parameters. Int J Ind Chem. 3:1

36. Talwar S, Krishnan AS, Hinestroza JP, Pourdeyhimi B, Khan SA (2010) Electrospun nanofibers with associative polymer-surfactant systems. Macromolecules 43:7650
37. Ki CS, Baek DH, Gang KD, Lee KH, Um IC, Park YH (2005) Characterization of gelatin nanofiber prepared from gelatin-formic acid solution. Polymer (Guildf) 46:5094
38. Rnjak-Kovacina J, Weiss AS (2011) Increasing the pore size of electrospun scaffolds. Tissue Eng—Part B Rev. 17:365
39. Zahmatkeshan M, Adel M, Bahrami S, Esmaeili F, Rezayat SM, Saeedi Y, Mehravi B, Jameie SB, Ashtari K (2018) Polymer based nanofibers: preparation, fabrication, and applications. Handb. Nanofibers 1
40. Jarusuwannapoom T, Hongrojjanawiwat W, Jitjaicham S, Wannatong L, Nithitanakul M, Pattamaprom C, Koombhongse P, Rangkupan R, Supaphol P (2005) Effect of solvents on electro-spinnability of polystyrene solutions and morphological appearance of resulting electrospun polystyrene fibers. Eur Polym J 41:409
41. Tao J, Shivkumar S (2007) Molecular weight dependent structural regimes during the electrospinning of PVA. Mater Lett 61:2325
42. Rivas L, Sanchez-Cortes S, García-Ramos JV, Morcillo G (2001) Growth of Silver colloidal particles obtained by citrate reduction to increase the Raman enhancement factor. Langmuir 17:574
43. Lee KJ, Jun BH, Kim TH, Joung J (2006) Direct synthesis and inkjetting of silver nanocrystals toward printed electronics. Nanotechnology 17:2424
44. Sun Y, Yin Y, Mayers BT, Herricks T, Xia Y (2002) Uniform Silver Nanowires synthesis by reducing $AgNO_3$ with ethylene glycol in the presence of seeds and poly(vinyl pyrrolidone). Chem Mater 14:4736
45. Zhang JP, Chen P, Sun CH, Hu XJ (2004) Sonochemical synthesis of colloidal silver catalysts for reduction of complexing silver in DTR system. Appl Catal A Gen 266:49
46. Darmanin T, Nativo P, Gilliland D, Ceccone G, Pascual C, De Berardis B, Guittard F, Rossi F (2012) Microwave-assisted synthesis of silver nanoprisms/nanoplates using a "Modified Polyol Process." Colloids Surfaces A Physicochem. Eng Asp 395:145
47. Kalimuthu P, John SA (2010) Studies on ligand exchange reaction of functionalized mercaptothiadiazole compounds onto citrate capped gold nanoparticles. Mater Chem Phys 122:380
48. Patel AS, Juneja S, Kanaujia PK, Maurya V, Prakash GV, Chakraborti A, Bhattacharya J (2018) Gold nanoflowers as efficient hosts for sers based sensing and bio-imaging. Nano-Struct Nano-Objects 16:329
49. Sonia K, Kukreti S, Kaushik M (2018) Exploring the DNA damaging potential of chitosan and citrate-reduced gold nanoparticles: physicochemical approach. Int J Biol Macromol 115:801
50. Philip D (2008) Synthesis and spectroscopic characterization of gold nanoparticles. Spectrochim Acta—Part A Mol Biomol Spectrosc 71:80
51. Radoń A, Łukowiec D (2018) Silver nanoparticles synthesized by UV-irradiation method using Chloramine T as modifier: structure, formation mechanism and catalytic activity. CrystEngComm 20:7130
52. Manikprabhu D, Lingappa K (2013) Microwave assisted rapid and green synthesis of silver nanoparticles using a pigment produced by streptomyces coelicolor Klmp33. Bioinorg Chem Appl 2013
53. Flores-Rojas GG, López-Saucedo F, Bucio E (2020) Gamma-irradiation applied in the synthesis of metallic and organic nanoparticles: a short review. Radiat Phys Chem 169
54. Wei L, Xiao L, He Y (2011) Synthesis of water soluble silver-nanoparticle-embedded polymer nanofibers with poly(2-ethyl-2-oxazoline) by a straightforward polyol process. J Mater Res 26:1614
55. Jin WJ, Lee HK, Jeong RH, Park WH, Youk JH (2005) Preparation of polymer nanofibers containing silver nanoparticles by using poly(n-vinylpyrrolidone). Macromol Rapid Commun 26:1903
56. Wang Y, Li Y, Yang S, Zhang G, An D, Wang C, Yang Q, Chen X, Jing X, Wei Y (2006) A convenient route to polyvinyl pyrrolidone/silver nanocomposite by electrospinning. Nanotechnology 17:3304

57. Aadil KR, Mussatto SI, Jha H (2018) synthesis and characterization of silver nanoparticles loaded poly(vinyl alcohol)-lignin electrospun nanofibers and their antimicrobial activity. Int J Biol Macromol 120:763

58. Pant B, Park M, Park SJ (2019) One-step synthesis of silver nanoparticles embedded polyurethane nano-fiber/net structured membrane as an effective antibacterial medium. Polymers (Basel) 11

59. Nguyen TH, Lee KH, Lee BT (2010) Fabrication of Ag nanoparticles dispersed in PVA nanowire mats by microwave irradiation and electro-spinning. Mater Sci Eng C 30:944

60. Dong G, Xiao X, Liu X, Qian B, Liao Y, Wang C, Chen D, Qiu J (2009) Functional Ag porous films prepared by electrospinning. Appl Surf Sci 255:7623

61. Tijing LD, Ruelo MTG, Amarjargal A, Pant HR, Park CH, Kim CS (2012) One-step fabrication of antibacterial (silver nanoparticles/poly(ethylene oxide))—polyurethane bicomponent hybrid nanofibrous mat by dual-spinneret electrospinning. Mater Chem Phys 134:557

62. Jiang C, Nie J, Ma G (2016) A polymer/metal core-shell nanofiber membrane by electrospinning with an electric field, and its application for catalyst support. RSC Adv 6:22996

63. Lakshman LR, Shalumon KT, Nair SV, Jayakumar R, Nair SV (2010) Preparation of silver nanoparticles incorporated electrospun polyurethane nano-fibrous mat for wound dressing. J Macromol Sci Part A Pure Appl Chem 47:1012

64. Dubey P, Bhushan B, Sachdev A, Matai I, Uday Kumar S, Gopinath P (2015) Silver-nanoparticle-incorporated composite nanofibers for potential wound-dressing applications. J Appl Polym Sci 132

65. Kowsalya E, MosaChristas K, Balashanmugam P, Tamil Selvi A, RaniIa JC (2019) Biocompatible silver nanoparticles/poly(vinyl alcohol) electrospun nanofibers for potential antimicrobial food packaging applications. Food Packag Shelf Life 21

66. Raj Kumar S, Gopinath P (2016) Dual applications of silver nanoparticles incorporated functionalized MWCNTs grafted surface modified PAN nanofibrous membrane for water purification. RSC Adv 6:109241

67. Shin JU, Gwon J, Lee SY, Yoo HS (2018) Silver-incorporated nanocellulose fibers for antibacterial hydrogels. ACS Omega 3:16150

68. Ravichandran R, Sridhar R, Venugopal JR, Sundarrajan S, Mukherjee S, Ramakrishna S (2014) Gold nanoparticle loaded hybrid nanofibers for cardiogenic differentiation of stem cells for infarcted myocardium regeneration. Macromol Biosci 14:515

69. Duan Z, Huang Y, Zhang D, Chen S (2019) Electrospinning fabricating Au/TiO$_2$ network-like nanofibers as visible light activated photocatalyst. Sci Rep 9

70. Wang P, Zhang L, Xia Y, Tong L, Xu X, Ying Y (2012) Polymer nanofibers embedded with aligned gold nanorods: a new platform for plasmonic studies and optical sensing. Nano Lett 12:3145

71. Jung D, Minami I, Patel S, Lee J, Jiang B, Yuan Q, Li L, Kobayashi S, Chen Y, Lee KB, Nakatsuji N (2012) Incorporation of functionalized gold nanoparticles into nanofibers for enhanced attachment and differentiation of mammalian cells. J Nanobiotechnol 10

72. Zhang H, Hu Z, Ma Z, Gecevičius M, Dong G, Zhou S, Qiu J (2016) Anisotropically enhanced nonlinear optical properties of ensembles of gold nanorods electrospun in polymer nanofiber film. ACS Appl Mater Interf 8:2048

73. Tanahashi I, Kanno H (2000) Temperature dependence of photoinduced au particle formation in polyvinyl alcohol films. Appl Phys Lett 77:3358

74. Bai J, Li Y, Yang S, Du J, Wang S, Zheng J, Wang Y, Yang Q, Chen X, Jing X (2007) A simple and effective route for the preparation of poly(vinylalcohol) (PVA) nanofibers containing gold nanoparticles by electrospinning method. Solid State Commun 141:292

75. Chen C, Tang Y, Vlahovic B, Yan F (2017) Electrospun polymer nanofibers decorated with noble metal nanoparticles for chemical sensing. Nanoscale Res Lett 12

76. Clinton Ifegwu O, Anyakora C, Torto N (2015) Nylon 6–Gold nanoparticle composite fibers for colorimetric detection of urinary 1-Hydroxypyrene. J Appl Spectrosc 82

77. Tang X, Yan X (2017) Dip-coating for fibrous materials: mechanism, methods and applications. J Sol-Gel Sci Technol 81:378
78. Xue J, Wu T, Dai Y, Xia Y (2019) Electrospinning and electrospun nanofibers: methods, materials, and applications. Chem Rev 119:5298
79. Celebioglu A, Ranjith KS, Eren H, Biyikli N, Uyar T (2017) Surface decoration of pt nanoparticles via ALD with TiO_2 protective layer on polymeric nanofibers as flexible and reusable heterogeneous nanocatalysts. Sci Rep 7
80. He M, Chen M, Dou Y, Ding J, Yue H, Yin G, Chen X, Cui Y (2020) Electrospun silver nanoparticles-embedded feather keratin/poly(vinyl alcohol)/poly(ethylene oxide) antibacterial composite nanofibers. Polymers (Basel) 12
81. Yu DG, Branford-White C, Bligh SWA, White K, Chatterton NP, Zhu LM (2011) Improving polymer nanofiber quality using a modified co-axial electrospinning process. Macromol Rapid Commun 32:744
82. Yu DG, Branford-White C, White K, Chatterton NP, Zhu LM, Huang LY, Wang B (2011) A modified coaxial electrospinning for preparing fibers from a high concentration polymer solution. Express Polym Lett 5:732
83. Loscertales IG, Barrero A, Márquez M, Spretz R, Velarde-Ortiz R, Larsen G (2004) electrically forced coaxial nanojets for one-step hollow nanofiber design. J Am Chem Soc 126:5376
84. Kim HM, Park JH, Lee SK (2019) Fiber optic sensor based on zno nanowires decorated by au nanoparticles for improved plasmonic biosensor. Sci Rep 9
85. Li C, Wang ZH, Yu DG, Williams GR (2014) Tunable biphasic drug release from ethyl cellulose nanofibers fabricated using a modified coaxial electrospinning process. Nanoscale Res Lett 9:1
86. Yu DG, Zhou J, Chatterton NP, Li Y, Huang J, Wang X (2012) Polyacrylonitrile nanofibers coated with silver nanoparticles using a modified coaxial electrospinning process. Int J Nanomed 7:5725
87. Wang J, Bin Yao H, He D, Zhang CL, Yu SH (2012) Facile fabrication of gold nanoparticles-poly(vinyl alcohol) electrospun water-stable nanofibrous mats: efficient substrate materials for biosensors. ACS Appl Mater Interf 4:1963
88. Dong H, Wang D, Sun G, Hinestroza JP (2008) Assembly of metal nanoparticles on electrospun nylon 6 nanofibers by control of interfacial hydrogen-bonding interactions. Chem Mater 20:6627
89. Son HY, Ryu JH, Lee H, Nam YS (2013) Bioinspired templating synthesis of metal-polymer hybrid nanostructures within 3D electrospun nanofibers. ACS Appl Mater Interf 5:6381
90. Chen D, Liu T, Zhou X, Tjiu WC, Hou H (2009) Electrospinning fabrication of high strength and toughness polyimide nanofiber membranes containing multiwalled carbon nanotubes. J Phys Chem B 113:9741
91. Xiao S, Shen M, Guo R, Huang Q, Wang S, Shi X (2010) Fabrication of multiwalled carbon nanotube-reinforced electrospun polymer nanofibers containing zero-valent iron nanoparticles for environmental applications. J Mater Chem 20:5700
92. Weng B, Xu F, Salinas A, Lozano K (2014) Mass production of carbon nanotube reinforced poly(methyl methacrylate) nonwoven nanofiber mats. Carbon N. Y. 75:217
93. Janjua MRSA (2019) Synthesis of CO_3O_4 nano aggregates by co-precipitation method and its catalytic and fuel additive applications. Open Chem 17:865
94. Unni M, Uhl AM, Savliwala S, Savitzky BH, Dhavalikar R, Garraud N, Arnold DP, Kourkoutis LF, Andrew JS, Rinaldi C (2017) Thermal decomposition synthesis of iron oxide nanoparticles with diminished magnetic dead layer by controlled addition of oxygen. ACS Nano 11:2284
95. Adschiri T, Hakuta Y, Arai K (2000) Hydrothermal synthesis of metal oxide fine particles at supercritical conditions. Ind Eng Chem Res 39:4901
96. Parashar M, Shukla VK, Singh R (2020) Metal oxides nanoparticles via sol-gel method: a review on synthesis, characterization and applications. J Mater Sci Mater Electron 31:3729
97. Hinman JJ, Suslick KS (2017) Nanostructured materials synthesis using ultrasound. Top Curr Chem 375

98. Annu AA, Ahmed S (2018) Green synthesis of metal, metal oxide nanoparticles, and their various applications. Handb Ecomater 1

99. Hernández-Hernández AA, Aguirre-Álvarez G, Cariño-Cortés R, Mendoza-Huizar LH, Jiménez-Alvarado R (2020) Iron oxide nanoparticles: synthesis, functionalization, and applications in diagnosis and treatment of cancer. Chem Pap 74:3809

100. Yasin AS, Obaid M, Mohamed IMA, Yousef A, Barakat NAM (2017) ZrO_2 nanofibers/activated carbon composite as a novel and effective electrode material for the enhancement of capacitive deionization performance. RSC Adv 7:4616

101. Prahsarn C, Klinsukhon W, Roungpaisan N (2011) Electrospinning of PAN/DMF/H_2O containing TiO_2 and photocatalytic activity of their webs. Mater Lett 65:2498

102. Taha AA, Na Wu Y, Wang H, Li F (2012) Preparation and application of functionalized cellulose acetate/silica composite nanofibrous membrane via electrospinning for Cr(VI) ion removal from aqueous solution. J Environ Manage 112:10

103. Yasin AS, Mohamed IMA, Park CH, Kim CS (2018) Design of novel electrode for capacitive deionization using electrospun composite titania/zirconia nanofibers doped-activated carbon. Mater Lett 213:62

104. Yasin SA, Abbas JA, Saeed IA, Ahmed IH (2020) The application of green synthesis of metal oxide nanoparticles embedded in polyethylene terephthalate nanofibers in the study of the photocatalytic degradation of methylene blue. Polym Bull 77:3473

105. Horzum N, Muñoz-Espí R, Glasser G, Demir MM, Landfester K, Crespy D (2012) Hierarchically structured metal oxide/silica nanofibers by colloid electrospinning. ACS Appl Mater Interfaces 4:6338

106. Drew C, Liu X, Ziegler D, Wang X, Bruno FF, Whitten J, Samuelson LA, Kumar J (2003) Metal oxide-coated polymer nanofibers. Nano Lett 3:143

107. Moallemian H, Kavosh M, Molaei H, Mehraniya H, Dehdashti ME (2014) Synthesis and characterization of nanoparticles Fe_2O_3 doped on nano fibers ZnO by electrospinning method. Synth React Inorganic Met Nano-Metal Chem 44:995

108. Lee JS, Kwon OS, Park SJ, Park EY, You SA, Yoon H, Jang J (2011) Fabrication of ultrafine metal-oxide-decorated carbon nanofibers for DMMP sensor application. ACS Nano 5:7992

109. Ding B, Ogawa T, Kim J, Fujimoto K, Shiratori S (2008) Fabrication of a super-hydrophobic nanofibrous zinc oxide film surface by electrospinning. Thin Solid Films 516:2495

110. Hang AT, Tae B, Park JS (2010) Non-woven mats of poly(vinyl alcohol)/chitosan blends containing silver nanoparticles: fabrication and characterization. Carbohydr Polym 82:472

111. Filip D, Macocinschi D, Paslaru E, Munteanu BS, Dumitriu RP, Lungu M, Vasile C (2014) Polyurethane biocompatible silver bionanocomposites for biomedical applications. J Nanoparticle Res 16

112. Au HT, Pham LN, Vu THT, Park JS (2012) Fabrication of an antibacterial non-woven mat of a poly(lactic acid)/chitosan blend by electrospinning. Macromol Res 20:51

113. Kim KI, Kim DA, Patel KD, Shin US, Kim HW, Lee JH, Lee HH (2019) Carbon nanotube incorporation in PMMA to prevent microbial adhesion. Sci Rep 9

114. Li J, Cao J, Wei Z, Yang M, Yin W, Yu K, Yao Y, Lv H, He X, Leng J (2014) Electrospun silica/nafion hybrid products: mechanical property improvement, wettability tuning and periodic structure adjustment. J Mater Chem A 2:16569

115. Jang W, Yun J, Jeon K, Byun H (2015) PVdF/graphene oxide hybrid membranes via electrospinning for water treatment applications. RSC Adv 5:46711

116. Choi SS, Lee YS, Joo CW, Lee SG, Park JK, Han KS (2004) Electrospun PVDF nanofiber web as polymer electrolyte or separator. Electrochim Acta 50:339

117. Kaur S, Ma Z, Gopal R, Singh G, Ramakrishna S, Matsuura T (2007) Plasma-induced graft copolymerization of poly(methacrylic acid) on electrospun poly(vinylidene fluoride) nanofiber membrane. Langmuir 23:13085

118. Kim GM, Lach R, Michler GH, Pötschke P, Albrecht K (2006) Relationships between phase morphology and deformation mechanisms in polymer nanocomposite nanofibres prepared by an electrospinning process. Nanotechnology 17:963

119. Wang X, Zhang K, Zhu M, Hsiao BS, Chu B (2008) Enhanced mechanical performance of self-bundled electrospun fiber yarns via post-treatments. Macromol Rapid Commun 29:826
120. Qu X, Alvarez PJJ, Li Q (2013) Applications of Nanotechnology in water and wastewater treatment. Water Res 47:3931
121. Elimelech M, Phillip WA (2011) The future of seawater desalination: energy, technology, and the environment. Science333(80–):712
122. Montgomery MA, Elimelech M (2007) Water and sanitation in developing countries: including health in the equation—millions suffer from preventable illnesses and die every year. Environ Sci Technol 41:17
123. Nasreen SAAN, Sundarrajan S, Nizar SAS, Balamurugan R, Ramakrishna S (2013) Advancement in electrospun nanofibrous membranes modification and their application in water treatment. Membranes (Basel) 3:266
124. Shen K, Cheng C, Zhang T, Wang X (2019) High performance polyamide composite nanofiltration membranes via reverse interfacial polymerization with the synergistic interaction of gelatin interlayer and trimesoyl chloride. J Memb Sci 588
125. Zhou T, Li J, Guo X, Yao Y, Zhu P, Xiang R (2019) Freestanding PTFE electrospun tubular membrane for reverse osmosis brine concentration by vacuum membrane distillation. Desalin Water Treat 165:63
126. Hu X, Yu Y, Zhou J, Wang Y, Liang J, Zhang X, Chang Q, Song L (2015) The improved oil/water separation performance of graphene oxide modified Al_2O_3 microfiltration membrane. J. Memb. Sci. 476:200
127. Dobosz KM, Kuo-Leblanc CA, Martin TJ, Schiffman JD (2017) Ultrafiltration membranes enhanced with electrospun nanofibers exhibit improved flux and fouling resistance. Ind Eng Chem Res 56:5724
128. Park MJ, Gonzales RR, Abdel-Wahab A, Phuntsho S, Shon HK (2018) Hydrophilic polyvinyl alcohol coating on hydrophobic electrospun nanofiber membrane for high performance thin film composite forward osmosis membrane. Desalination 426:50
129. Ma Z, Kotaki M, Ramakrishna S (2005) Electrospun cellulose nanofiber as affinity membrane. J Memb Sci 265:115
130. Gopal R, Kaur S, Feng CY, Chan C, Ramakrishna S, Tabe S, Matsuura T (2007) Electrospun nanofibrous polysulfone membranes as pre-filters: particulate removal. J Memb Sci 289:210
131. Gopal R, Kaur S, Ma Z, Chan C, Ramakrishna S, Matsuura T (2006) Electrospun nanofibrous filtration membrane. J Memb Sci 281:581
132. Kaur S, Rana D, Matsuura T, Sundarrajan S, Ramakrishna S (2012) Preparation and characterization of surface modified electrospun membranes for higher filtration flux. J Memb Sci 390–391:235
133. Moradi G, Zinadini S (2020) A high flux graphene oxide nanoparticles embedded in pan nanofiber microfiltration membrane for water treatment applications with improved antifouling performance. Iran Polym J (English Ed) 29:827
134. Aussawasathien D, Teerawattananon C, Vongachariya A (2008) Separation of micron to submicron particles from water: electrospun nylon-6 nanofibrous membranes as pre-filters. J Memb Sci 315:11
135. Li M, Wang D, Xiao R, Sun G, Zhao Q, Li H (2013) A novel high flux poly(trimethylene terephthalate) nanofiber membrane for microfiltration media. Sep Purif Technol 116:199
136. Fauzi A, Hapidin DA, Munir MM, Iskandar F, Khairurrijal K (2020) A superhydrophilic bilayer structure of a nylon 6 nanofiber/cellulose membrane and its characterization as potential water filtration media. RSC Adv 10:17205
137. Wang X, Fang D, Yoon K, Hsiao BS, Chu B (2006) High performance ultrafiltration composite membranes based on poly(vinyl alcohol) hydrogel coating on crosslinked nanofibrous poly(vinyl alcohol) scaffold. J. Memb. Sci. 278:261
138. Yoon B, Lee S (2011) Designing waterproof breathable materials based on electrospun nanofibers and assessing the performance characteristics. Fibers Polym. 12:57
139. Fan Z, Wang Z, Duan M, Wang J, Wang S (2008) Preparation and characterization of polyaniline/polysulfone nanocomposite ultrafiltration membrane. J Memb Sci 310:402

140. Yang Z, Zhou Y, Feng Z, Rui X, Zhang T, Zhang Z (2019) A review on reverse osmosis and nanofiltration membranes for water purification. Polymers (Basel) 11

141. Tian M, Qiu C, Liao Y, Chou S, Wang R (2013) Preparation of polyamide thin film composite forward osmosis membranes using electrospun polyvinylidene fluoride (PVDF) nanofibers as substrates. Sep Purif Technol 118:727

142. Bui NN, Lind ML, Hoek EMV, McCutcheon JR (2011) Electrospun nanofiber supported thin film composite membranes for engineered osmosis. J Memb Sci 385–386:10

143. Botes M, Cloete TE (2010) The potential of nanofibers and nanobiocides in water purification. Crit Rev Microbiol 36:68

144. Abdel-Fatah MA (2018) Nanofiltration systems and applications in wastewater treatment: review article. Ain Shams Eng J 9:3077

145. Ritcharoen W, Supaphol P, Pavasant P (2008) Development of polyelectrolyte multilayer-coated electrospun cellulose acetate fiber mat as composite membranes. Eur Polym J 44:3963

146. Kaur S, Barhate R, Sundarrajan S, Matsuura T, Ramakrishna S (2011) Hot pressing of electro-spun membrane composite and its influence on separation performance on thin film composite nanofiltration membrane. Desalination 279:201

147. Yung L, Ma H, Wang X, Yoon K, Wang R, Hsiao BS, Chu B (2010) Fabrication of thin-film nanofibrous composite membranes by interfacial polymerization using ionic liquids as additives. J Memb Sci 365:52

148. Liu W, Herrmann AK, Bigall NC, Rodriguez P, Wen D, Oezaslan M, Schmidt TJ, Gaponik N, Eychmüller A (2015) Noble metal aerogels-synthesis, characterization, and application as electrocatalysts. Acc Chem Res 48:154

149. Medici S, Peana M, Nurchi VM, Lachowicz JI, Crisponi G, Zoroddu MA (2015) Noble metals in medicine: latest advances. Coord Chem Rev 284:329

150. Taurozzi JS, Arul H, Bosak VZ, Burban AF, Voice TC, Bruening ML, Tarabara VV (2008) Effect of filler incorporation route on the properties of polysulfone-silver nanocomposite membranes of different porosities. J. Memb. Sci. 325:58

151. Gunawan P, Guan C, Song X, Zhang Q, Leong SSJ, Tang C, Chen Y, Chan-Park MB, Chang MW, Wang K, Xu R (2011) Hollow fiber membrane decorated with ag/mwnts: toward effective water disinfection and biofouling control. ACS Nano 5:10033

152. Zodrow K, Brunet L, Mahendra S, Li D, Zhang A, Li Q, Alvarez PJJ (2009) Polysulfone ultrafiltration membranes impregnated with silver nanoparticles show improved biofouling resistance and virus removal. Water Res 43:715

153. Dasari A, Quirós J, Herrero B, Boltes K, García-Calvo E, Rosal R (2012) Antifouling membranes prepared by electrospinning polylactic acid containing biocidal nanoparticles. J Memb Sci 405–406:134

154. Sui C, Wang C, Wang Z, Xu Y, Gong E, Cheng T, Zhou G (2017) Different coating on elec-trospun nanofiber via layer-by-layer self-assembly for their photocatalytic activities. Colloids Surf A Physicochem Eng Asp 529:425

155. Pant HR, Pandeya DR, Nam KT, Baek W, Hong ST, Kim HY (2011) Photocatalytic and antibacterial properties of a TiO_2/Nylon-6 electrospun nanocomposite mat containing silver nanoparticles. J Hazard Mater 189: 465

156. Chen L, Peng X (2017) Silver nanoparticle decorated cellulose nanofibrous membrane with good antibacterial ability and high water permeability. Appl Mater Today 9:130

157. Sambhy V, MacBride MM, Peterson BR, Sen A (2006) Silver bromide nanoparticle/polymer composites: dual action tunable antimicrobial materials. J Am Chem Soc 128:9798

158. Lv Y, Liu H, Wang Z, Liu S, Hao L, Sang Y, Liu D, Wang J, Boughton RI (2009) Silver nanoparticle-decorated porous ceramic composite for water treatment. J Memb Sci 331:50

159. Dotzauer DM, Dai J, Sun L, Bruening ML (2006) Catalytic membranes prepared using layer-by-layer adsorption of polyelectrolyte/metal nanoparticle films in porous supports. Nano Lett 6:2268

160. Huang X (2009) Fabrication and properties of carbon fibers. Materials (Basel) 2:2369

161. Wangxi Z, Jie L, Gang W (2003) Evolution of structure and properties of pan precursors during their conversion to carbon fibers. Carbon NY 41:2805

162. Inagaki M, Yang Y, Kang F (2012) Carbon nanofibers prepared via electrospinning. Adv Mater 24:2547

163. Ren T, Si Y, Yang J, Ding B, Yang X, Hong F, Yu J (2012) Polyacrylonitrile/Polybenzoxazine-Based Fe_3O_4@carbon nanofibers: hierarchical porous structure and magnetic adsorption property. J Mater Chem 22:15919

164. Park C, Engel ES, Crowe A, Gilbert TR, Rodriguez NM (2000) Use of carbon nanofibers in the removal of organic solvents from water. Langmuir 16:8050

165. Noamani S, Niroomand S, Rastgar M, Sadrzadeh M (2019) Carbon-based polymer nanocomposite membranes for oily wastewater treatment. NPJ Clean Water 2

166. Shim WG, Kim C, Lee JW, Yun JJ, Jeong YI, Moon H, Yang KS (2006) Adsorption characteristics of benzene on electrospun-derived porous carbon nanofibers. J Appl Polym Sci 102:2454

167. Cuervo MR, Asedegbega-Nieto E, Díaz E, Vega A, Ordóñez S, Castillejos-López E, Rodríguez-Ramos I (2008) Effect of carbon nanofiber functionalization on the adsorption properties of volatile organic compounds. J Chromatogr A 1188:264

168. Maitra T, Sharma S, Srivastava A, Cho YK, Madou M, Sharma A (2012) Improved graphitization and electrical conductivity of suspended carbon nanofibers derived from carbon nanotube/polyacrylonitrile composites by directed electrospinning. Carbon N. Y. 50:1753

169. Prilutsky S, Zussman E, Cohen Y (2008) The effect of embedded carbon nanotubes on the morphological evolution during the carbonization of poly(acrylonitrile) nanofibers. Nanotechnology 19

170. Singh G, Rana D, Matsuura T, Ramakrishna S, Narbaitz RM, Tabe S (2010) Removal of disinfection byproducts from water by carbonized electrospun nanofibrous membranes. Sep Purif Technol 74:202

171. Tai MH, Gao P, Tan BYL, Sun DD, Leckie JO (2014) Highly Efficient and flexible electrospun carbon-silica nanofibrous membrane for ultrafast gravity-driven oil-water separation. ACS Appl Mater Interf 6:9393

172. Teng M, Qiao J, Li F, Bera PK (2012) Electrospun mesoporous carbon nanofibers produced from phenolic resin and their use in the adsorption of large dye molecules. Carbon NY 50:2877

173. Faccini M, Borja G, Boerrigter M, Morillo Martín D, Martìnez Crespiera S, Vázquez-Campos S, Aubouy L, Amantia D (2015) Electrospun carbon nanofiber membranes for filtration of nanoparticles from water. J Nanomater 2015

174. Ji L, Lin Z, Medford AJ, Zhang X (2009) Porous carbon nanofibers from electrospun polyacrylonitrile/SiO_2 composites as an energy storage material. Carbon NY 47:3346

175. Kim C, Jeong Young Il, Ngoc BTN, Yang KS, Kojima M, Kim YA, Endo M, Lee JW (2007) Synthesis and characterization of porous carbon nanofibers with hollow cores through the thermal treatment of electrospun copolymeric nanofiber web. Small 3:91

176. Lee BS, Park KM, Yu WR, Youk JH (2012) An effective method for manufacturing hollow carbon nanofibers and microstructural analysis. Macromol Res 20:605

177. Liu H, Cao CY, Wei FF, Huang PP, Bin Sun Y, Jiang L, Song WG (2014) Flexible macroporous carbon nanofiber film with high oil adsorption capacity. J Mater Chem A 2:3557

178. Al-Anzi BS, Siang OC (2017) Recent developments of carbon based nanomaterials and membranes for oily wastewater treatment. RSC Adv 7:20981

179. Barsan MM, Ghica ME, Brett CMA (2015) Electrochemical sensors and biosensors based on redox polymer/carbon nanotube modified electrodes: a review. Anal Chim Acta 881:1

180. Meyyappan M (2016) Carbon nanotube-based chemical sensors. Small 12:2118

181. Sui X, Wagner HD (2009) Tough nanocomposites: the role of carbon nanotube type. Nano Lett 9:1423

182. Joly S, Kane R, Radzilowski L, Wang T, Wu A, Cohen RE, Thomas EL, Rubner MF (2000) Multilayer nanoreactors for metallic and semiconducting particles. Langmuir 16:1354

183. Ji J, Sui G, Yu Y, Liu Y, Lin Y, Du Z, Ryu S, Yang X (2009) Significant improvement of mechanical properties observed in highly aligned carbon-nanotube-reinforced nanofibers. J Phys Chem C 113:4779

184. Srivastava A, Srivastava ON, Talapatra S, Vajtai R, Ajayan PM (2004) Carbon nanotube filters. Nat Mater 3:610
185. Salipira KL, Mamba BB, Krause RW, Malefetse TJ, Durbach SH (2007) Carbon nanotubes and cyclodextrin polymers for removing organic pollutants from water. Environ Chem Lett 5:13
186. Wang J, Zhang P, Liang B, Liu Y, Xu T, Wang L, Cao B, Pan K (2016) Graphene oxide as an effective barrier on a porous nanofibrous membrane for water treatment. ACS Appl Mater Interf 8:6211
187. Wang C, Shao C, Liu Y, Li X (2009) Water—dichloromethane interface controlled synthesis of hierarchical rutile TiO_2 superstructures and their photocatalytic properties. Inorg Chem 48:1105
188. Zhang L, Wang W, Yang J, Chen Z, Zhang W, Zhou L, Liu S (2006) Sonochemical synthesis of nanocrystallite Bi_2O_3 as a visible-light-driven photocatalyst. Appl Catal A Gen 308:105
189. Li L, Chu Y, Liu Y, Dong L (2007) Template-free synthesis and photocatalytic properties of novel Fe_2O_3 hollow spheres. J Phys Chem C 111:2123
190. Smith YR, Kar A, Subramanian V (2009) Investigation of physicochemical parameters that influence photocatalytic degradation of methyl orange over TiO_2 nanotubes. Ind Eng Chem Res 48:10268
191. Guan H, Shao C, Chen B, Gong J, Yang X (2003) A novel method for making CuO superfine fibres via an electrospinning technique. Inorg Chem Commun 6:1409
192. Lee SS, Bai H, Liu Z, Sun DD (2013) Novel-structured electrospun TiO_2/CuO composite nanofibers for high efficient photocatalytic cogeneration of clean water and energy from dye wastewater. Water Res 47:4059
193. Zhao Y, He X, Li J, Gao X, Jia J (2012) Porous CuO/SnO_2 composite nanofibers fabricated by electrospinning and their H_2S sensing properties, sensors actuators. B Chem 165:82
194. Jang M, Cannon FS, Parette RB, Joung Yoon S, Chen W (2009) Combined hydrous ferric oxide and quaternary ammonium surfactant tailoring of granular activated carbon for concurrent arsenate and perchlorate removal. Water Res 43:3133
195. Mu J, Shao C, Guo Z, Zhang Z, Zhang M, Zhang P, Chen B, Liu Y (2011) High photocatalytic activity of ZnO-carbon nanofiber heteroarchitectures. ACS Appl Mater Interf 3:590
196. Li CJ, Li YJ, Wang JN, Cheng J (2013) PA6@FexOy nanofibrous membrane preparation and its strong Cr (VI)-removal performance. Chem Eng J 220:294
197. Wang Q, Gao D, Gao C, Wei Q, Cai Y, Xu J, Liu X, Xu Y (2012) Removal of a cationic dye by adsorption/photodegradation using electrospun PAN/O-MMT composite nanofibrous membranes coated with TiO_2. Int J Photoenergy 2012
198. Liu Z, Miao YE, Liu M, Ding Q, Tjiu WW, Cui X, Liu T (2014) Flexible polyaniline-coated TiO_2/SiO_2 nanofiber membranes with enhanced visible-light photocatalytic degradation performance. J Colloid Interf Sci 424:49
199. Joo Kim H, Raj Pant H, Hee Kim J, Jung Choi N, Sang Kim C (2014) Fabrication of multi-functional TiO_2-fly ash/polyurethane nanocomposite membrane via electrospinning. Ceram Int 40:3023
200. Taha AA, Qiao J, Li F, Zhang B (2012) Preparation and application of amino functionalized mesoporous nanofiber membrane via electrospinning for adsorption of Cr^{3+} from aqueous solution. J Environ Sci 24:610
201. Wu S, Li F, Wu Y, Xu R, Li G (2010) Preparation of novel poly(vinyl alcohol)/SiO_2 composite nanofiber membranes with mesostructure and their application for removal of Cu^{2+} from waste water. Chem Commun 46:1694
202. Dastbaz A, Keshtkar AR (2014) Adsorption of Th^{4+}, U^{6+}, Cd^{2+}, and Ni^{2+} from aqueous solution by a novel modified polyacrylonitrile composite nanofiber adsorbent prepared by electrospinning. Appl Surf Sci 293:336
203. Teng M, Li F, Zhang B, Taha AA (2011) Electrospun cyclodextrin-functionalized meso-porous polyvinyl alcohol/SiO_2 nanofiber membranes as a highly efficient adsorbent for indigo carmine dye. Colloids Surf A Physicochem Eng Asp 385:229

204. Hu Z, Ma Z, He X, Liao C, Li Y, Qiu J (2015) Preparation and characterization of flexible and thermally stable CuO nanocrystal-decorated SiO_2 nanofibers. J Sol-Gel Sci Technol 76:492
205. Afkhami A, Saber-Tehrani M, Bagheri H (2010) Simultaneous removal of heavy-metal ions in wastewater samples using nano-alumina modified with 2,4-dinitrophenylhydrazine. J Hazard Mater 181:836
206. Mahanta N, Valiyaveettil S (2013) Functionalized poly(vinyl alcohol) based nanofibers for the removal of arsenic from water. RSC Adv 3:2776
207. Teo WE, Ramakrishna S (2009) Electrospun nanofibers as a platform for multifunctional, hierarchically organized nanocomposite. Compos Sci Technol 69:1804
208. Patel S, Hota G (2016) Iron oxide nanoparticle-immobilized pan nanofibers: synthesis and adsorption studies. RSC Adv 6:15402
209. Han C, Ma Q, Yang Y, Yang M, Yu W, Dong X, Wang J, Liu G (2015) Electrospinning-derived $[C/Fe_3O_4]$@C coaxial nanocables with tuned magnetism, electrical conduction and highly efficient adsorption trifunctionality. J Mater Sci Mater Electron 26:8054
210. Horzum N, Demir MM, Nairat M, Shahwan T (2013) Chitosan fiber-supported zero-valent iron nanoparticles as a novel sorbent for sequestration of inorganic arsenic. RSC Adv 3:7828
211. Zhu HY, Jiang R, Xiao L (2010) Adsorption of an anionic azo dye by chitosan/kaolin/γ-Fe_2O_3 composites. Appl Clay Sci 48:522

Functionalized Natural Polymer-Based Electrospun Nanofiber

Yuanfang Cheng, Xiaoxiao Ma, Weiting Huang, and Yu Chen

Abstract Electrostatic spinning technology has already been proved to be a passable and effective method for manufacturing one-dimensional nanomaterials, which has received special interest of researchers. Electrospinning is currently the only technique that allows the fabrication of continuous fibers with diameters down to a few nanometers. In this chapter, nanofiber materials and electrospinning are summarized, especially the nanofiber prepared from natural polymer was emphasized. The basic process and forming mechanism of electrospinning nanofibers were theoretically explained, and the factors and device types affecting the morphology and state of the fibers were analyzed. Finally, the application of electrospinning nanofibers, especially the naturally derived nanofibers, in many aspects were summarized.

Keywords Electrospinning · Nanofiber · Natural polymer · Application

1 Introduction

Nanofiber material is a new kind of material. Nanofibers were initially defined with a diameter of less than 1000 nm. With the further improvement of science and technology, nanofibers with a diameter of 1–100 nm have been steadily put into use. It can be prepared from many kinds of materials. Synthetic nanofibers are made of nylon 6, nylon 66, ethylene terephthalate, acrylic acid and other polymers. Carbon nanofibers are carbonized at high temperatures in an inert atmosphere, which were made by removing the carbon element in polymer. It also combined the advantages of graphene and nanofibers. Have excellent mechanical properties, good electrical conductivity, specific surface area, porosity, etc. In the capacitor electrode, adsorption filtration, catalyst, etc., all have great potential. Ceramic nanofibers are made from salts of inorganic compounds, such as silicates, which are then fired to form ceramic nanofibers [1]. The present silicate nanofibers are nanofibers from natural mg-Aluminosilicate acicular silicate and layered silicate nanofibers from turgor rock.

Y. Cheng · X. Ma · W. Huang · Y. Chen (✉)
School of Material Science and Engineering, Beijing Institute of Technology, Beijing 100081, China
e-mail: cylsy@163.com

© The Author(s), under exclusive license to Springer Nature Switzerland AG 2021
S. K. Tiwari et al. (eds.), *Electrospun Nanofibers*, Springer Series on Polymer and Composite Materials, https://doi.org/10.1007/978-3-030-79979-3_11

The natural nanofiber materials are made of cellulose, chitosan, polylactose, and other natural polymers, which have excellent biocompatibility and degradability. Nanofibers composed of natural macromolecules (chitosan, gelatin, alginate, gelatin, chitin, etc.) are not only widely available and relatively cheap, but also can be modified and designed in various ways. Compared with petroleum-based polymer nanomaterials, natural polymer nanofibers not only have unique nanometer size effect, but also have unique advantages in biocompatibility, degradability, and environmental friendliness. In recent years, the preparation of polymer nanofibers has attracted much attention. This is because when the diameter of the fiber is reduced from micron (10–100 m) to nanometer (10–100 nm), the fiber will have some special properties, such as large specific surface area, super strong mechanical properties, etc. Electrospinning is a direct and relatively easy method to prepare nanofibers. Natural polymers are spun by conventional dry or wet processes. To get a finer diameter of natural polymer fibers, researchers began experimenting with electrospinning using natural polymers. However, because the natural polymers are mostly polyelectrolytes, it is more difficult to prepare the nanofibers of natural polymers by static spinning than to synthesize them. Therefore, in recent years, its application scope has expanded from traditional architecture, clothing, and decoration to new high-tech materials with high added value, such as transparent display panels, separation membranes, energy storage devices and catalytic carriers. Natural polymer fibers spun by electrospinning have many potential applications in medical fields such as trauma dressings, tissue engineered scaffolds and drug delivery systems.

2 Overview of Nanofiber Materials

2.1 Properties of Nanofiber Materials

Nanoscale materials have the following characteristics.

2.1.1 Size Effect

The surface area per unit volume of fiber increases, while the volume decreases, so that the reactivity and selectivity can be significantly improved, and the ultra-low consumption energy can be materialized. Toray's successful nanofiber nylon consists of more than 1.4 million nanoscale filaments [2]. The nanofibers have a very large surface area compared with conventional fibers, and give the nanofibers filaments their shape. For conventional nylon fibers, the absorptivity on the surface of the fiber is only one thousandth of that inside the fiber, that is, the absorptivity on the surface is negligible. However, the surface area of nylon nanofibers is 1000 times that of conventional polymers, and the surface absorption effect becomes very obvious.

2.1.2 Self-assemble of Supramolecules

The supramolecular nanofibers are arranged in a regular way and have a unified function through self-assemble. For example, through periodic placement and directional dynamic bond synthesis, polymers and other materials are arranged into larger bundles to form supramolecular nanofibers [3]. Especially for natural polymer materials, due to its rich functional groups, it is suitable to form nanofibers through self-assembly. Flexible polymer chains are assembled into supramolecular nanofibers collectively. The formation of these nanofibers leads to a coordinated, long-term polyvalent interaction between the chains that can improve the tensile strength or thermal stability of the material.

2.1.3 Recognition of Cells and Biomaterials

Nanofibers have unique structures to recognize and adhere to cells, which can well simulate extracellular matrix, and provide a site for cell adsorption. By combining cellular knowledge makes specific structure nanofibers. For example, hyaluronic acid is a kind of linear macromolecular mucopolysaccharide composed of glucuronic acid and N-acetylglucosamine disaccharide connected alternately and repeatedly. It has good biocompatibility, biodegradability and high viscoelasticity, and can specifically bind to some specific receptors on the cell surface. With its unique molecular structure and physical and chemical properties, hyaluronic acid has many important physiological functions in the body, such as lubricating joints, regulating vascular wall permeability, regulating proteins, promoting wound healing, etc. It has special water-holding function and can improve skin nutrition metabolism.

2.1.4 Hierarchical Effects

Nanofibers have a large specific surface area and high strength, but due to the impact of their huge specific surface area, they cause strong self-aggregation, which makes it difficult to disperse, stabilize and combine with other base materials [4].

3 Electrostatic Spinning Technology

3.1 Concepts of Electrospinning

Electrospinning refers to the process in which a polymer solution is used for spinning by deforming in the electrostatic field of a high-voltage electrostatic charged polymer solution or melt flow, and then by cooling and solidifying the polymer solution or melt to obtain the fibrous material. Electrospinning can be considered a variant of electrospray technology, both of which rely on the use of high pressure to spray liquids. The main differences between electrostatic spinning and electrostatic spray are the viscosity and viscoelasticity of the liquid involved and the behavior of the spray. Electrostatic spray is a spray method in which spray droplets are charged by a high-voltage electrostatic generator. In electrospinning, the spray can be kept in a continuous state to produce fibers instead of breaking down into droplets (used to form particles) as electrospinning does. The principle is to make the polymer solution or melt with thousands of volts of high-voltage static electricity, the charged polymer droplets under the action of electric field force in the capillary Taylor cone tip to accelerate the movement.

3.2   Influencing Factors of Electrospinning Nanofibers

The final morphology of electrospinning nanofibers is affected by some factors.

(1)   Polymer solution concentration. The higher the concentration of polymer solution is, the greater the viscosity is and the greater the external tension is. However, only with the increase of external tension can the droplet fragmentation be weakened after it is removed from the nozzle. Generally, the fiber diameter increases with the increase of concentration under other conditions.

(2)   Electric field intensity. With the increase of electric field intensity, the jet of polymer electrospinning liquid has greater external charge density and electrostatic repulsion. At the same time, the higher electric field intensity makes the jet get more acceleration. Both of these two factors can cause greater tensile stress of the jet and the fiber, resulting in a higher tensile strain rate, which is conducive to the preparation of finer fibers.

(3)   Spacing between capillary opening and collector. After being ejected through the capillary, the polymer droplets are volatilized in the air with the solvent, and the polymer condenses and solidifies into the fiber, which is eventually accepted by the acceptor. As the spacing between the two increases, the diameter decreases.

(4)   Activity rate of electrostatic spinning fluid. When the nozzle aperture is fixed, the average jet velocity is obviously proportional to the fiber diameter.

(5)   Status of collector. When fixed collectors are used, the nanofibers appear random irregularities. When the rotary disk collector is used, the nanofibers appear in parallel arrangement. Thus, different equipment conditions produced different fibrous omentum.

3.3   Electrostatic Spinning Device

The structure of the electrostatic spinning device has a decisive role in the morphology of the formed nanofibers. Different nozzle or the introduction of appropriate template agent can be used to prepare fibers with different morphologies, including honeycomb-like nanofibers, porous nanofibers, hollow nanofibers, and multi-cavity nanofibers.

The electrostatic spinning process is completed under the action of a liquid supply pump, a spinneret, a high-voltage power supply, and a receiver. The liquid supply pump is used to provide a spinning solution with a stable flow rate, and a spinneret is installed on it. The two electrodes of the high-voltage power supply are connected to the spinneret and receiver, respectively. The purpose is to create a high-voltage electric field in the spinning area. Generally, the electrospinning liquid (polymer solution or dissolved matter) is injected into a syringe with a small nozzle with an inner diameter of 100 μm–1 mm. The nozzle ACTS as an electrode, and the collecting end is connected to another electrode, providing a high-voltage electric field of 100–3000 KV m^{-1} between the two. Under most experimental conditions, the distance between the nozzle and the collector plate is 5–25 cm. After starting the high-voltage power supply, charged ions or polar molecules in the solution polymerize towards the tip of the spinel under the action of electrostatic force to form a cone structure, namely, Taylor cone. When the voltage exceeds the critical value, the electrostatic force on the liquid breaks through the surface tension, and the solution is excited from the Tip of the Taylor cone to form a jet stream. The jets are stable near the Taylor cone, but they soon undergo high-frequency bending, stretching, and splitting, and the jet diameter decreases, eventually forming nanoscale or submicron scale filaments [5–8]. At the same time, the solvent evaporates, the jet solidifies, and the fiber ends up deposited on the receiver [9].

3.3.1 Classification by Spinneret

Single Sprinkler Electrospinning Device

The simplest single nozzle electrospinning device consists of a glass tube with a very fine tip, which holds the polymer liquid and controls the flow rate by adjusting its tilt Angle with the horizontal plane. This is often used in cases where spinning requirements are not very high. The relatively standard electrospinning device uses a syringe instead of a glass tube as the spinneret. During spinning, spinning liquid is loaded into the syringe and an injection pump is connected to the end of the syringe to regulate the flow of liquid. The advantage of this device is that the flow rate of liquid can be controlled at will, so that it can be stabilized at a certain value, which also provides a specific value of liquid flow rate for theoretical study of electrostatic spinning process. The spinneret can be placed horizontally, vertically or at an angle without much effect on the spinning result.

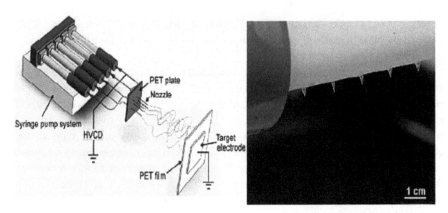

Fig. 1 An experimental setup for multijet electrospinning [11]

Multi-nozzle Electrospinning Device

The electrospinning process of multi-nozzle is better than single nozzle. Because each jet has the same electric charge, this device overcomes the shortcoming of low spinning efficiency of single nozzle and provides the possibility for future mass production. At the same time, the polymer nanofibers that cannot be dissolved in the same solvent or processed under the same conditions can be uniformly mixed together, so that the same fiber membrane has multiple functions. Ding's team used this multi-nozzle electrospinning device to prepare biodegradable nanofibers composed of PVA(polyvinyl alcohol)/CA(acetyl cellulose), and the results showed that the two fibers were mixed very evenly [10] (Fig. 1).

Coaxial Electrospinning Device

The coaxial electrospinning device shown in Fig. 2 consists of two syringes with different inner diameters. Two polymer solutions were added to two syringes inside and outside, and an electrode and an injection pump were used to generate electric field and electric field, respectively. Control the flow rate. Coaxial nanofibers can be obtained by adjusting the electric field intensity and liquid flow.

Multi-channel Sprinkler Electrospinning Device

The multi-channel electrospinning technology has successfully prepared bionic micron fibers with multi-channel, which provides a new idea for the preparation of bionic multi-channel micronanotubes (Fig. 3).

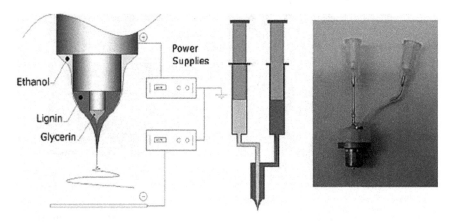

Fig. 2 A setup for coaxial electrospinning

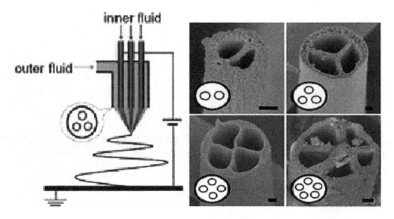

Fig. 3 Schematic illustration of the three-channel tube fabrication system and SEM images of multichannel tubes with variable diameter and channel number

Scanning Probe Electrospinning Device

Kameoka et al. [12] designed a possible method for one-dimensional assembly of nanofibers, scanning probe electrospinning. They used a specially fabricated scanning probe as a nozzle for electrostatic spinning.

The specific process of preparing nanofibers by this method can be roughly divided into three steps: (a) Contact the probe with the polymer solution surface; (b) A drop of liquid is absorbed as the raw material for electrospinning and then leaves the surface of the liquid; (c) By connecting to the probe by a piece of gold on the droplet applied voltage, the voltage size reaches a certain value to overcome surface tension, the spinning on the probe tip droplets form Taylor cone, when the voltage is increased

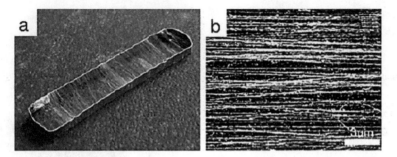

Fig. 4 **a** Frame electrode used for obtaining parallel fiber, **b** obtained oriented fibers [13]

further, fiber bundle will be out of droplet surface spray, collected through high-speed rotating electrode, the structure of the fiber will be in a specific orientation on the surface.

3.3.2 Classification by Receiving Device

The traditional electrostatic spinning device is evolved from the electrostatic spray device. The basic unit of electrospinning consists of three parts: high-voltage power supply, spinneret, and collector. The traditional electrostatic spinning method usually produces fiber felt in the form of nonwoven fabric with a disorderly arrangement of fibers. With the continuous development of the research on electrostatic spinning devices, many kinds of electrostatic spinning devices have been developed, which can produce core–shell structure, parallel arrangement of fibers, hollow structure, and internal multi-channel structure.

Metal Plate Collection Device

The most common receiver used in electrospinning is a metal plate. The fibers collected by this device are of one kind. The advantage of an amorphous nonwoven fiber membrane is that the fiber product is a naturally porous structure.

Metal Frame Collection Device

As shown in Fig. 4, in 2003, Wendorff from Germany [12] invented a metal frame with length and width of 6 cm and 2 cm, respectively as a negative pole to receive fibers, and obtained nanofibers with the same orientation.

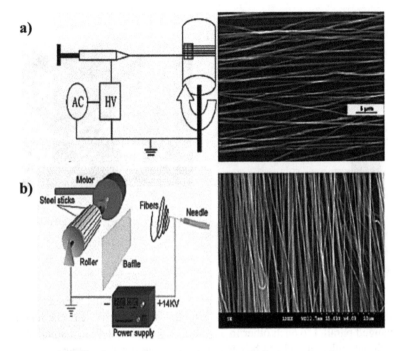

Fig. 5 **a** Schematic diagram of setup for electrospinning with a rotary drum, **b** Schematic diagram of modified setup for electrospinning [14]

Drum-Type Collection Device

Figure 5 is a schematic diagram of the drum collection device, which adopts a drum with a speed of several thousand revolutions per minute to receive electrostatic spinning.

When collecting, the collected fibers can have a certain orientation. In 2001, researchers at Virginia Commonwealth University successfully obtained well-oriented nanofibers using this technique. Yao [14] designed to add a baffle between the spinneret and the roller, which greatly improved the orientation of the fiber.

High-Speed Runner Collecting Device

Yarin [15] combined electrostatic spinning technology with electric field assembly technology, and adopted high-speed rotating wheel to collect nanofibers. The nanofibers with better orientation were obtained.

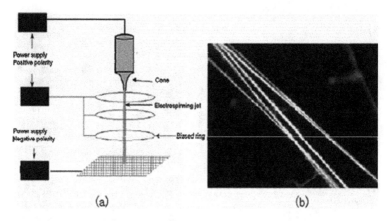

Fig. 6 a A multiple field technique, **b** aligned PEO fiber yarn obtained

Parallel Electrode Collection Device

The collector of the plate receiving device consists of two parallel conductive silicon plates spaced at certain distances from each other to form grooves. The nanofibers are arranged parallel to each other in an additional electric field in a small slot on a conventional collector and can be easily transferred to other substrates.

Multiple Electric Field Orientation Collection Device

Figure 6 is a schematic diagram of a multi-field oriented collection electrospinning device. The device comprises three high-voltage power sources, one of which is connected to the spinneret to provide the process voltage during spinning. The high-voltage power supply connected with eight copper rings in parallel provides the ring voltage that causes the nanofibers to align, and the high-voltage power supply connected with the receiving device provides the voltage of the receiving electrode. The fiber bundles prepared by this device have good orientation.

Tip Oriented Collection Device

The device is shown in Fig. 7. Its collector consists of two parts: a stainless steel tip as a negative pole and the nozzle to form an electric field; Nanofibers are collected using a polyester film mounted on a drum. The method has two characteristics: one is to use a needle tip instead of a large metal plate as a counter electrode, and spray from the spinneret. The fibers will converge around the tip of the needle. The other is to use a rotating cylinder with high speed to collect the fibers, which will be wound around the rotation. The continuous nanofibers with the same orientation were obtained on the tube.

Fig. 7 A steel pin aligned exactly with the needle of syringe is used as counter electrode

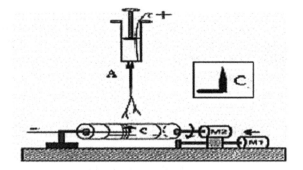

Collection Device with Magnetic Field

Two magnets placed parallel to the collected aluminum were used to electrospin a polymer solution containing magnetic nanoparticles, resulting in extremely directional nanofibers.

3.4 Formation Mechanism of Electrospinning Nanofibers

The whole process of preparing nanofibers by electrospinning is closely related to the hydrodynamics properties. When the applied electric field force is greater than the surface tension of the spinning liquid, the spinning liquid with static electricity is ejected from the spinneret as a bundle of fibers and flies rapidly towards the collecting plate. This process can be roughly divided into three stages: (1) forming fiber bundle and extending along its axis; (2) The bending movement and splitting refinement of fiber bundles; (3) The solvent continuously volatilizes, causing the solidification of the fiber bundle, and finally forming nanofibers on the collection plate [16].

The motion process of a typical electrostatic spinning jet in the air is a very complex electronic hydrodynamics process, that is, the spinning solution flows out of the jet nozzle to the fiber and finally deposits to the collection plate, which can be roughly divided into the following three processes:

(1) The formation of Taylor conical droplets and the elongation of spinning jet;
(2) Folds and instability of the efflux;
(3) Volatilization of solvent and movement of forming fibers.

The theoretical mechanism involved in the preparation of polymer nanofibers by electrospinning is complicated. It involves static, electricity, Science, rheology, electrohydrodynamics, aerodynamics, turbulence, solid–liquid surface charge transport, mass transport, and heat transfer. Some researchers have found that cellulose solution system of high solution viscosity and surface tension is not conducive to continuous and stable fiber spinning process, and a ene propyl, 3 in 1 methyl chloride imidazole salt (AMIMC1) as solvent, by adding cosolvent dimethyl sulfoxide (DMSO), spinning liquid surface tension and viscosity can be greatly reduced, so that the cellulose solution electrostatic spinning jet flow tends to be stable, even further. The addition of DMSO improves the chain motion capability to a certain extent, so that stable and continuous jet flow can be obtained under the action of electric field force.

4 Fabrication of Electrospun Nanofibers

Electrospinning can be divided into three categories, i.e., solution electrospinning, melt electrospinning and emulsion electrospinning according to the polymer state used to form fibers (Fig. 8).

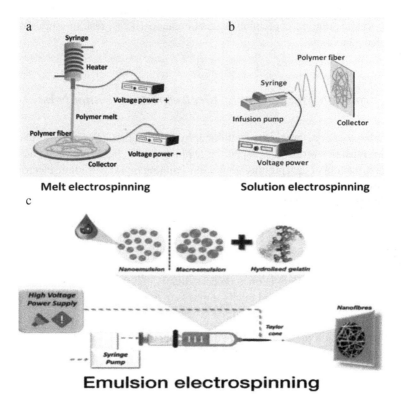

Fig. 8 (**a**) Melt electrospinning apparatus, (**b**) solution electrospinning [17], (**c**) emulsion electrospinning of gelatin-based nanofibres process [29]

4.1 Solution Electrospinning

The process of solution electrospinning is as follows. Form a uniform solution by dissolving a given polymer in a compatible solvent, and then apply high voltage to form a superfine jet flow from the charged solution. Later, the jet solidifies due to the rapid evaporation of the solvent, after which solid nanofibers are deposited on the collector.

Diversified types of materials can be prepared by solution electrospinning, including natural biopolymers, such as chitin, silk fibroin, alginate, DNA, dextran, chitosan fibrinogen, collagen, and gelatin, as well as a great quantity of biodegradable and biocompatible synthetic polymers, such as polylactic acid (PLA), polycaprolactone (PCL), and polylactic co glycolic acid (PLGA) [18].

Solution electrospinning can produce continuous nanofibers in several biomedical fields such as drug delivery and tissue engineering as the most traditional and commonly used method to. Qin et al. [19] successfully used solution electrospinning to prepare the self-supporting, fast dissolvable, and oral nanofiber membrane by mixing chitosan and pullulan polysaccharide in different proportion without using any inedible polymer matrix. The test results showed that the film has fast solubility and good encapsulation property, and confirmed to be anticipated for wider use in oral mucosa release system by further research. Scanning electron microscopy (SEM) results demonstrated that the diameter of nanofibers increased first and then decreased when chitosan content in solutions increased. Kishimoto et al. [20] fabricated silk fibroin (SF) nonwoven mat, which can be potentially useful as medical materials (such as tissue engineering cell scaffold) by using solution electrospinning technology It is noteworthy that this technology did not use any co-existing water-soluble polymers, such as poly (ethylene oxide), while prepared products in <10 wt% aqueous solutions. The experimental results suggested that the average tensile stress and strain of the high molecular weight SF non-woven mat is 0.83 ± 0.05 MPa / (g / m^2) and $12 \pm 5.6\%$, which had stable morphology without beads. Dodero et al. [21] prepared ZnO nanoparticles/alginate-based composite membranes by solution electrospinning, which can be used as biosorption materials. The mats are characterized by uniform texture, nanofibers interconnected voids with an average size of 140 nm and with an average diameter of 100 nm. Due to its excellent biocompatibility and antibacterial activity, the electrospun nanocomposite mats can be used as renewable and biocompatible adsorbent systems, materials to stabilize the long-lasting process of water environment, drug delivery, and passivation processes. Owing to the high cost-effectiveness of the development of the electrospun nanomaterials, it has broad application prospects in widespread domains.

Polymer, solvent, polymer solution, environmental conditions, and processing parameters determine the success of electrospinning polymer solution, and have impact on the structure and morphology of nanofibers. However, the application of electrospinning will be limited, mainly because the electrospinning solution contains a lot of toxic solvents, which has the risk of polluting the environment and human health, so that limits the application of electrospinning fiber in biomedical related

fields. Moreover, electrospinning is not suitable for natural polymers and some functional polymers [22]. However, suitable electrospinning solution can be obtained by
an efficient way which is mixing one functional polymer with another electrospinning polymer. Dai et al. [23] proposed a method to construct curcumin/gelatin-blend
nanofiber mats (NMS) via solution electrospinning. In vitro experiments demonstrated that this bionic structure had the necessary properties and activities for fibroblast growth, similar to the role of extracellular matrix (ECM) of granulation tissue in
physiological trauma. We observed that nanopores in the electrospun NMs avoided
cell infiltration and tissue endogenesis. Therefore, the prepared NMS can be used as
a mechanical active carrier, which can deliver bioactive curcumin to wounds locally,
so as to improve wound healing.

4.2 Melt Electrospinning

Solvent-free makes melt electrospinning is a safer, more reliable, and more environmentally friendly method, which is to prepare nanofibers with high yield, less defects,
and better physical and mechanical properties by mixing salt with molten polymer,
reducing melt viscosity by adding plasticizer, or reinforcing by adding assisted wind.

The polymer used for melt electrospinning needs to melt above glass transition
temperature and below thermal degradation temperature. Therefore, thermally
unstable polymers, proteins and thermosetting polymers are not applicable to this
method. The most commonly used polymers are thermoplastic plastics, such as and
PLGA, polypropylene (PP), and polyester (e.g., polyurethane, PLA, PCL). Notably,
it is very suitable for low melting point polymers with good thermal stability
and processability, such as PCL. Melt electrospinning is also used for polymers
that are only soluble in certain solvents, such as polyamides and polyolefins. It is
also applicable for some industrial polymers which are commonly used, such as
polyethylene terephthalate (PET) nylon-6, polymethylmethacrylate (PMMA), and
polyethylene [18].

Melt electrospinning has been studied and used in certain fields of biomedical
engineering, such as regenerative medicine, tissue engineering, biosensors, and drug
delivery. The research of Lian et al. [24] provideed the basis for the application of
melt electrospinning technology in the field of fusible drug loading. In this study,
PCL fiber membrane loaded with daunorubicin hydrochloride was prepared by melt
electrospinning and used as an antitumor drug delivery system for local drug delivery
of tumor therapy. It was indicated that the drug was encapsulated inside the fibers
which appeared curled and had smooth surfaces without the appearance of drug
crystals or other impurities. The study of Oh et al. [25] confirmed that with the
increase of draw ratio and temperature, the crystal phase transformation of poly(l-
lactic acid) (PLLA)/BaTiO3 fiber would be enhanced, thus affected its structural
development. Based on this property, they mass-produced PLLA/BaTiO3 fibers via
melt electrospinning technology and applied them to cotton fabrics in the form of
embroidery. The flexible and stretchable electrospun fiber can also be employed

as an environment-friendly sensor in the field of wearable electronic products to promote the development of electronic textile industry. Eichholz et al. [26] first studied the effect of melt electrospun fibers with controllable architectures on the behavior and differentiation of stem cells, and predicted the impact of this study on tissue engineering and disease progression. They fabricated cell micro-environments with various fiberous structures to study their effects on the behavior of human stem cells. The fiber diameter was 10.4 ± 2 μm, and the angles between adjacent layers were random, which might be $10°$, $45°$, $90°$. Su et al. [27] successfully fabricated 3D coil compression PCL scaffold with variable short-range adjustable coil density and long-range pattern. This study confirmed that the melt electrospinning technology could be programmed to prepare three-dimensional electrospun scaffolds with hierarchical structure, whose orderliness and biocompatibility made it great potential material in tissue engineering.

However, various problems need to be considered in the process of melt electrospinning, such as how to improve the accuracy of the preparation process of 3D multi-layered porous scaffolds, how to ameliorate the stability of melt electrospinning, and how to reduce the thermal damage of material properties of melt electrospun fibers and their additive agent, such as drugs or bioactive molecules. Nguyen et al. [28] proposed a beam bridge test method to determine the sagging behavior of melt spun microfibers in order to prepare a 3D lattice architecture with controllable structure and clear transverse pores. In order to improve the production efficiency of nanofibers, reduce the manufacturing cost of nanofibers, and expand the large-scale production line, Chen et al. [22] proposed the polymer melt differential electrospinning (PMDES) method, using umbrella spinneret equipment, setting the minimum jet spacing of 1.1 mm to generate multiple jets, and established a 300–600 g/h PMDES large-scale production line to promote the development of nanofibers industrialization and commercialization.

4.3 Emulsion Electrospinning

Emulsion electrospinning fabricates the core–shell structure electrospun fiber with unique morphology, large specific surface area, and high porosity through the following process. Apply high pressure on the feed emulsion, and then the droplet extends into the core–shell fiber to emitt the charged polymer jet, which is ejected from the conductive capillary and collected into the collector with a grounded capillary pump. This technology is taken advantage of preparing core–shell structured micro nanofibers from oil-in-water (O/W) or oil-in-water (W/O) emulsions. Various biopolymers (such as polysaccharides, proteins) and biocompatible polymers (such as PCL, polyoxyethylene (PEO), polyvinyl alcohol (PVA)), and other renewable resources new materials are suitable for emulsion electrospinning technology [30].

Emulsion electrospun fibers show attractive potential of application in biomedical engineering and food fields, such as tissue engineering scaffolds, drug delivery, wound dressings, and food applications, owing to the excellent performance of the emulsion electrospun nanofibers in improving the stability, biocompatibility, and entrapment efficiency of bioactive compounds, which can be used for targeting drug delivery and controlling drug release, so as to provide new ideas for improving its barrier properties. Ricurtea et al. [29] prepared gelatin-based nanofibers (NF) containing a high-oleic palm oil nanoemulsion or macroemulsions with smooth surfaces and uneven oil distribution via emulsion electrospinning. The emulsion had good physical interaction with the gelatin solution, which could be used as an excellent encapsulated aliphatic compound wall material for edible packaging of fat bread, cheese or meat substitute. Zhang et al. [31] successfully fabricated uniform core–shell nanofibers by electrospinning by adding anticharged gum arabic into gelatin-stabilized emulsions to form bilayer emulsions. It was found that the morphology and diameter of the electrospun fibers were appreciably influenced by the viscosity of the bilayer emulsions, which could be achieved by increasing the weight ratios of gum arabic and gelatin. This work opened up a new way for designing emulsion-based electrospun fibers for the encapsulation of hydrophobic bioactive compounds. Basar et al. [32] developed an electrospun material that would be employed as a new wound dressing for drug release with excellent control ability. By comparison of product properties, it was confirmed that emulsion electrospun fibers had advantages over some properties of solution electrospun fibers. PCL (single) and PCL/ gelatin (binary) electrospinning fibers containing Ketoprofen were prepared by electrospinning of solution and emulsion, respectively. According to in vitro release studies, compared with the single PCL electrospun mat, the binary PCL/ gelatin mat significantly inhibited the sudden release of ketoprofen and showed the drug release capacity lasted for four days. The study of Tao et al. [33] proved that the emulsion electrospinning technology can be applied to the preparation of transplantable biomimetic scaffolds. They successfully fabricated a PCL/carboxymethyl chitosan (CMCs)/sodium alginate (SA) micro-fiber biomimetic periosteum. The results showed that the micron-sized fiber immobilized CMCs and SA well, and had smooth surface morphology with an average diameter of $2.381 \pm 1.068 \, \mu m$ which was similar to the fiber network in natural periosteal ECM and had excellent tensile strength, so it can be used to repair large bone defects. Moydeen et al. [34] prepared a promising green drug delivery system via emulsion electrospinning technology, that is, the core–shell type nanofibers containing PVA/ sulfate dextran (Dex) loaded with ciprofloxacin (Cipro) nanofibers which acted as a carrier for drug delivery.

However, there are still some limitations that restrict production and applications of emulsion electrospinning. Relatively low productivity is one major limitation of the electrospinning of emulsion, and the usual solution is to modify the structure of the electrospinning device. The high requirements of traditional equipments for processing natural materials is another challenge. In addition, severe plasticization resulting from the hydrophilicity of the materials, and the decrease of barrier performance caused by the increase of relative humidity are also a major obstacle to the application of emulsion electrospinning. Therefore, it is urgent to improve the overall functionality and barrier properties of natural biopolymer materials.

5   Application and Functional Research of Electrospinning Nanofibers

Electrospinning is a commonly used way for preparing nano-sized or micro-sized fibers from other materials. Electrospun nanofibers have been used as catalysts, energy storage devices, biomedical applications, and aerospace applications due to their large porosity and high specific surface area [35].

5.1   Application of Electrospun Fiber as Polymer Film

With the rapid development of cities, the water crisis has become a new challenge in development. Membrane distillation (MD) [36] is an unique type of non-isothermal membrane method, in which the vapor is transported to the condensate through the hydrophobic membrane in the heat source solution by the difference of vapor pressure, and there is no water pressure or high-temperature vapor. Compared with other means, MD technology has significant advantages such as 100% theoretical oil displacement capability, high saltwater resistance capability, and solar energy utilization. Membrane is the core part of MD. In order to give full play to the advantages of MD, it is necessary to design MD reasonably. The ideal MD film should meet the requirements of moisture resistance, high permeability, and high thermal stability. However, currently MD films mainly originate from ultrafiltration membranes, such as polypropylene(PP), polyvinylidene fluoride(PVDF), polytetrafluoroethylene(PTFE), etc., and its performance is difficult to meet the requirements. Therefore, it is of great significance to search for ideal films with high permeability and stability for MD applications. Nanofiber electrospinning film (ENM) has developed into an excellent membrane distillation (MD) material in recent years due to its unique advantages, such as adjustable porosity, narrow pore size distribution, low bending degree, controllable thickness and flexible structure, etc.

In recent years, new ENMs, such as double-layer or multi-layer electrospun nanofiber films, have shown better permeability in MD than traditional films. However, due to the lamination of the film layer that the performance loss after a long run is still not cheerful. Based on this, inspired by the branching and interwoven nestlike structure, An et al. [37] made a solid and stable interwoven layered fiber composite (iHFC) membrane composing of interconnected polyethylene terephthalate (PET) microfibers and polyvinylidene-hexafluoropropylene (PH) nanofibers by electrospinning, as shown in Fig. 9. PET microfiber can not only reduce the mass transfer resistance, but also improve the adiabatic performance. PH nanofibers have good hydrophobicity, moisture resistance, and high salt repellency. Membrane structure optimization focuses on the composition and film thickness of customized PH nanofibers and PET superfine fibers, which have different contributions to the quality and heat transfer in the MD process. The results show that the prominent iHFC film

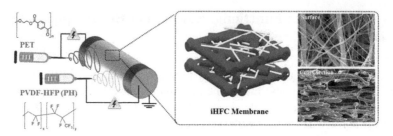

Fig. 9 Schematic process of iHFC membrane fabrication from PET and PH solutions [37]

with a PH/PET (1.5/0.8) ratio and a film thickness of 80 m has a good permeability of 65 LMH and a stable performance when worked for 60 h at a temperature change of 40°C. The interleaved structure makes the iHFC film have good long-term stability and strong mechanical strength. It is believed that the development of iHFC membrane with high separability and high separation performance and its unique method can support the way for the industrialization of MD membrane.

Polyimide (PI) is an engineering polymer, which has been generously applied in many high-tech fields in virtue of its excellent mechanical performance, excellent thermal stability, low dielectric constant, solvent and radiation resistance, etc. It is of great significance to prepare high-performance multifunctional materials with unique functions such as excellent mechanical performance and polyamide. Since the solubility of PI precursors allows them to electrospun from solution, various PI nanomaterials can be made by electrospun. In recent years, polyimide nanofibers have been widely studied due to their various structures and high degree of arrangement. Among them, polyaniline is easy to polymerize, low cost, good environmental stability, and high conductivity, so it has attracted wide attention. However, the polyaniline blend is actually a conductive material dispersed in a three-dimensional insulating matrix, so it usually has a high conductive permeability threshold. Nanometer polyaniline is widely concerned for its nanometer size, large specific surface area, three-dimensional network, and high pore structure.

Until now, it had been possible to manufacture different types of polymer nanofibers whose membranes via electrostatic radiation, and most of the membranes consisted of randomly distributed nanofibers [38]. Nanofiber membrane can be used as a homogeneous reinforcing material to improve the mechanical properties of polyimide matrix. As electrospun polyimide nanofiber membrane has a large specific surface area, Chen et al. [39] used electrostatic spinning to prepare high-orientation polyimide (PI) nanofiber membrane [40] taking PI fiber membrane as template and $FeCl_3$ as oxidant to grow polyaniline in situ. The resulting composite films have not only excellent thermal performance and mechanical performance, but also pH sensitivity and excellent electrical conductivity. The electromagnetic impedance is significantly improved. Due to the high arrangement of PI nanofibers, PANI/PI composite films show anisotropic conductivity in both parallel and vertical arrangement. PANI/PI can rapidly and effectively conduct protonation and deprotonation, showing sensitive reactions to different pH environments.

In food processing and transportation, contamination, and packaging of pathogens often arise on the surface of food contamination and transmission. Therefore, packaging plays a vital role in increasing the selling life of food and reducing cross-contamination. Electrospinning technology can produce nanofibers with a diameter of a few millimeters from polymer solutions containing different substrates by using electrostatic force so that nanofibers can form homogeneous nanoscale structural films. It is important to select food-class biopolymers containing natural antibacterial agents to make nanofiber packaging films. Chitosan and gelatin have good biocompatibility and biodegradability. They are two main natural biomolecular materials used in food packaging films.

Liu et al. [41] used the electrostatic spinning technology adding ε-polylysine(ε-PL) into gelatin/chitosan-based mixture which can get a kind of effective antibacterial nanofiber membrane. All nanofiber films have uniform disordered fiber structures with well-distributed diameter. Gelatin, chitosan, and ε-PL polymers interact with each other by hydrogen bonds, and the electric cycle lowers the crystal degree of the nanofiber film. On the one hand, the addition of combustible material improves the thermal stability of the film, on the other hand, it reduces the oxygen permeability and permeability of the film, and the combustible material is not easy to be separate from the fiber film. The weight ratio of gelatin/chitosan/ε-PL(G/CS/P) affected solution properties and nanofiber morphology. Antibacterial tests showed that G/CS/P(6:1:0.125) membrane was more capable of controlling six foodborne pathogens by destroying bacterial membranes than G/CS nanofiber membrane. Therefore, G/CS/P nanofibers can be applied as food packaging materials to weaken the harm of foodborne pathogenic bacteria.

5.2 Application of Electrospinning Nanofibers in Intelligent and Flexible Electronics

How to store and maximize energy conversion is a long-term problem facing our society. In order to face these problems, supercapacitor, secondary battery, fuel cell, and solar cell are highly sought after by various researchers which electron or ion conduction is particularly important in the application of energy storage devices [42]. Nanometer electrodes are considered as ideal materials in the field of flexible electronics in virtue of their high specific surface area and short diffusion distance [43]. In addition, devices made from electrically spun nanofibers are light and flexible, which makes them attractive in the energy field.

5.2.1 Application of Electrospinning Fiber in Supercapacitors

Supercapacitors (SCs), as an electronic device that can be charged and discharged rapidly, have a growing demand in high-power density equipment. Therefore, it is important to find sustainable, smaller, and more flexible polymers [44]. Polyaniline (PANI) is a traditional conductive polymer (ICPs). Because of its high stability, easy preparation, cheap monomer, and changing conductivity by adding, it has become one of the most popular conductive polymers. Bhattacharya et al. [45] have designed a method for solvent electric spin polymers from high boiling point at relatively low voltage. Utilizing this way, the camphor sulfonic acid (CSA) as the main additives, m-cresol, and chloroform as mixed solvent while chloroform and dimethyl formamide (DMF) in the mixed solvent to compare of polyaniline electrospinning fiber. This method was the first to apply the concept of secondary mixing to the electrodeposition fiber, forming a multi-branched well-arranged conductive fiber grid to facilitate electron conduction. The electrical conductivity of polyethylene oxide (PEO) fiber is 1.73 S/cm, more than twice as high as the average conductivity reported in the literature. When the fiber network is applied to the supercapacitor, its specific capacitance achieves 3121 F g^{-1} at 0.1 A g^{-1}, realizing higher capacitance of PANI-PEO electrospun fiber reported so far. The prepared fiber felt electrode has a good internal porosity, which promotes the diffusion of ions at the active site, thus increasing the electrochemical performance of the composite.

While applying SCs adequately, researchers are also exploring ways to improve its electrochemical properties. Because of its DC path, one-dimensional nanomaterials can not only promote electron transfer, but also shorten the ion diffusion distance, thus maximizing the electrode performance. At the same time, it can alleviate the volume expansion of electrode materials and weaken their mechanical degradation, thus obtaining a longer cycle life. Diatomic doping is beneficial to synergistic effect, electron conduction, electrolyte infiltration, and introduction of more pseudocapacitors.

The preparation of porous carbon nanofibers (PNFs) of biomass materials is still a challenge for SCs due to carbon yield and molecular flexibility limitations. Chen et al. [46] prepared P and N double-carried one-dimensional PNFs using chitosan (CS)/polyvinylpyrrolidone (PVP) phytic acid solution as electrospinning solution by a controllable method, as shown in Fig. 10. When using it as the SCs electrode, the specific capacitance of 0.2-PNF can reach 358.7 F g^{-1} at the 1 A g^{-1}. When the current density is 0.5 A g^{-1}, the specific capacitance of 0.2-PNF reaches 94F g^{-1}, and the energy density is 13.1 W h kg^{-1}, and the power density is 248 W kg^{-1}. After 10,000 cycles, the capacitance is 87.5% of the original. The performance is higher than that most reported nitrogen carried carbon nanofiber SCs.

Carbon-based materials are largely applied in SCs electrode materials because of their remarkable electrochemical properties, abundant structure, and large resource reserves. Electrospun carbon nanofibers are used in flexible and wearable electronic devices on account of outstanding peculiarities such as high aspect ratio, large special surface area, and self-supporting film structure. To ameliorate the electrochemical performance of electrospun carbon nanofibers as SCs electrodes, a number of methods have been drawing forth to enhance the specific surface area and the number of pores. Impurity atoms can improve the electrochemical performance of the fiber, especially the fabricability of the fiber. Heteroatoms of organic salts are

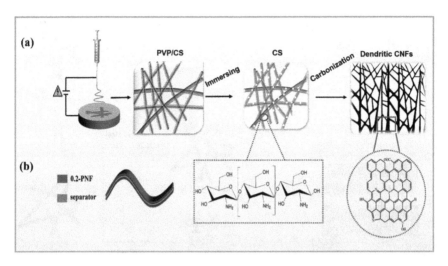

Fig. 10 **a** PNF preparation flow chart. **b** Symmetrical SC structure [46]

usually preserved after carbonization, in which metal ions, especially alkali metal ions, may act as a distal activator for the formation of pores.

Zhang et al. [47] prepared carbon nanofibers (HPCNFs) with O and N composition changes by electrospinning and pyrolysis using polyacrylonitrile and polymethyl methacrylate as raw materials and sodium alginate (SA) as additive. It has an outstanding specific capacitance ($253.2 \ F \ g^{-1}$ at $0.25 \ A \ g^{-1}$ in 1 M H_2SO_4 aqueous electrolyte) and a projecting cycle life (99.8% capacity after 10,000 cycles) when applied to supercapacitors. By changing SA content, the morphology of carbon nanofibers would be easily adjusted, while the layered porous nanostructure and rich nitrogen and oxygen functions play an important role in the bilayer structure charge, thus increasing the charge density.

SCs electrodes based on electrospun carbon fiber have token on satisfactory energy production/reserve performance. The introduction of additives (such as silver, nickel or carbon nanotubes) and the application of conductive polymers (such as polypyrrole) or transition metal oxides on the fiber surface can also improve its capacitance. However, most carbon fibers produced by electrospinning are made of polyacrylonitrile, polybenzimidazole or asphalt as carbon sources, which will release toxic substances during carbonization. Cellulose is composed only of carbon elements, oxygen, hydrogen whose carbonation reaction releases environmentally friendly components.

Cai et al. [48] deacetylated cellulose acetate by electrostatic spinning and polymerized polypyrrole to obtain polypyrrole coated cellulose, and preparation of nitrogen sensorized carbon nanofibers (N-CNFs) by carbonization, as shown in Fig. 11. The asymmetric SCs were further assembled with N-CNFs and N-CNFs/Ni(OH)$_2$ as negative and positive poles, respectively. SCs operated in KOH solution (6.0 M) with a voltage of 1.6 V, an energy density up to $51 \ W \ h \ kg^{-1}$, and a maximum

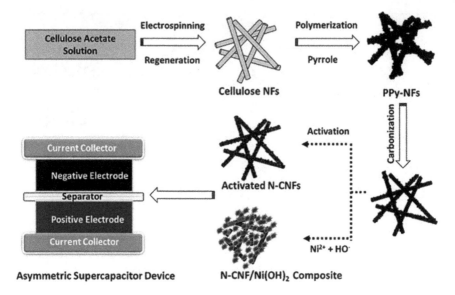

Fig. 11 Schematic illustration of NF preparation process and supercapacitor setup [48]

power density of 117KW kg^{-1}. The device has a good cycle life, after 5000 cycles, its specific capacitance reserve of 84%. N-CNFs extracted from electrospinning cellulose can be applied as electrode materials for energy storage devices such as SCs.

Deng et al. [49] prepared MWNT/cellulose acetate blends by electrospinning method, and MWNT/cellulose composite nanofibers were made after deacetylation. The results showed that the oxidative stability activation energy of cellulose nanofibers decreased from 230 kJ mol^{-1} to 180 kJ mol^{-1} after adding MWNTs. At the same time, the addition of nanotubes greatly increases the surface area of the composite material. The nanotubes protrude from the fibers and the surface becomes rough. When used as an electrode for supercapacitors, the original ACNF and 6% MWNT/ACNF ratios are 105 F g^{-1} and 145 F g^{-1}, respectively, at a current density of 10A g^{-1}.

5.2.2 Electrospinning Nanomaterials Used for Sodium and Lithium Electric Electrode

Electrostatic spinning nanofibers hold higher specific surface area and larger porosity, which can increase the reaction site of active materials and improve the performance of energy storage devices.

In recent years, lithium-ion battery (LIBs) has been developing rapidly all the way to satisfy the need of carriable electronic devices and mixed-power electric vehicles [50]. Sodium-ion battery (SIBs) is one of the most likely alternatives to lithium-ion batteries because of their high sodium content and lower cost [51]. Zhou et al. [52] satisfactorily prepared ZnSe nitrogen-mixed carbon composite nanofibers suitable for SIBs and LIBs by using electrostatic spinning and simple seleniza-tion. One-dimensional nanostructure allows faster ion diffusion, which improves the cycle stability of ZnSe caused by amorphous crystallization during charging and discharging. The synergistic effect of ZnSe and carbon greatly improves the energy storage capacity of anode. The capacity of ZnSe@N-CNFs for SIBs and LIBs can reach 455.0 and 1226.1 mA h g^{-1}, respectively. When at the 2 A g^{-1}, ZnSe@N-CNFs maintained excellent stability (LIBs after 600 times of 701.7 mA h g^{-1} and SIBs after 200 times of 368.9 mA h g^{-1}). The ZnSe@N-CNFs in SIBs and LIBs retain a one-dimensional structure which not have additional stacking by a long period at a high current density, while the large-period stability of ZnSe@N-CNFs puts down to the excellent microstructural stability (Fig. 12).

Cellulose as a reproducible resource is an affluent semifinished material compared with fossil polymer. Cellulose acetate (CA), a widely used cellulose derivative, is soluble in common solvents. It is easy to electrospin and prepare nanofibers, which can be carbonized to generate CNFs and regenerated into cellulose. Han et al. [53] acetic acid-soluble cellulose composite Fe_3O_4 to obtain mesoporous Fe_3O_4@CNFs nanomaterials for LIBs. The carbon substrate can not only buffer the volume change caused by Fe_3O_4, but also accelerate the soaking and electron shuttle. Carbon in the outer layer can form a stable SEI membrane, thus isolating active materials and electrolytes. The Fe_3O_4@CNFs electrode shows a high reversible capacity. When at the 2 A g^{-1}, the specific capacity can reach 596.5 mAh g^{-1}, that the capacitance holds to 99% original after 300 cycles while the current density is 1 A g^{-1}, the specific capacity is 773.6 mAh g^{-1} and the capacity retention rate is 98% after 300 cycles.

Electrostatic spinning is the best technology to prepare long nanofibers. Low-cost carbon nanofibers derived from lignin which is the second most abundant material in the fibers and then undergoing proper heat treatment. Among all kinds of polymer precursors, PAN is the best to prepare carbon nanofibers with good mechanical

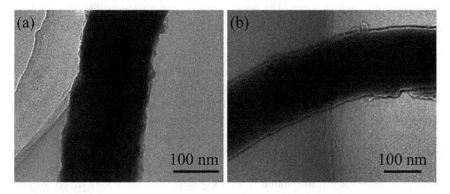

Fig. 12 (a) TEM images of ZnSe@N-CNFs after 600 cycles at 2 A g^{-1} in LIBs; (b) after 200 cycles at 2 A g^{-1} in SIBs [52]

strength. Choi et al. [54] prepared carbon nanofibers using polyacrylonitrile and lignin as raw materials through electrostatic spinning. The thickness changes of carbon nanofibers and their capacitance changes when applied to LIBs anode materials were observed by adjusting the ratio of PAN and lignin. With the increase of precursor lignin content, carbon nanofibers became thinner and thinner. When the PAN/lignin mass ratio is 50:50, the discharge capacity of the carbon nanofibers prepared by lignin precursor system is 150 mAh g^{-1} under the condition of rapid charge and discharge (7 min), which shows a high rate and cycling performance similar to that of conventional polyacrylonitrile spinning. Therefore, lignin-based carbon nanofiber is a promising low-cost high-power LIBs anode material.

5.3   Electrospinning Fibers for Biomedical Field

In the biomedical ground, nanofibers have a diameter smaller than cells, so they can biological function of natural extracellular matrix and mimic the structure. Most organs and tissues of human are alike to nanofibers in structure and shape, which makes it possible for nanofibers to be used in tissue and organ repair. Some electrospinning material has favorable biological compatibility and biodegradable properties which is used as a transport to enter human body, and easy to be taken in, combined with the electrostatic spinning of nanofibers and large specific surface area, excellent properties such as porosity, so it caused the researchers in the field of biomedical continuous attention, and has set up a wound repair, in biological tissue engineering applications are well [55].

In recent years, electrostatic spinning has been widely used for the preparation of tissue engineering scaffolds because of its simple preparation process. Ngadiman et al. [56] mixed magnetite (Fe_2O_3) with polyvinyl alcohol (PVA) and obtained mats with high porosity and biodegradability by changing the parameters such as rotation speed, voltage, nanoparticle content, and rotation distance. Figure 13 shows the electrospinning principle diagram. The results showed that PVA/γ-Fe_2O_3 with the best porosity (90.85%) could be obtained when the nanoparticle content was 5% W/V, the voltage was 35 kV, the spinning distance was 8 cm, the solution volume flow was 1.1 mL/h, and the rotation speed was 2455 rpm. The nanofibers own excellent degradation rate, lower mechanical properties, and good cellular compatibility, which are satisfactory to soft tissue engineering scaffolds. Electrospun nanofiber mats with porosity of more than 80% had higher cell survival rate. Electrospinning nanofiber mats have high porosity and support cell proliferation and adhesion. Therefore, scaffold production in this way can provide an appropriate place to represent the "state" for cells to grow and new tissue to be created.

As the first barrier to the outside world, skin plays a significant part of role in the protection of the body. If the wound is not adequately repaired, it may lead to wound infection or even cause disease response, which can endanger life when serious. Generally speaking, suitable wound dressings should be non-toxic, actively healing, and have good biocompatibility. In addition, it should insulate bacteria, remove extra

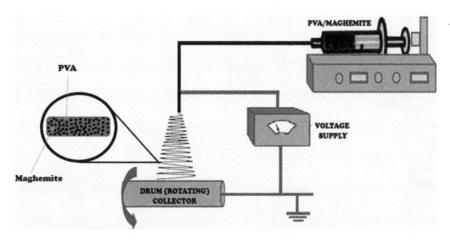

Fig. 13 Electrospinning process schematic diagram [56]

secretions, and maintain air permeability to the injured area. Among the available wound dressings, hydrogels and electrospinning films have attracted more and more attention in virtue of their excellent performance in wound dressings.

Electrostatic spinning has been widely used in membrane and tissue engineering in recent years. Xia et al. [57] prepared a super-transparent porous cellulose membrane with nanofibers composite chitosan by one-step electrostatic spinning and used it as a wound dressing. cellulose membrane- chitosan (CM-CS) was prepared as shown in Fig. 14. Firstly, dissolve pure cellulose membranes (CMs) and then regenerate them in an acid bath using the casting way. The chitosan solution is polarized into nanofibers to form continuous fiber pads on CMs by charge repulsion between molecules. As a result, the CM-CS has larger porosity and can maintain good air permeability to the greatest extent, as well as excellent light transmittance and mechanical properties. Because of the influence of hydrogen bond between van der Waals force and cellulose molecules between hydroxyl groups, cellulose can adsorb and easily form a relatively stable three-dimensional porous fiber membrane, which makes it the preferred material for wound dressing. Chitosan is largely applied in biomedical wound dressing. Chitosan can not only interfere with the function of macrophages, but also irritate cell proliferation and accelerate the granules of tissues, which can promote wound healing and have higher angiogenesis ability. Through a series of biological tests, CM-CS not only has good biocompatibility but also has good antibacterial activity against Staphylococcus aureus and Escherichia coli. CM-CS can be served as a potential dressing to promote wound healing.

Tissue engineering, as a kind of functional tissue manufacturing or restoration means, requires three-dimensional (3D) stents to provide a structural basis for the attachment and proliferation of cells. Among the variable ways of preparing 3D scaffolds, electrostatic spinning has attracted great attention because it can

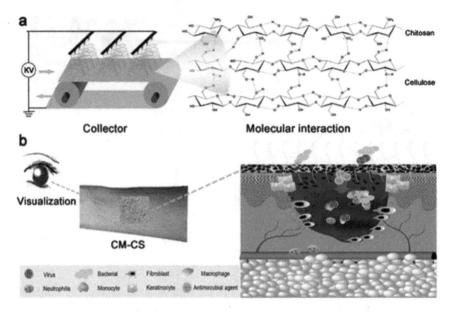

Fig. 14 Schematic diagram of CM-CS preparation (**a**) and covering skin wound (**b**) [57]

produce porous nanocrystals with specific properties. The structure of electro-spinning nanocrystal fiber mesh is similar to the size range of natural extracel-lular matrix (ECM) fibers. In recent years, polymer mixtures have attracted much interest in the preparation of poly (blend) nanofibers, which exhibit combinations of properties associated with each polymer. Polyblend nanofiber scaffolds have hydrophilic/hydrophobic properties, surface charge, mechanical strength of synthetic polymers, and biochemical characteristics of natural fibers.

Ajalloueian et al. [58] prepared a novel nanofiber blended with polylactic acid glycolic acid (PLGA) and chitosan by electrostatic spinning method. The spin-ning liquid used polyvinyl alcohol (PVA) as emulsifier. Finally formed scaffolds composed of a mixture of PLGA and chitosan. Compared with pure PLGA, the electro-spinning acid/chitosan blanket has stronger hydrophilicity, which is benefi-cial to improve the interaction of the cellular scaffold. Under dry and wet conditions, the tensile strength of PLGA/chitosan was 4.94 and 4.21 MPa, respectively, which showed sufficient strength for biomedical application. Cell culture studies showed that compared with PLGA membrane, PLGA/Chitosan nanofibers could promote the adhesion and proliferation of fibroblasts.

6   Conclusions and Future Prospects

Nanofiber materials prepared by electrostatic spinning are widely used, including polymers, ceramics, small molecules, and their complexes. In addition to preparing smooth surface nanofibers, electrostatic spinning can also produce nanofibers with secondary structures, including stomata, cavities, and nuclear sheath structures. Many different types of materials have been successfully used in the direct production of nanofibers by electrospinning. For example, synthetic polymers such as polystyrene and polyvinyl chloride have been electrospun into nanofibers for commercial applications related to environmental protection. Many biocompatible and biodegradable synthetic polymers, such as poly (lactic acid), it has been directly electrospun into nanofibers and have been further developed as scaffolds for biomedical applications. Natural biopolymers, such as DNA, silk protein, fibrinogen, glucan, chitin, chitosan, alginate, collagen, and gelatin, can also be electrospun into nanofibers from solution. As a new processing technology, electrospinning is a simple and efficient way to produce polymer nanofibers. Due to its unique advantages, electrospinning has been widely concerned in the field of functional materials. The preparation of biodegradable polymer fiber materials with low modulus, high flexibility, and high strength is an important subject in material science. In the future, the research direction to be strengthened is to use the principle of electrostatic spinning, improve the electrostatic spinning device and control the parameters of electrostatic spinning, so as to obtain continuous, uniform size, controllable defect, and regular arrangement of polymer nanofibers. On the basis of understanding the process of electrostatic spinning and the properties of the electrostatic spinning polymer nanofibers, the polymer nanofibers with controllable degradation rate, good physical and mechanical properties, and biocompatibility were developed to realize the practical application of the nanofibers. Recently, the development of self-ejecting multi-strand nanofiber electrospinning technology with high production efficiency has made electrospinning step towards industrialization. The electrospinning of natural polymers will also develop towards large-scale industrialization.

Due to its safety, biocompatibility, and biodegradability, natural polymers have attracted more and more attention from electrospinning researchers. The innovation of electrospinning technology is also a necessary step to promote the industrial application of nanofiber membranes. Generally, the mechanical properties and solvent stability of natural polymer nanofiber membranes are poor, which need to be improved by means of mixed electrospinning or chemical cross-linking with synthetic polymers. In recent years, the safe and non-toxic crosslinking of natural polymers has also been studied. Directions such as cross-linking by polyphenols, Maillard reactions, etc. Electrospinning as a simple and efficient polymer production. The new processing technology of nanofiber, due to the unique advantages of the polymer nanofiber.it has been widely paid attention to in the field of functional materials. Preparation of degradable polymer fiber materials with low modulus, high flexibility, and high strength is an important subject in material science. If successful, these materials can be used to make highly flexible monofilament surgical sutures and wound dressings to promote wound healing, as well as to make flexible catheters and other tissue engineering scaffolds, which will greatly promote the progress of surgery and the development of the medical fiber industry. Therefore, the future research direction should be strengthened as follows: using the principle of electrospinning, by improving the electrospinning device and controlling the parameters of electrospinning, to obtain continuous, uniform size, controllable defects, and regular arrangement of polymer nanofibers; based on the understanding of the electrospinning process and the properties of electrospinning polymer nanofibers, the polymer nanofibers with controllable degradation rate, good physical and mechanical properties, and biocompatibility were developed to realize the practical application of nanofibers. Synthesize suture material which can maintain a long time and strength in the body and be absorbed in a short time after wound healing.

References

1. Matthews A, Wnek E, Simpson G (2002) Electrospinning of collagen nanofibers. Biomacromol 3(2):232–238
2. Taylor GL (1969) Electrically drive jet. Proc R Soc Acad Ser A Math 313(1515):453–475
3. Gobius SG, Ivana V, Zorica V (2011) Nanoporous network channels from self-assembled triblock copolymer supramolecules. Macromol Rapid Commun 32(4):366–370
4. Ge C, Yang X, Li C (2012) Synthesis of polyaniline nanofiber and copolymerization with acrylate through in situ emulsion polymerization. Appl Polymer 123(1):627–635
5. Yarin A, Zussman E (2004) Upward needleless electrospinning of multiple nanofibers. Polymer 45(9):2977–2980
6. Kameoka J, Orth R, Yang Y (2003) Scanning tip electrospinning source for deposition of oriented nanofibres. Nanotechnology 14(10):1124
7. Yang QB, Li Z, Hong Y (2004) Influence of solvents on the formation of ultrathin uniform poly(vinyl pyrrolidone) nanofibers with electrospinning. J Polym Sci Part B Polym Phys 42(20):3721–3726
8. Zhao Y, Yang B, Lu X (2005) Study on correlation of morphology of electrospun products of polyacrylamide with ultrahigh molecular weight. J Polym Sci Part B Polym Phys 43(16):2190–2195
9. Li D, McCann T, Xia Y (2006) Electrospinning: a simple and versatile technique for producing ceramic nanofibers and nanotubes. Am Ceram Soc 89(6):1861–1869
10. Ding B, Kimura E, Sato T (2004) Fabrication of blend biodegradable nanofibrous nonwoven mats via multi-jet electrospinning. Polymer 45(6):1895-1902P
11. Dosunmu G, Chase W, Kataphinan G (2006) Electrospinning of polymer nanofibres from multiple jets on a porous tubular surface. Nanotechnology 17(4):1123-1127P
12. Kameoka J, Orth R, Yang Y (2003) A scanning tip electrospinning source for deposition of oriented nanofibres. Nanotechnology 14(10):1124-1128P
13. Dersch R, Liu T, Schaper A (2003) Electrospun nanofibers: Internal structure and intrinsic orientation. J Polym Sci Part A Polym Chem 41(4):545–553
14. Yao F, Gu Z, Zhang J (2007) Fiber-oriented liquid crystal polarizers based on anisotropic electrospinning. Adv Mater 19(21):3707–3711
15. Theron E, Zussman L (2001) Electrostatic field-assisted alignment of electrospun nanofibers. Nanotechnology 12(3):384–390
16. Taylor GL (1969) Electrically driven jet. Proc R Soc Ser A Math 313(1515):453–475
17. Lian H, Meng Z (2017) Melt electrospinning vs. solution electrospinning: a comparative study of drug-loaded poly (ε-caprolactone) fibres. Mater Sci Eng C 74:117–123.

18. Xue J, Wu T, Dai Y, Xia Y (2019) Electrospinning and electrospun nanofibers: methods, materials, and applications. Chem Rev 119:5298–5415. https://pubs.acs.org/doi/pdf/10.1021/acs.chemrev.8b00593
19. Qin Z, Jia X, Liu Q, Kong B, Wang H (2019) Fast dissolving oral films for drug delivery prepared from chitosan/pullulan electrospinning nanofibers. Int J Biol Macromol 137:224–231
20. Kishimoto Y, Morikawa H, Yamanaka S, Tamada Y (2017) Electrospinning of silk fibroin from all aqueous solution at low concentration. Mater Sci Eng C 73:498–506
21. Dodero A, Vicini S, Lova P, Alloisio M, Castellano M (2020) Nanocomposite alginate-based electrospun membranes as novel adsorbent systems. Int J Biol Macromol 165:1939–1948
22. Chen M, Zhang Y, Li H, Li X, Ding Y, Bubakirb MM, Yang W (2019) An example of industrialization of melt electrospinning: polymer melt differential electrospinnin. Adv Ind Eng Polym Res 2:110–115
23. Dai X, Liu J, Zheng H, Wichmann J, Hopfner U, Sudhop S, Prein C, Shen Y, Machens HG, Schilling AF (2017) Nano-formulated curcumin accelerates acute wound healing through Dkk-1-mediated fibroblast mobilization and MCP-1-mediated anti-inflammation. NPG Asia Mater 9:e368
24. Lian H, Meng Z (2017) Melt electrospinning of daunorubicin hydrochloride-loaded poly (ε-caprolactone) fibrous membrane for tumor therapy. Bioact Mater 2:96–100
25. Oh HJ, Kim DK, Choi YC, Lim SJ, Jeong JB, Ko JH, Hahm WG, Kim SW, Lee Y, Kim H, Yeang BJ (2020) Fabrication of piezoelectric poly(l-lactic acid)/BaTiO3 fibre by the melt-spinning process. Sci Rep 10:16339
26. Eichholz KF, Hoey DA (2018) Mediating human stem cell behaviour via defined fibrous architectures by melt electrospinning writing. Acta Biomater 75:140–151
27. Su Y, Zhang Z, Wan Y, Zhang Y, Wang Z, Klausen LH, Huang P, Dong M, Han X, Cui B, Chen M (2020) A hierarchically ordered compacted coil scaffold for tissue regeneration. NPG Asia Mater 12:55
28. Nguyen NT, Kim JH (2019) Identification of sagging in melt-electrospinning of microfiber scaffolds. Mater Sci Eng C 103:109785
29. L.R. Patricio, R.S. Luis, E.D. Maria, X.Q. Carvajala, Edible gelatin-based nanofibres loaded with oil encapsulating high-oleic palm oil emulsions, Colloids and Surfaces A: Physicochemical and Engineering Aspects, 595(2020), 124673.
30. Zhang C, Feng F, Zhang H (2018) Emulsion electrospinning: Fundamentals, food applications and prospects. Trends Food Sci Technol 80:175–186
31. Zhang C, Li Y, Wang P, Zhang A, Feng F, Zhang H (2019) Electrospinning of bilayer emulsions: the role of gum Arabic as a coating layer in the gelatin-stabilized emulsions. Food Hydrocoll 94:38–47
32. Basar AO, Castro S, Torres-Ginerb S, Lagaron JM, Sasmaze HT (2017) Novel poly(ε-caprolactone)/gelatin wound dressings prepared by emulsion electrospinning with controlled release capacity of Ketoprofen anti-inflammatory drug. Mater Sci Eng C 81:459–468
33. Tao F, Cheng Y, Tao H, Jin L, Wan Z, Dai F, Xiang W, Deng H (2020) Carboxymethyl chitosan/sodium alginate-based micron-fibers fabricated by emulsion electrospinning for periosteal tissue engineering. Mater Design 194:108849
34. Moydeen AM, Padusha MSA, Aboelfetoh EF, Al-Deyab SS, El-Newehy MH (2018) Fabrication of electrospun poly(vinyl alcohol)/dextran nanofibers via emulsion process as drug delivery system: kinetics and in vitro release study. Int J Biol Macromol 116:1250–1259
35. Moheman A, Alam MS, Mohammad A (2016) Recent trends in electrospinning of polymer nanofibers and their applications in ultrathin layer chromatography. Adv Colloid Interfac 229:1–24
36. Kiss AA, Kattan Readi OM (2018) An industrial perspective on membrane distillation processes. J Chem Technol Biotechnol 93:2047–2055
37. An X, Bai Y, Xu G, Xie B, Hu Y (2020) Fabrication of interweaving hierarchical fibrous composite (iHFC) membranes for high-flux and robust direct contact membrane distillation Desalination 477:114264

38. Yan G, Niu H, Zhao X, Shao H, Wang H, Zhou H, Lin T (2017) Improving nanofiber production and application performance by electrospinning at elevated temperatures. Ind Eng Chem Res 56:12337–12343
39. Chen D, Miao Y, Liu T (2013) Electrically conductive polyaniline/polyimide nanofiber membranes prepared via a combination of elecrospinning and subsequent in situ polymerization growth. Acs Appl Mater Inter 5:1206–1212
40. Jiang S, Hou H, Agarwal S, Greiner A (2016) Polyimide Nanofibers by "Green" Electrospinning via Aqueous Solution for Filtration Applications Acs Sustain Chem Eng 4:4797–4804
41. Liu F, Liu Y, Sun Z, Wang D, Wu H, Du L, Wang D (2020) Preparation and antibacterial properties of ε-polylysine-containing gelatin/chitosan nanofiber films. Int J Biol Macromol 164:3376–3387
42. Li X, Zhi L (2018) Graphene hybridization for energy storage applications. Chem Soc Rev 47:3189–3216
43. Nioradze N, Chen R, Kim J, Shen M, Santhosh P, Amemiya S (2013) Origins of nanoscale damage to glass-sealed platinum electrodes with submicrometer and nanometer size. Anal Chem 85:6198–6202
44. Zhu Q, Zhao D, Cheng M, Zhou J, Owusu KA, Mai L, Yu Y (2019) A new view of supercapacitors: integrated supercapacitors. Adv Energy Mater 9:1901081
45. Bhattacharya S, Roy I, Tice A, Chapman C et al (2020) High-conductivity and high-capacitance electrospun fibers for supercapacitor applications. Acs Appl Mater Inter 12:19369–19376
46. Chen K, Liu J, Bian H, Wang W, Wang F, Shao Z (2020) Dexterous and friendly preparation of N/P co-doping hierarchical porous carbon nanofibers via electrospun chitosan for high performance supercapacitors. J Electroanal Chem 878:114473
47. Zhang R, Wang L, Zhao J, Guo S (2019) Effects of Sodium Alginate on the Composition, Morphology, and electrochemical properties of electrospun carbon nanofibers as electrodes for supercapacitors. Acs Sustain Chem Eng 7:632–640
48. Cai J, Niu H, Wang H, Shao H et al. (2016) High-performance supercapacitor electrode from cellulose-derived, inter-bonded carbon nanofibers. J Power Sources 324:302–308
49. Deng L, Young RJ, Kinloch IA, Abdelkader AM, Holmes SM, De Haro-Del Rio DA, Eichhorn SJ (2013) Supercapacitance form cellulose and carbon nanotube nanocomposite fibers. Acs Appl Mater Inter 5:9983–9990
50. Manthiram A (2017) An outlook on lithium ion battery technology. Acs Central Sci 3:1063–1069
51. Dunn B, Kamath H, Tarascon JM (2011) Electrical energy storage for the grid: a battery of choices. Science 334:928–935
52. Zhou P, Zhang M, Wang L, Huang Q et al (2019) Synthesis and electrochemical performance of ZnSe electrospinning nanofibers as an anode material for lithium ion and sodium ion batteries. Front Chem 7:1–10
53. Han W, Xiao Y, Yin J, Gong Y, Tuo X, Cao J (2020) Fe_3O_4 @Carbon nanofibers synthesized from cellulose acetate and application in lithium-ion battery. Langmuir 36:11237–11244
54. Choi DI, Lee J, Song J, Kang P, Park J, Lee YM (2013) Fabrication of polyacrylonitrile/lignin-based carbon nanofibers for high-power lithium ion battery anodes. J Solid State Electr 17:2471–2475.
55. Karayeğen G, Koçum IC, Çökeliler Serdaroğlu D, Doğan M (2018) Aligned polyvinylpyrroli-done nanofibers with advanced electrospinning for biomedical applications. Bio-Med Mater Eng 29:685–697
56. Ngadiman NHA, Yusof NM, Idris A, Misran E, Kurniawan D (2017) Development of highly porous biodegradable γ-Fe_2O_3/polyvinyl alcohol nanofiber mats using electrospinning process for biomedical application. Mater Sci Eng C 70:520–534
57. Xia J, Zhang H, Yu F, Pei Y, Luo X (2020) Superclear porous cellulose membranes with chitosan-coated nanofibers for visualized cutaneous wound healing dressing. Acs Appl Mater Inter 12:24370–24379
58. Ajalloueian F, Tavanai H, Hilborn J, Donzel-Gargand O, Leifer K, Wickham A, Arpanaei A (2014) Emulsion electrospinning as an approach to fabricate PLDA/chitosan nanofibers for biomedical applications. Biomed Res Int 2014:1–13

Surface-Functionalized Electrospun Nanofibers for Tissue Engineering

Raunak Pandey, Ramesh Pokhrel, Prabhav Thapa, Sushant Mahat, K. C. Sandip, Bibek Uprety, and Rahul Chhetri

Abstract Electrospun nanofibers have been investigated for applications in diverse fields of tissue engineering such as degradable polymers, bioactive inorganics and nano-composites/ hybrids. Poly (ε-caprolactone) (PCL), poly (L-lactide-co-3-caprolactone) (PLLACL) and poly(lactic co-glycolic acid) (PLGA) electrospun nanofibers have been reported to be an effective scaffold for tissue engineering and drug delivery due to high surface-to-volume ratio, tunable porosity, cell affinity, hydrophilicity and ease of surface functionalization. In particular, electrospun fibrous scaffolds prepared by coaxial and co-electrospinning showed promising applications in adhesion, proliferation, elongation, cell growth and apoptosis which is highly desired for human body applications in tissues such as bone, nerve, ligament along with bio-artificial bone graft mimicking and bio-mineralization. Different characterization methods such as FESEM, SEM, FTIR, XRD and wet chemical precipitation have been used for these studies. Furthermore, a wide range of materials suitable for extracellular matrix scaffold has been prepared by electrospinning technique. This review summarizes preparation methods, functionalization and characterization techniques of nanofibers by electrospinning and their wide application in the field of tissue engineering. In addition, challenges pertaining to cell infiltration, low-density growth and inadequate mechanical strength of nanofibers as well as suggestions to mitigate these problems are also pointed out.

Keywords Electrospun nanofibers · Surface functionalization · Electrospinning · Tissue Engineering · Cell infiltration

Full Forms

FESEM	Field Emission Scanning Electron Microscope
SEM	Scanning Electron Microscope

R. Pandey (✉) · R. Pokhrel · P. Thapa · S. Mahat · K. C. Sandip · B. Uprety · R. Chhetri
Department of Chemical Science and Engineering, Kathmandu University, Dhulikhel, Kavre, Nepal

© The Author(s), under exclusive license to Springer Nature Switzerland AG 2021   315
S. K. Tiwari et al. (eds.), *Electrospun Nanofibers*, Springer Series on Polymer and Composite Materials, https://doi.org/10.1007/978-3-030-79979-3_12

FTIR   Fourier Transform Infrared Spectroscopy
XRD   X-ray Diffraction

1 Introduction

Tissue engineering applies the knowledge of nanotechnology, engineering and biology to engineer cells or a combination of cells for replacing the biological tissues that have been damaged due to natural conditions or injuries [1]. The process begins by seeding cells into a scaffold followed by incubation in a growth medium that helps the cells to multiply and grow across the scaffold forming a substitute tissue. The cells forming the substitute tissue are taken from the stem cells of the human body. It is vital for the substitute tissue to be able to mimic the native fibers of the site and to provide essential topographical and chemical cues for regeneration [2, 3]. The scaffolds used for tissue generation are either extracted from donor organs or artificially developed from biodegradable polymers. The lack of sufficient donor organs and the difficult extraction process have necessitated the development of artificial techniques for fabricating scaffolds [4, 5]. Several well-known methods have been developed for producing nanofibrous scaffolds such as self-assembly, solvent casting, gas foaming, phase separation, melt blowing and electrospinning. The scaffolds made from these techniques have good porosity, small diameter, superior mechanical integrity, good biocompatibility (with native tissues at the site) and low cost for application in tissue engineering [6, 7].

Among the different techniques, the electrospinning process produces high surface area fibrous structures with necessary cell attachment, proliferation, differentiation and stiffness properties required for growing a substitute tissue [8]. The electrospinning process uses an electric field to generate nanofibers from a solution of polymer using a setup shown in Fig. 1. In general, a polymer solution is held at the tip of the capillary tube forming the moderate viscous Taylor cone at the tip of the plunger. This polymeric solution is subjected to an electric field which causes a jet of the solution to eject onto the collector metal screen (Fig. 1). As the solution travels through the air onto the collector screen, the solvent evaporates producing a continuous layer of polymer fiber. The orientation and the alignment of the evaporated nanofibers are based on the behavior and type of collector used in the electrospinning process. The magnitude of voltage also determines the orientation. Nanofibers of polyesters [1, 3, 7, 9–17], poly-anhydrides [18–22], polycarbonate [10, 23], poly (ethylene glycol) [24] and natural biopolymers (collagen, gelatin, fibrin, lipids) [25–27] have been produced using the electrospinning process [6]. Synthetic polymers offer easier process-ability compared to natural polymers and provide more controllable nanofibrous morphology. These nanofibers have been investigated widely for use in different sectors such as drug release, bone tissue engineering, skin tissue engineering, bio-sensing, cardiac tissue engineering and other applications [28, 29].

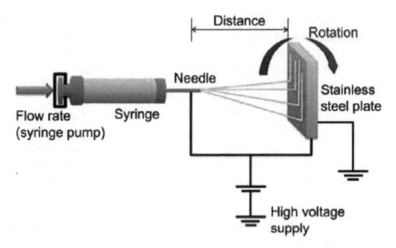

Fig. 1 Schematic illustration of the electrospinning setup. The mandrel can be rotated at various speeds to achieve different fiber orientations. Reprinted with permission from [2]. Copyright 2008, Elsevier

Although electrospun fibers are able to mimic the native tissues of the site, the fibers require surface treatments to match the bio-properties of the native site and to ensure the safety of the host site. In general, hydrophilicity, biocompatibility, stiffness, cell proliferation and antibacterial properties of the electrospun nanofibers are modified using the surface treatment methods. Surface treatment or functionalization is a common method that keeps the bulk properties of the material intact while promoting the biocompatibility of the scaffolds. To this end, plasma treatment, wet chemical approach, surface graft polymerization and surface coating are popular approaches for surface functionalization. Plasma treatment followed by wet chemical methods such as treatment with sodium hydroxide, potassium hydroxide increases hydrophilicity and improves biocompatibility. Wet chemical approach of alkaline hydrolysis and subsequent chlorination of nanofibers produce antibacterial polyacrylonitrile (PAN) nanofibers which enhances the cell attachment, hydrophilicity and mechanical properties [18, 19, 22]. Surface graft polymerization employs coating second material by physical adsorption. This method provides better stability by increasing cell attachment, fiber diameter and proliferation and by enhancing the hydrophilic nature of the cell. To improve their usability and efficiency, the nanofibers are often grafted before the electrospinning process to incorporate biological cues in the scaffolds [30]. The surface coating is usually coating the surface of nanofibers by molten or semi-molten substrates without any action of radiation or plasma sources. This modification enhances attachment, proliferation and other properties which helps to tune the scaffolds. Toward this end, Calcium Phosphate (CaP) coated Nylon-6 nanofibers have been produced which have been shown to promote cell proliferation [23]. Similarly, Polycaprolactone (PCL) fibers doped with silver nanoparticles have been shown to be highly beneficial in reducing bacterial infections in the implanted devices. Thus, surface treatment is observed to be an effective method to engineer the surfaces of electrospun nanofibers in two and three-dimensional structures to promote cell infiltration and growth [31]. Structural changes of nanomaterials are perceived after the surface functionalization. Increase of hydrophilicity is the main changes that is the nanomaterials will have more affinity toward water. This increase in hydrophilicity will enhance structural changes leading to an increase in water contact angle and wettability, proliferation, biocompatibility and so on. Mechanical strength is also increased, giving certain nanofibers a suitable candidate for use in grafts which are durable and strong. Improvement of covalent cross-linking after functionalization is also seen. Cell adhesion and attachment is an important characteristic which is obtained after surface functionalization. These help the nanofibers to remain at a fixed position. High surface area and increase in fiber diameter after functionalization ensures more contact to the surface of the cell leading to various applications of functionalized nanofibers in bone grafts, cardiovascular systems, cartilages, wound healing and so on.

Electrospun poly (lactide-co-glycolide) (PLGA) scaffolds are commonly used in biomedical engineering and drug delivery fields due to their improved cell attachment, proliferation and adhesion [28]. Anterior Cruciate Ligament (ACL) replacement parts are being developed with the recent advancements in electrospun nanofibers and the surface grafting [32]. Polycaprolactone nanofibers have applications in skin tissue engineering along with bone tissue engineering. Poly-L-lactic acid can help in cartilage tissue regeneration and Poly (L-lactide-co-3-caprolactone) in neural tissue engineering. G-Rg3/PLLA electrospun fibrous scaffolds can be used in wounded skin for patients with deep trauma, severe burn injury and surgical incision having potential application in human skins. Likewise, tendon, muscles, blood vessels and so on also have potential applications of tissue engineering.

The nanofibers are often characterized for their morphological and mechanical properties and product consistency throughout the synthesis process. Scanning Electron Microscope (SEM), Atomic Force Microscope (AFM) and Field Emission Scanning Electron Microscope (FESEM) are used to characterize morphologic properties of nanofibers. Similarly, the chemicals used in producing the electrospun nanofibers and their surface treatments are characterized by Fourier Transform Infrared Spectroscopy (FTIR), Nuclear Magnetic Resonance (NMR) and Raman Spectroscopy techniques. The bending test and Young's modulus test are performed using the AFM, SEM and TEM techniques to characterize the mechanical characteristics of the nanofiber. X-ray diffraction (XRD) is also used to analyze the structure of nanofibers [33–38].

Thus, the objective of this review is to discuss electrospinning, surface treatment methods, characterization techniques and the applications of electrospun nanofibers in tissue engineering. The future aspects, along with challenges faced, is also pointed out.

2 Electrospinning

Electrospinning is the most widely used process for creating scaffolds for tissue engineering applications among various methods such as solvent casting, gas foaming and phase separation. In particular, the electrospinning process is inexpensive, produces high aspect ratio materials and provides the ease of combining different biomolecules for tailoring the properties of the final structure. Electrospinning is a method of synthesizing nanofibers from a polymer solution by using the electric field. The solution is held at the tip of a capillary tube by surface tension and subjected to an electric field which induces a charge on the liquid surface. The induced charge develops a force directly opposing the surface tension in the liquid, and when this force overcomes the surface tension, a charged jet of the polymer solution is ejected from the capillary tube end towards the metal screen (Fig. 1). The trajectory can be controlled with the help of an electric field. The solvent evaporates while the jet travels through the air producing a continuous layer of polymer fiber that lay randomly in the collector metal screen [39]. There are various types of electrospinning processes such as uniaxial, uniaxial with emulsion, side by side, coaxial, nozzle and melt electrospinning.

Coaxial electrospinning is a leading technology to surface modify the nanofiber scaffolds using multiple feed needles instead of a single feed needle. The two feed needles are connected to the common spinneret that helps to produce the scaffolds of different mixtures. In the two feed systems, one feed usually contains a hydrophobic polymer and second feed contains water-soluble bioactive solutions [1]. Using this technique, substrates of heparin contained poly (L-lactic acid-co-ε-caprolactone) (PLLA-CL), Poly(lactic-co-glycolic) acid (PLGA) and polyurethane nanofibers have been produced for tissue engineering applications [1, 40, 41].

The polymer solution in the emulsion electrospinning process is usually polymers dissolved in an organic solvent. Emulsion electrospinning enables loading of drug solutions into polymer solutions for spinning into nanofibers. The emulsion solution for electrospinning is prepared by mixing drug solution and polymer solution with the help of an emulsifier. The rest of the apparatus set up is the same as the electrospinning process (Fig. 1). Using this method nanofibers with Poly (ethylene oxide)-Fluorescein iso-thiocyanate (PEO-FITC) inside copolymer of Poly (ethylene glycol) / Poly (lactic acid) (PEG/PLA), Doxorubicin hydrochloride (Dox) an anticancer agent inside copolymer of Poly (ethylene glycol) / Poly (lactic acid) (PEG/PLA), proteins in Poly (l-lactide-co-caprolactone) (PLCL) has been prepared [42–44].

Melt electrospinning is the process of electrospinning which uses polymer melt instead of the polymer solution for generating the nanofibers. The polymer melt is generated from the polymer rod by melting the polymer rod. These rods should be non-conductive and viscous [45]. Gases and radiations are mainly used to melt the rod. Radiation is generally not preferred as it leads to poisoning of the polymer and is also economically infeasible. Gases can be used either directly or through steel jackets. Use of gas is convenient and abundant to melt the polymer rod. Also, the diameter of the nanofibers after electrospinning can be controlled by controlling the gas flow rate, which is advantageous in surface modifications. Nanofibers of PCL (Poly(ε-caprolactone)) and PLA (Poly-lactic acid) have been fabricated using the melt electrospinning [46, 47]. The primary advantage of this method is the exclusion of the use of harmful organic solvents as used in other electrospinning techniques.

3 Components of the Electrospinning Process

3.1 Apparatus

Major components of the electrospinning process include syringe pump, voltage supply, needle and collector. A syringe pump is usually a plunger and plays a vital role to maintain the proper flow rate of the polymer solution. Generally used syringe pump is a single needle that creates core /shell fiber structure [15]. A programmable syringe pump is also available [24]. The voltage supply is between the collector and the needle. Change in voltage will change the acceleration of the electrospinning jet resulting in the change in the volume of the solution. Morphological changes in the scaffolds are experienced if the voltage is manipulated while electrospinning. The process includes DC focused potential (external voltage) to alter the mean free path of electrospun nanofibers, but in some cases, AC potential is also used giving advantages of less accumulation of like charges on the polymer surface [48]. A time-varying electric field can be used when deflection of the electric field is needed to produce nanofibers [49]. The needle is the release point of the polymer solution. For the industrial production, the apparatus is similar, but with little adjustment in the number of needles. Experiments in labs generally use a single needle, while industries use multiple needles to allow scale-up [50]. The collector plays an essential role in the alignment of fibers. The types of collectors are traditional (stationary plate), rotatory drum, parallel disc, rotating flat disc plate, rotating disc sharp and flat edge and conveyer collectors [15, 51]. The materials generally used for making collectors are (Aluminum foil [29, 52]), tin-foil [53] and stainless steel plate [54]. Collectors can also be modified to improve mechanical strength and cell infiltration of the nanofibers by employing drum, disc shape, and sharp edges collectors and so on. Modifications of the collectors can lead to the formation of nano-yarns, which are nanofibers formed from powdered polymers. Two collector system consists of a metal rod as the first collector and hollow metal hemisphere as the second collector. The surface of the collectors is insulated, resulting from deposition of electrospun nanofibers at the tip and the edge [55]. Porous nanofibers can be obtained with the help of non-conductive collectors [51].

3.2 Parameters Affecting Fiber Quality

The characteristics of the nanofibers can be controlled by the flow rate of the solution and applied voltage. In addition, polymer solution properties such as viscosity, conductivity, elasticity, concentration, environmental factors such as temperature and humidity also play an essential role. Voltage is known to affect the diameter of the nanofiber. The average diameter of the fiber initially decreases as the voltage applied is increased. The diameter reduces to the lowest point after which any further increase in the voltage increases the diameter of the nanofiber due to the increase in bead defects. However, some studies have shown that both the high and low voltage can lead to decrease in the diameter of nanofibers [15, 51]. The solution flow rate affects the pore size and the diameter of the nanofibers. Increase in flow rate leads to a rise in both pore size and diameter [51]. Low flow rate is generally preferred because the polymer solution will have more time for evaporation, and uniform nanofibers are formed. Any change in the flow rate is followed by the change in the voltage to obtain uniform nanofibers [56]. Collector distance plays a vital role in the size of nanofibers. Distance between needle and collector should be chosen carefully because too-far can lead to large diameter formation, and too-near can result in inadequate drying [51]. Increase in viscosity causes increases in intermolecular forces leading to an increase in fiber diameter [51]. These forces are challenging to break and hence to draw small nanofibers; the viscosity should be moderate. Volatility is an important factor to produce nanofibers as a low volatile solution cannot evaporate quickly and thus not suitable for forming nanofibers. A more volatile solution leads to an increase in micropores in the nanofiber structure. The increase in micropores leads to an increase in the surface area of the nanofibers, which is desired for increasing the reactivity [51].

4 Importance of Surface Functionalization for Electrospun Nanofibers

Surface modification techniques are the change in physical/surface properties of desired nanoparticles. As multiple processes have been described in the above section, the main goal is to alter the surface chemistry of the desired substrate. The modified surface of biomaterial is the first layer that connects with the body. For instance, in the dialysis process dialyzer will not work if a clot is formed in the surface of dialyzer [57]. The connection of biomaterial with body surface is mainly determined by the surface modification techniques. Where adhesiveness and connectivity is the main factor, the modification should be precise and accurate. In this scenario, the ion beam deposition technique is used. In the case of bone damage, the modified surface of nanomaterials containing osteogenic cell should be precise and accurate. This will lead to the formation of an adequate structure of bone and will give dynamic support to the body. This can be achieved only through perfect adhesiveness of the foreign material. In this scenario, the ion beam deposited nano-materials will be a suitable choice. Not only adhesiveness surface modification also enhances the biocompatibility of nanomaterials. The human body can contain a lot of compounds within a small vicinity. There may be lipids, fatty acids, cholesterol and so on. In the case of phospholipids of cell membranes, it contains the phosphate head group and two fatty acid chains. For instance, pure graphene with neutral charge can interact hydrophobically with lipid tails. This will extract all the cholesterol from the cell membrane and damage it. Similarly, it will affect the cytoplasm by penetrating due to their small sizes and sharp edges. But using the right amount and altering the surface chemistry will maintain the cell viability. For instance, it has been reported using graphene oxide with polyethylene glycol, poly (acrylamide) cell viability can be maintained up to 100% at 100 μg/ml [58]. Implanting different functional groups in graphene nanomaterials can be done by different surface modification techniques.

Polymeric nanoparticles are widely used in drug delivery systems due to their stability and ease of surface modification. To enhance the drug delivery, opsoniza-tion and prolongs circulation should be minimized. Opsonization refers to the coating of dangerous antigens. The target material such as microbes, molecules are modi-fied and develop a stronger interaction with surface receptors. It invades the specific antigens that can be done only by specific components called as opsonins. Simi-larly, in vivo circulation of nanoparticles is equally important. It enhances the non-uniform distribution of the deliverance system throughout the target vicinity. To prevent this phenomenon, nanoparticle should be coated with polymer surfactants with biodegradable copolymers. The presence of charge in the surface of nanoparti-cles agglomerates the particles. In drug delivery system, aggregation of particles should be strictly prohibited. It has been reported, modified nanoparticles with zeta potential will develop stable suspension [59]. Along with the targeted drug delivery, modified nanoparticles should be of good mechanical strength. Large vari-ation between untreated and treated nanofibers has been reported. Increase in tensile strength and young's modulus of sisal/polyester composites nanofibers has been reported after each modification step [60].

5 Surface Functionalization/Modification Techniques for Nanofiber Scaffolds

Nanofibers are incapable of activating specific cell functions and responses such as cell adhesion and proliferation until their surface is modified. These cell functions and responses are important for application in tissue engineering. An ideal scaf-fold must mimic the chemical and physical properties of the tissues for application in tissue engineering [26]. Surface functionalization/ modifications is a technique which involves changes or addition of some reagents in the surface of the nanofiber so that the fibers can bio mimic the microenvironments surrounding the cells and tissues. The surface is modified physically or chemically according to the properties of the scaffold after electrospinning. Modification enhances hydrophilicity, mechanical properties, chemical properties, biocompatibility and fiber diameter of the scaffold. Modifications do not compromise the initial physical properties and structures of the electrospun scaffold. Surface treatments are mainly performed after the electrospinning is done. It also points out to those methods which will lead to the improvement of adhesion, biocompatibility, strength and so on which can be used in tissue engineering for regenerative medicines. Surface functionalization generally comprises physical approach and chemical approach. Some of the techniques used for surface modification/functionalization are discussed below.

5.1 Physical Approach

Surface modifications via physical approach method have been widely adopted in the field of nano-engineering. This approach is practical, feasible and simple. Modifications in the surface of nanoparticle are done by bringing physical changes in the substrate. Deposition and layering are the major principles of physical approach. Some of the techniques for the physical approach are given below: (Fig. 2).

5.1.1 Ion Beam Deposition

Ion beam deposition techniques is one of the automated process used for surface modification. It works on the principle of ionization of the target material. It involves ionized inert gases, substrate and target material. Initially, a substrate is placed inside the chamber where low energy inert gas source is used for cleaning of the substrate. A high-intensity ionized inert gas is bombarded to the target material. The ionized inert gas transfers its energy to the target surface. This energy melt vaporizes and alters the target surface. The vaporized surface is deposited on the substrate and forms a new modified substrate's surface [62, 63]. An accelerator is used to accelerate the ionized particle. The ionized particle travels through the mass analyzer. Here, only selective ionized particle with certain mass is passed. In this way, process of deposition is achieved.

Ion beam deposition can be both assisted, induced and sputtered. They can be further classified as Ion Beam Induced Deposition (IBID), Ion Beam Sputtering Deposition and Ion Beam Assisted Deposition (IBAD). Among them, IBAD is most common because it omits the oxide formation in substrate surface and produces compact coating [62, 63]. Ion beam deposition has been widely used because of control over stoichiometric parameters such as sputtering rate and ion current density. It produces uniform deposition with high precision coatings that have low absorption and scatterings over other methods. It improves wearability, adhesive properties and biocompatibility of the nanomaterial. It has been reported that coatings with high densities, high refractive indices, low extinction coefficients and good mechanical

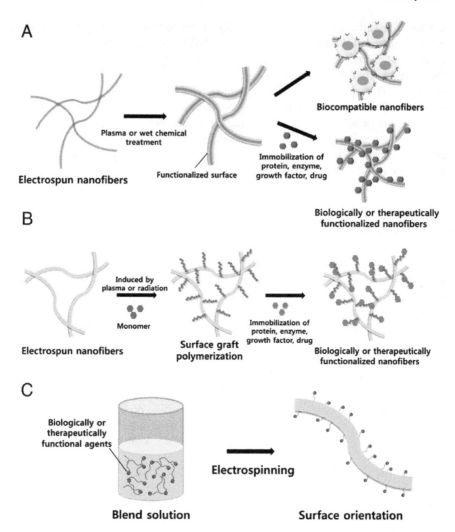

Fig. 2 Surface functionalization techniques of electrospun nanofibers. **a** Plasma treatment or wet chemical method. **b** Surface graft polymerization. **c** Co-electrospinning. Adapted with permission from [61]. Copyright 2009, Elsevier

properties for nanomaterial have been produced [62, 64]. Similarly, deposition of 10 nm INAs quantum dots in gallium arsenide substrate through this process has also been reported [64]. The enhancement of properties related to higher energy promoting rapid cellular adhesion and spreading in polyetheretherketone (PEEK) by ion beam deposition can help in biomedical applications [62]. Ion beam deposition created biomimetic apatite precipitation layers by emerging calcium phosphate thin film on pure titanium, helping to modulate precipitation processes and bioactivity enhancement [63].

5.1.2   Layer by Layer Deposition

Layer by layer deposition process is based on the principle of electrostatic inter-actions. Initially, a dipping of the charged substrate into a dilute aqueous solution of anionic polyelectrolyte is done. After a certain time, the polymer absorbs the charged particle. Then another layer of cationic polyelectrolyte is placed on the top of anionic polyelectrolyte. This process is repeated from time to time, and the layer of anionic polyelectrolyte followed by cationic polyelectrolyte is achieved. At each step, a new layer is formed, and the modified surface is achieved [65]. The result is the self-assembled layer of opposite charge. In this manner, the chemical stability of the film is achieved due to covalent cross-linking [63].

This deposition method is used where the chemically well-defined structure is not important [65]. Similarly, control over stoichiometric parameters such as depo-sition rate, deposition thickness cannot be achieved [65]. Though it cannot produce uniform coating, this process is cheap, less time-consuming and practical approach [63, 65]. The working principle is simple and widely used in MSC cells treat-ment therapy [63]. Deposition of 5 nanometer (nm) nanoparticle quantum dots on a gold substrate have been done by this method [66]. Similarly, deposition of tita-nium oxide (TiO_2) nanoparticles on multifunctional woven fabrics has also been reported [67]. Collagen/hyaluronic acid (Col/HA) polyelectrolyte multilayer film on titanium-based implants, ultrathin anionic polysaccharide support was created using the layer by layer deposition helping in proliferation, hydrophilicity, increased cross-linking properties on nanofibers [63]. Stainless steel stent with chitosan/heparin (CH/HE) multilayer functionalized with poly l-lysine (PLL), ultra-strong and stiff multilayered polymer nano-composites layer by layer coating on β-tricalcium phos-phate/polycaprolactone porous scaffolds (3DP bTCP/PCL) fabricated by 3D printing, antibacterial coating through the deposition of negatively charged gold nanoparti-cles (GNPs) and positively charged lysozyme (Lys) on highly porous electrospun cellulose nanofibrous mats have been used as functionalized nanofibers using layer by layer deposition [68].

5.1.3   Langmuir-Blodgett Method

Langmuir–Blodgett method is one of the commonly used methods for the deposition of the nanoparticles into a solid substrate. The substance to be deposited in the substrate is in solid-state in other modifications techniques, while in this method it is in a liquid state. Initially, a thin film of required deposition is developed. The development of film is done by adding a few drops of desired film substance in a liquid layer. The desired deposit and liquid makes a bond and film is developed over the liquid surface. A vertical substrate is placed in the middle of the liquid pool. After the complete immersion of the vertical substrate, the distance of barriers within the sides of the pool is decreased. This increases the pressure within the film surface. The substrate is gradually pulled out of the pool. The liquid film is then transferred to the substrate in this process. After repeating this process for a few times, a homogeneous layer of film is developed in the substrate [63, 69].

This method is based on the absorption of the film in the substrate. It is a powerful method for depositing amphiphilic molecules into a substrate at the air–water interface. Deposition of the film on the substrate depends upon the affinity of the film to the substrate [63, 70]. This method can be used in nano-medicines, nanoelectronics and so on. Formation of carbonated hydroxyapatite films on the metallic surface has been reported [70]. Similarly, fabrication of multifunctional nanoparticle has been done by this method [69]. This method is simple and can be widely used for flat substrate deposition.

5.1.4   Rapid Prototyping

Rapid prototyping is a fabrication method which is fully computerized and based on developing a template. Initially, a computer-aided design is drawn and fed into a system that can prototype without design error. The target metallic powders are melted by high-intensity laser beam and fabricated over the focused laser beam spot. The template is placed over the substrate, and target materials are passed over the substrate. This process is done in multiple types, and the desired prototype is fabricated all over the substrate [63]. It is a simple, inexpensive and rapid method. Nanoparticles between 10 and 100 nm can be fabricated easily. The templates used in this approach can be reusable. It is widely used in the field of nanoelectronics. The nanomaterials developed are uniform, compact and mechanically sound. Scaffold developed in this manner possesses a structured matrix with controlled porosity [63]. Deposition of silver (Ag) nanoparticles on TiO_2 precursors has been reported [71].

5.1.5   Surface Graft Polymerization

Surface graft polymerization is a simple, useful and versatile approach to improve surface properties by introducing radicals or groups like peroxide groups into polymer surface and then covalently bonding and polymerizing the monomers on the surface of scaffolds [48]. Surface graft polymerization improves hydrophilicity by decreasing water contact angle and exposing the peroxides to the polymer surface, increases cell attachment by chemically modifying the interior of porous scaffolds and improves the proliferation rate of scaffolds by instigating the polymer by free radicals through radiation or plasma [12, 18]. Surface modification tunes wettability, changes the chemical composition and creates a suitable environment for adhesion [18]. Surface collagen grafting has also become popular along with surface graft polymerization where collagen is added to the surface of the polymer after the action of plasma or radiation.

Radiation is often used to assist in surface grafting as it helps in the generation of reactive sites to polymer surfaces. This method is relatively simple and used to initiate the polymerization as it does not require any chemical additives or catalyst on the polymer surface. Penetration depth and graft yield can be easily controlled by gamma-ray irradiation. Radiation method successfully cultivated human mesenchymal stem cells (hMSCs) by grafting acrylic acid on Poly (lactide-co-ε-caprolactone) (PLCL) film [1]. Structural changes may occur, but inner surface modifications are difficult due to variability in penetration depth [1]. Surface graft polymerization, when used on PCL surface by air plasma treatment physically, absorbed the gelatin to PCL surface [12]. The process maintained the expression of key markers typical for gelatin, improved cell attachment and growth of fibers after culturing with endothelial cells [12].

Radiation is a common way of surface penetration followed by grafting but krypton radical (Kr+) irradiation, when done to PLLA nanofibers, improves cell attachment with the cost of cell damage to make the surface thin [8]. However, discarding the damaged cells can lead to a refined larger surface area and spaces for cell attachment [8]. Cell adhesion can also be seen by the induction of new functional groups and carbonization [8]. Also, when argon radical (Ar+) irradiation is done to PLLA, fibronectin adsorption is increased along with an increase in cell attachment and excellent biocompatibility of PLLA nanofiber scaffolds [8]. When ultraviolet (UV) light is passed to nanofiber mats, cell cytotoxicity was decreased, and positive control for cell viability was seen with no morphological changes [72].

Adding collagen and solvents after electrospinning has also become an unvarying method for surface modification techniques. Residual solvents can also affect the biological performance of scaffold. Biological and structural changes during processing are observed in natural collagen, although being relatively non-immunogenic. Type-I collagen, when dissolved in hexafluoropropanol (HFP) has a negative impact on cell culture [25]. Biocompatibility and hydrophilicity of the scaffolds may be increased when collagen is added. The unrestricted somatic stem cells (USSCs) prefer extracellular matrix (ECM) for attachment than in the surface of polyethersulfone (PES). Also, better morphological characteristics can be seen at these collagen sites than plasma-treated sites. Proliferation and attachment of USSCs on natural microenvironment is also managed by the collagen [25, 26]. Grafting of poly trimethylsilyl-propargyl-hexaethylene glycol methacrylate-co-OEGMA (TMS-Prg-HEGMA-co-OEGMA) brushes from styrene-based copolymer surfaces was done for the surface graft polymerization of protected alkyne monomers. The polymerization of these alkyne monomers was through atom transfer radical polymerization (ATRP). The copolymerization of 37, 47 TMS-Prg-HEGMA with OEGMA475 formed polymer brush coating on fiber surface for effective SI-ATRP. The grafting of fiber and de-protection of the collagen was due to the formation of the polymer brush, and this formation confirmed that fiber structure was not altered by the reaction. The observation of minimal contrast between solid electrospun core and polymer brush coating was seen which was not improved by attachment of azide-functional gold nanoparticles and staining protocols [21]. After surface modification using collagen and surface graft polymerization, hydrophilicity, cell attachment and proliferation will increase to help make the scaffolds to be used in tissue engineering [26].

5.1.6 Surface Coatings

Surface coating is a process of reinforcing the surface functions of the substrate by depositing the layer of molten, semi-molten or chemical materials onto the surface [73]. Surface coatings with compounds enhance the physical and chemical properties such as the increase in fiber diameter, proliferation and attachment of the electrospun nanofibers. Hydrophilicity is also enhanced when the polymer surface is coated. The surface coating does not require penetration by plasma or radiation, unlike the surface graft polymerization technique.

Mussel inspired coatings to produce proteins have applications of adhesives in spite of wet conditions. Also, cell attachment and proliferation of endothelial cells were increased when PCL was coated with polydopamine [1]. Cell adhesion rate and proliferation of hMSCs were increased when polydopamine was coated to PLLA electrospun nanofibers [1]. This coating further increased osteogenic differentiation and calcium mineralization of hMSCs [1]. Calcium phosphate (CaP) coating on nylon-6 (N6) fibers also increased the amount of hydroxyapatite (HAp) which enhanced surface roughness and coated layer thickness [23]. These coating increased the fiber diameter and improved the crystalline phase of N6. CaP coatings increases the hydrophilicity of the nanofibers, enhances bone tissue growth and proliferation with an increase in cellular attachment and maintenance of surface morphology [23]. Surface chemical composition, surface roughness, topography and wettability enhances the biocompatibility and biological, environmental functions on Nylon-6 nanofiber when coated with CaP [23]. Tensile strength is increased after coating with CaP on Nylon-6, but the nanofibers become brittle when the HAp layer is increased [23]. For coating poly (lactide-*co*-glycolide) (PLGA)/gelatin nanofiber with CaP coating, soaking time of the nanofiber was increased along with agglomeration [27]. The hydrophilicity was also enhanced when this coating is applied [27]. When N6 nanofibers are coated with hydroxyapatite, homogeneous nucleation, and fast precipitation of hydroxyapatite is experienced, which increases layer thickness resulting from extended reaction time and pressure [10]. The coating of poly-pyrrole on electrospun PLGA nanofibers has an application on neural tissues. These coatings make the nanofiber non-cytotoxic and influence uniformity, conductance and morphologies of the nanofibers. In addition, the fiber diameter is increased [11].

5.1.7 Electrospinning

Electrospinning is a process of synthesizing the nanofiber scaffolds from the polymer solution by using an electric field. Electrospinning is the basic requirement to produce nanofibers for tissue engineering applications, but by different methods of electrospinning, the physical and chemical changes in the nanofibers can be altered and modified. Coaxial electrospinning, co-electrospinning, blend electrospinning are some of the modified techniques of electrospinning to modify the surface of nanofibers. Coaxial electrospinning uses multiple feed system, electrospinning two or more polymer solutions from coaxial capillaries [74]. Co-electrospinning is a process of simultaneously collecting scaffolds onto a single collection device [18]. Blending drugs with a polymeric solution before electrospinning is blend electrospinning [18]. Localization of biomolecules by co-electrospinning allows sustained release profiles. This method bio-functionalizes the fiber in situ and the interspersed fibers from different molecules are produced from concomitant electrospinning from two or more spinnerets. Co-electrospinning combines the natural and synthetic polymer which has advantages of modulating the chemical, physical, mechanical, bio-resorb ability profile and degradation properties.

Composite scaffolds are obtained from co-electrospinning Poly (L-lactic acid) and gelatin from two distinct spinnerets which improves morphology and mechanical properties [7]. Loss of biological activity of biomolecules by denaturation or aggregation leads to a conformational change in organic environment so, preservation of the protein activity must be done for successful growth factor delivery, and appropriate concentration of growth tissue is needed [1]. Coaxial electrospinning is used to functionalize nanofiber surfaces by generating core-shell structured nanofibers. The good properties of cell attachment and proliferation of smooth muscle cells and endothelial cells are shown by coaxial electrospinning of Poly (L-lactide-co-3-caprolactone) (PLLACL) solution as shell and the phosphate-buffered saline solution containing protein as a core. The average strength of PLLACL was, however, higher than the nanofibers with Bovine Serum Albumin (BSA) at the core [14]. Poly (glycerol sebacate) (PGS) based electrospun nanofibers was produced from coaxial electrospinning by blending PGS with poly (ε-caprolactone) (PCL) [3]. When electrospun fibers containing PGS and poly (L-lactic acid) (PLLA) were produced from coaxial electrospinning, fiber diameter increased having a higher concentration of PGS [3]. The use of PLCL/HA core-shell matrices from coaxial electrospinning reduced the biocompatibility of the scaffolds [17]. The HA concentrations should be higher to enable smooth electrospinning. From coaxial electrospinning, hydrophilicity was improved, which aids on cell attachment, proliferation and differentiation by decreasing the water contact angle of the scaffolds.

5.2 Chemical Approach

Different chemicals require the activation of the surface, i.e., synthesizing the functionality of nano biomaterials on the surface. The amine-functionalized graphene oxide (AGO) provides improved modulus, increased proliferation of hMSCs, increased osteogenesis of stem cells and inhibited biofilm formation compared to PCL graphene oxide composites and reduced graphene oxide. Alkaline hydrolysis, covalent adsorption and the wet chemical process are some of the most common methods for chemically modifying substrates surfaces [63].

5.2.1   Adsorption via Covalent Bonding

In comparison to the physical adsorption process, the substrate and adsorbed molecules are binding in the covalent bonding method or by electrostatic interactions. This approach is more surface-resistant than physical adsorption due to its covalent bonds or electrostatics. In terms of cell growth and bodily fluid movement, it is also strongly bio-monitored.

The covalent linkage between the reactive groups on the surfaces is the primary determinant for the physical and chemical properties of the biomaterial changed by the surface. There have been experiments in which covalently adsorbed peptides incorporated in human vitronectin with oxidized titanium substrate has improved endogenous implant inclusion in the process of bone tissue engineering. Also, it has the ability to regulate the actions of neuronal stem cells for therapy applications by adaptation with different signaling molecules by covalent bonding of conducting PEDOT (poly (3, 4-ethylene dioxythiophene)/GO (Graphene oxide) scaffold. In combination with other polymers for tissue engineering purposes, new biomaterials can also be created through the enzymatic functions of natural polymers, including chitosan. Another study has also shown that in the presence of porcine pancreatic lipase, copolymer (poly-lactic acid chitosan) can be produced that enhances the adhesiveness of cells and can be used as a scaffold in tissue engineering.

5.2.2   Alkali Acid Hydrolysis

In Alkali hydrolysis process, ester bonds are cleaved to cause the formation of carboxyl and hydroxyl functional group by diffusing protons between the polymer chains, which induces hydrolysis of the surface. The hydrophilicity of Poly-Based Scaffolds can be improved by this process. When citric acid is washed, the water contact angle on the PLLA film surface decreases and surface roughness increases significantly.

5.2.3   Self-Assembled Monolayer

Self-assembling is an energy-efficient method used to prepare modeled monolayers on biomaterial surfaces on the substrate. Intermolecular or inter-particle forces at a critical point facilitate a normal molecular arrangement that minimizes the total free energy of the entire surface. In an area from the nanoscale to the micro-scale, molecules or nanoparticles can self-assemble. Many material forms, such as block copolymers, nanospheres, nanoparticles and biomolecules, have already been used to auto assume patterns.

Endothelial cells cultured on a gold substrate covalently adsorbed by the base fibroblast growth factor (bFGF) achieved the highest rate of proliferation when grown on carboxylic-ended self assembled monolayer (SAMs). In stem cell, research nanoparticles also have been utilized that functioned with monolayer SAMs. Also recently developed, the nanotopography-mediated reverse uptake (NanoRU) delivery system successfully transfers siRNA to neural stem cells (NSCs). The surface charge on SAMs could therefore be concluded to play an important part with respect to the binding characteristics of various substrates and of ligands/biomolecules.

5.2.4 Plasma Treatment

Plasma treatment is the process of immobilizing polar groups such as carboxyl, hydroxyl or amine moieties on the surface of the scaffold by the action of plasma. High energy plasma which impinges on the surface of the scaffold alters the wettability and hydrophilicity of the scaffolds by exposing the polar groups of the polymer and decreasing the water contact angles [15]. Hydrophilicity enhances fibroblast adhesion, proliferation and biocompatibility of the scaffold allowing steady seeding process of the scaffolds [1, 18]. The surface morphology of plasma-treated fibers becomes rough and is able to control surface roughness and to crosslink during the process of graft polymerization and thin-film coating [75]. Plasma treatment with air, oxygen, argon, nitrogen and acetylene is mostly common.

Argon plasma treatments have been performed to remove surface contaminants and produce ex-situ oxygen surface functionalities for poly(L-Lactide) (PLLA) nanofibers [13, 75]. Surface wettability and surface roughness properties of the nanofiber were found to increase with the plasma power [13, 75]. The decrease in average fiber diameter, increase in cell attachment, wettability and modification of fiber morphology to increase biocompatibility were experienced after plasma treatment with argon oxide (ArO_2), nitrogen and hydrogen ($N_2 + H_2$), and ammonia and oxygen ($NH_3 + O_2$) to polycaprolactone (PCL) nanofibers [75]. Oxygen plasma treatment was able to remove residual organic impurities and weakly bound organic contamination on the anatase titania nanofiber mats/ indium tin oxide (TiO_2-NF/ITO) surface. Ultra-clean surface for biomolecules loading was also offered after the plasma treatment with oxygen [76]. Thus, modification of surface hydrophobic or hydrophilic characteristics, etching and nano-texturing of polymer surface and improving mechanical properties by treatment conditions are prominent advantages of plasma treatment [16].

5.2.5 Wet Chemical Method

Wet chemical method involves generating reactive functional groups on random chemical excision of ester linkages on polymer backbone by treatment with mild acids or bases. With the wet chemical method, chemical functional groups are not generated, but carboxylic and hydroxyl groups on conventional hydroxyl-esters are normally seen. Surface-modified nanofibers with minimal change in bulk properties are produced by determining the duration of hydrolysis and concentration of hydrolyzing agent [1]. The wet chemical approach provides flexibility for surface modification and helps to enhance hydrophilicity by decreasing the water contact angle. Biocompatibility through surface immobilization and an increase in mechanical strength is experienced when electrospun nanofibers are immersed into the acidic or basic medium.

When poly (ε-caprolactone) (PCL) /polyacrylonitrile (PAN) is electrospun to form highly aligned nanofibers while engineering the parallel aligned myoblasts and myotubes, PAN modified the surface of PCL. After modifications, the alignment of the topography of nanofibrous scaffolds on cell viability was prevalent on the myoblasts tissues [18]. An acidic and alkaline solution such as sodium hydroxide (NaOH) can create nanotopography by modifying the surface wettability of electrospun polyesters. Surface wettability is induced by partial surface hydrolysis of the nanofibers. In the case of poly (ε-caprolactone), NaOH hydrolyzes the ester bond by random chemical scission to change the surface morphology in the more agglomerated form which influences cellular attachment and proliferation [9, 12]. This scission exposes hydroxyl and carboxyl groups to the polymer surface, which improves the hydrophilicity and decreases the water contact angle [12, 77]. Polyacrylonitrile (PAN) nanofibers modified by loading ferric oxide (Fe_2O_3) were found to have increased the mechanical strength and uniformity [19]. However, the challenges of agglomeration and particle dispersion were also prevalent [19]. The mechanical integrity of scaffold is also influenced by the concentration of hydrolyzing agents and submersion time [12]. In the case of chitosan/ hydroxyapatite nanofibers, the chemical method showed the spindle-like morphology, which delayed cell attachment and proliferation [78]. When cellulose acetate is hydrolyzed with NaOH better biocompatibility was observed, which resulted in enhanced cell adhesion and proliferation along with improved cellular response [79]. The degree of crystallinity and hydrophilicity of poly (vinyl alcohol) (PVA) increases when treated with methanol, but the reduction of cell proliferation was not seen [72]. Poly (ethylene glycol) (PEG) loading in Poly(lactic acid) (PLA) nanofibers showed an increase in diameter and uniform morphology of PLA [24]. Hydrolysis by potassium hydroxide (KOH) and chlorine loading in polyacrylonitrile (PAN) nanofiber was done where less number of chlorine exhibited effective biocidal functionality, and more number of chlorine resulted in slow cell inactivation [22]. Hence, immobilization of bioactive molecules is mainly done by applying chemical modifications where immobilized molecules are covalently attached to nanofibers and aren't easily leached out.

5.2.6   Click Chemistry

A recent approach in surface modification by click chemistry has been of interest to many scientists in tissue engineering. Click chemistry is 1,3-dipolar cyclo-addition of azide and alkyne to form 1,2,3-triazole applied to functionalize nanomaterials to rapidly create scaffolds necessary for tissue engineering [80]. Click chemistry induces click reactions which can be used as a powerful method for surface modifications due to the high efficiency and excellent selectivity under mild reaction conditions.

4-dibenzocyclooctynol (DIBO)-terminated PCL nanofiber is modified by performing conjugation of azide-containing peptide and azide-containing gold nanoparticles onto DIBO-terminated poly (γ- benzyl-L-glutamate) (DIBO-PBLG) nanofibers. Conjugation is done via the strain promoted azide–alkyne cyclo-addition (SPAAC) where SPAAC provided efficient conjugation of the nanoparticles [20]. Utilization of photo-initiators and UV-irradiation for functionalization of alkene bearing nanofibers in click reactions are also prominent. Enhancement of hydrophilicity and anti-biofouling characteristics are provided by TEG groups, and reactive handles are provided by polymeric precursors of polylactide-based copolymer with furan groups when nanofibers are formed. When nanofibers are conjugated with cell adhesive peptide, cRGDfK provided cell adhesion and prolif-eration. A poly (styrene-co-VBC) macro-initiator was synthesized with smooth morphology, and fiber size also increased to allow cell infiltration into scaffolds [20, 21] (Table 1).

6 Characterization of Nanofibers

Nanofibers are differentiated in the synthesis process by correlating test methods with functional properties of the materials and confirming the product quality. Basic data and information can be obtained from the single fiber characterization for under-standing the structural relationship and properties of nanofibers. The method of char-acterization is obviously still in its initial phase, and the demand for the production of effective characterization techniques has steadily increased. The study focuses on chemical characterization, as well as the physical characterization of nanofibers.

6.1 Physical Characterization of Nanofibers

Physical evaluation of nanofiber scaffolds after surface modification is done to observe the morphological changes in the scaffolds. Morphological properties such as porosity, fiber diameter and length of the scaffold are usually analyzed. A scanning electron microscopy (SEM) scans a focused electron beam to form the image of the sample by producing various signals. The surface of the sample in SEM is sputtered and coated with a thin layer of gold, platinum-coated aluminum or coated palladium-platinum-gold or palladium-platinum or aluminum/palladium [11, 14, 17, 23, 24, 81–87]. Field emission electron microscopy (FeSEM) is applicable to analyze the morphological behavior of the samples at fairly high resolution and provides surface or whole small particles topographical information to determine the elemental composition of the surface of the sample [34–36, 76, 88]. Furthermore, the SEM and FeSEM were used to analyze the structure and morphology of nanofiber for PLA [89], PCL/PEI and PLC [2, 90, 91], PES [26], PLLA [8], PGS-PMGA/gelatin

Table 1 Advantages and disadvantages of functionalized nanofibers [57–70]

S.N		Approach	Advantages	Disadvantages
1	Physical	Ion beam deposition	Good adhesiveness and biocompatibility of deposited material, packing with high densities, high refractive indices, low extinction coefficients, Compact packing and good mechanical properties. A fully computerized process with Easy monitoring of stoichiometric parameters	A complex process with a time-consuming, expensive and lower spatial resolution
2		Layer by Layer Deposition	Simple working chemistry, cheap, Deposition of multiple layers and Less time-consuming	No control over stoichiometric parameters such as deposition rates, deposition thickness with a less chemically well-defined structure
3		Langmuir- Blodgett	Control over film thickness, substrate deposition and barrier pressure with simple, easy, less time-consuming with the cheap process	Mixing of other phases while multiple layering, resulting dis-oriented structure
4		Rapid prototyping	A simple, easy, cheap computerized process with less design error and time Good control over structural integrity and porosity of the matrix	Frequent monitoring of vibration suppression and fiber solidification
5		Surface graft polymerization	Simple, easy and versatile approach for incorporating polymer in solid nanoparticles, with good adhesion of the attached graft. Control over-penetration	Poor graft density in grafting to and grafting through method due to the limited number of polymer chain and slower chain reaction
6		Surface coating	Simple and easy and cheap process with an increase in hydrophilicity. No use of harmful radiations or plasma	Tends to increase the stress in the outer side of nanoparticles that may lead to internal stress

(continued)

Table 1 (continued)

S.N		Approach	Advantages	Disadvantages
7	Chemical	Adsorption via covalent bonding	Provides strong bonding to biomolecules	Loss of functional conformation of biological molecules
8		Alkali acid hydrolysis	Increases roughness of the surface with a short time of functionalization	Requires exact conditions, permanent surface functionalization
9		Self-assembled monolayer	Limits to small nanofibers from nm to microns for fabrication	The high cost of synthesis, complex process
10		Plasma treatment	Low-temperature treatment, low operating costs, environmentally friendly	Adaptation mechanism is effected, depth of plasma penetration is increased.
11		Wet chemical	Versatile and economical, easy scale-up	Difficult to achieve a uniform layer system, the external stimulus is required
12		Click chemistry	High selectivity, efficiency, biocompatibility and stability	Expensive substrate material, the problem of heating requirement issues in certain cases

[3], PLGA/gelatin [92], HEC/PVA [93], PCL-MWCNTs [94], DTX/PDLLA [95] and PLO/PLEY [96].

Transmission electron microscopy (TEM) is technique of microscopy in which a beam of energetic electrons is transmitted through a sample and an image is formed by electrons interacting with the sample. TEM has been widely used in characterizing nanofibers core-shell structure, internal morphology, crystalline structure and membrane elemental information [17, 36, 40, 76, 97]. The samples for TEM observation were prepared by depositing the nanofibers onto a copper mesh or grid [17, 38, 78, 81, 97].

Atomic force microscopy (AFM) is known as a widely used technique for the characterization of nanomaterials which gives information about topography, morphology and particle/grain distribution from the surface of the sample. The topology of nanofibers including polycaprolactone [98] and morphology of Poly (l-lactide) (PLLA) [13], polyvinyl alcohol (PVA) [99] and polycaprolactone [100] has been investigated through AFM images.

Mercury intrusion porosimetry is the technique based on the property of mercury that does not wet the surface of solid materials [41]. The analysis done by the mercury porosimetry indicated that the porosity of the electrospun copolymer nonwoven was more than 80% with a median size of 8 micrometer (mm), total pore area of $5m^2/g$ and pore size of (10–200) mm [101]. Similarly, another study showed porosity ratios, and pore size PHB ranges varied from 62 to 83% and from 0.4 to 8 μm, respectively [81].

Capillary flow porometry is a well-known process for measuring pore size distributions in polymer membranes and fibrous media. It offers a simple and non-destructive technique that enables rapid and accurate pore size and distribution measurements [102]. Structural pore properties of electrospun PCL/gelatin [102] and PCL with natural polymers [103] were done by capillary flow porometry.

6.2 Chemical Characterization of Nanofibers

The chemical composition, functional groups and chemical bonding of the nanofibers are analyzed to investigate the creation of scaffolds having these characteristics. The Fourier Transform Infrared Spectroscopy (FT-IR) is a method of investigating the composition of organic and inorganic materials based on the chemical bonds and functional groups that have different characteristics of energy levels. FTIR collects and transforms interference pattern data to a spectrum that offers information about molecular transmission and absorption that gives the materials a molecular fingerprint. The FTIR spectroscopy were used to analyze the functional groups of many nanofibers such as PCL [38, 91, 98, 102, 104], H-PCL, PCL-matrigel [91], PVA, PVA/γ-Fe$_2$O$_3$ [34], Nylon 6 [23, 35], non irridated and Kr$^+$ irridated [8], PLLA [3, 83, 105], PLLA/HA, PLLA/collagen/HA [83], (PANI-CNT)/PNIPAm-co-MAA [86], HEC [93, 106], HEC(PVA) [106], TA-g-PCL [105], PLA/GO, PLA/GO-g-PEG [107], mineralized PLGA [27, 36], PCL/P3ANA [108], PLCL and PLCL/HA [17], PLGA/gelatin [27, 92], PANi-CSPA, pure PCL, PCL/PANi [55], PET, PVA/PAA [109], silica/PVP [110], PCL/gelatin [102], CA [109, 111], CA-DA [111], PCL/SF, MeOH PCL/SF [38], PAN, PAN Fe$_2$O$_3$ [19], CA/CS/AG, CA/CS/MWCNTs [79], γ-PGA [112], raw DEX, PCL NFMs [113], HA-NPs [6], PolyNaSS, ungrafted PCL [114], AM/ST/PEO [115] and PLO, PLEY [96].

X-ray photoelectron spectroscopy (XPS) is one of the most powerful techniques that measure the chemical composition, elemental and molecular distribution of the material's surface in the range of (5–10) nm. The x-ray in the source stimulates the emission of photoelectrons and photoelectron kinetic energy is measured by the analyzer, and the detection of emitted electrons provides the information for quantitative and qualitative analysis. XPS was used to investigate the chemical composition of the core-shell structure of electrospun nanofibers of PCL, H-PCL, PCL-matrigel [91], SA/PEO [85], Hap /CTS [78], pure PCL and blended PCL/PANi [55] and PPy-PLGA [11] by measuring the binding electrons associated with atoms [85]. The C 1 s core-level spectra of PCL fitted into three peaks components about 284.6, 286.1 and 288.5 eV ascribed to C–H, C–O, O = C–O species, respectively.

Raman spectroscopy is a technique to inspect the vibrational, rotational and other low-frequency modes. From the analysis of vibrations, useful information such as functional group, chain orientation, interfacial properties of polymeric composition can be revealed. Electrospun cellulose [116], carbonization [8], N6 fiber [23], graphite, synthesized GO, electrospun fibers PLA, PLA/GO, PLA/GO-g- PEG [107], TiO_2-NF [76] and CaP [117] were characterized using Raman Spectroscopy.

6.3 Mechanical Characterization of Nanofibers

Mechanical strength for the durability of the scaffolds to be used in tissue engineering must be tested and analyzed. The tensile test is mainly performed where electrospun nanofibrous scaffolds are cut into rectangular samples [85, 91, 94, 104, 118] to perform the tensile testing. Instron tensile tester [91], universal testing machine [55], tabletop tensile tester [118] are applied to determine the tensile properties such as tensile strength [55, 118], elongation [55, 104, 118], yield point [104] and Young Modulus [55] from the stress–strain curves of electrospun nanofibers.

6.4 Structural and Thermal Evaluation of Nanofibers

Techniques are provided to determine thermodynamic data and the evaluation of whether the nanofibers can endure the thermodynamic changes when applied in tissue engineering. X-ray diffraction (XRD) is a non-destructive technique for characterizing the crystalline materials and also the chemical composition and physical properties of materials. XRD has been used to study the crystal structure in several nanofibers including TiO_2 [33], PVA-γ-Fe_2O_3 [34], N 6 [35], HAp/CTS [78], HEC/PVA [52], PLGA [36], PA-6,6 [37], PCL [38, 98, 100, 104], PCL/SF [38], HA [10, 36], HA/chitosan [88], PCL-MNA [119], HA-NPs [6], TiO_2-NF [76], Fe_3O_4, PUA/ Fe_3O_4, PANI/PUA/ Fe_3O_4 [120], PLLA [121], PCL@MS [104] and AM/ST/PEO, AM/ST/PVA [115]. The X-ray spectrum was performed using Cu Kα radiation with recommended operating conditions [33, 37, 76, 88, 115].

Differential scanning calorimetry (DSC) is a thermal technique in which the difference in the amount of heat required to increase the temperature of a sample and reference is measured as a function of temperature. The thermal and crystalline properties (melting temperature [34, 94, 100], crystallization [34, 94, 100, 119] and glass transition [94]) for different electrospun nanofibers such as PVA, PVA-γ-Fe_2O_3 [34], HEC/PVA [93], PCL-MWNTs [94], PCL NFMs[113], PCL-MNA [119] and PCL [100] have been assessed by DSC process.

Thermo gravimetric analysis (TGA) is especially useful for identification of compositions in materials by decomposing all organic polymer content at high temperatures. Properties of decomposition temperature [37, 116], thermal stability [106, 107, 122] and residual weight [35] of electrospun nanofibers such as cellulose [116], Nylon-6 [10, 35], CNC-ZnO [123], PGS-PMMA [3], HEC, PVA, HEC/PVA [106], PLA, PLA/GO, PLA/GO-g-PEG [107], PA-6,6 [37], pristine PVP, precalcined silica/PVP [110], CS/MWNTs, CS/ALG [79], HEC/PVA [93], PCL [122], PAN [22] and PCL@MS [104] have been analyzed by TGA.

7 Applications of Electrospun Nanofibers in Tissue Engineering Scaffolds

Various studies on tissue regeneration with the use of electrospun scaffolds have been performed. These scaffolds are first surface-functionalized and characterized before applications in the human body mainly in bones, cartilages and neural tissues. Some of the applications are discussed in detail below.

7.1 Bone Tissue Engineering

There has been wide research in bone tissues using electrospun nanofibers. PCL nanofibers have been extensively used for bone tissue engineering [38, 98, 104, 124]. It was seen that the core-sheath structured PCL/silk fibroin (SF) nanofibrous scaffolds provide desirable mechanical, biological properties and is potential of controlled drugs release [38].

In addition, the currently developed MS-shelled PCL hybrid nanofiber has also demonstrated outstanding properties for bone [99] favorable such as vitro bone bioactivity, mechanical functionality, osteogenic stimulation of stem cells and the loading and delivery capacity. These outcomes are considered to be useful as nano-bio matrix platforms therapies for the repair and regeneration though the further studies of bone regenerative ability of scaffolds in vivo are still remained [104]. Similarly, electrospun cross-linked HA containing chitosan nanofibrous scaffolds subsequently cross-linked with genipin are potential candidates for cranial and maxillofacial reconstruction [88].

The PANI nanofibers have been found to increase the biocompatibility of PES [125]. Similarly, the nanofibrous membrane of chitosan/PEO with a ratio of 9:1 has retained good structural integrity in water and exhibited better adhesion of chondrocytes [87]. It is also anticipated that the natural bone ECM in terms of nanoscale structure and chemical composition is a good choice of biomimetic bone tissues [36]. Biological in vitro cell culture with human fetal osteoblast (hFOB) cells for up to 15 days has demonstrated that the incorporation of HAp nanoparticles into chitosan nanofibrous scaffolds led to significant bone formation compared to that of the pure electrospun CTS scaffolds [78].

Other electrospun biomimetic PLLA/coll/HA nanofibers [83], PLLA/HAP/PCL nanofibers [124], the addition of gelatin in nanofibers [92, 110] and PLLACL–collagen (3:1) nanofibrous mats [97] have been examined. They have shown excellent bioactivity [124], adhesion [83, 92], hydrophilicity [97], proliferation [83, 92, 97, 98, 124, 125], mineralization of osteoblasts [83], osteogenic differentiation of MC3T3-E1 cells by increasing cell numbers, ALP activity, osteocalcin concentration [98, 124], for bone tissue regeneration. It is worth noting that mineralized

PLGA/gelatin nanofibers are a suitable choice because of their high efficiency of mineralization [27].

7.2 Neural Tissue Engineering

One of the most promising methods to restore nerve systems in human health care is nerve tissue engineering. The role of scaffold is significant in nerve tissue engineering. Many studies of nanofibers are done in nerve tissue engineering.

The Nerve Growth Factor (NGF) released from coaxial electrospun PLLACL nanofibers was studied by observing the differentiation of PC12 cells into neurons, in the presence of the supernatant obtained from the electrospun NGF which revealed that the PLLACL fibers retained at least some degree of bioactivity for up to 10 days [14]. In another study, rat PC12 cells seeded onto PGS-PMMA/gelatin nanofibers showed the potential to induce the differentiation of PC12 cells into neuron-like cells even in the absence of any nerve growth factor or chemical treatment [3]. Likewise, the TS PCL-NFs nanofibers, compared to ES PCL-NFs, have shown a positive impact on the development of neurons due to the enhanced tensile properties [100].

In vitro studies using C17.2 nerve stem cells on aligned PCL/gelatin, nanofibrous scaffolds have shown that the direction of nerve cell elongation and neurite outgrowth is parallel to the direction of fiber alignment. It is seen that the fiber alignment of PCL/gelatin nanofibrous scaffolds is less than the PCL nanofiber scaffolds and also the proliferation of C17.2 was higher on aligned nanofibrous scaffolds in comparison to random nanofibrous scaffolds for both PCL and PCL/gelatin [102]. Similarly, in vitro nerve stem cells culture and the electrical stimulation of Rat nerve stem cells (C17.2) on PLLA/PANi nanofibers revealed that the electrical stimulation of nerve cells could stimulate the differentiation or neurite elongation [118].

Furthermore, in vitro culture of PC12 cells and hippocampal cells on the PPy-PLGA meshes demonstrated that the electrical stimulation of PC12 cells on the conducting nanofiber scaffolds improved the neurite outgrowth compared to the non-stimulated cells [11]. These studies reveal increases in neurite length and percentage of neurite bearing cells will aid to design neuronal tissue interfaces integrated with topographical and electrical cues for use in nerve tissue scaffolds and for neural interfacing.

7.3 Cartilage Tissue Engineering

Bi-layer scaffold of collagen and electrospun poly-L-lactic acid nanofibers (COL-nanofiber) bi-layer in which mesenchymal stem cells were cultured on the bi-layer scaffold revealed that the implantation of COL-nanofiber scaffold seeded with cells induced more rapid subchondral bone emergence, and better cartilage formation, which led to the better functional repair of osteochondral defects as manifested by histological staining, biomechanical test and micro-computed tomography data [54].

Similarly, the study of PLLA/gelatin nanofibrous scaffolds showed that the chondrocyte growth and differentiation markers emphasized the cartilage tissue growth with PLLA50GEL50 and PLLA70GEL30, providing the best cellular response. Furthermore, the mineralization experiment too suggests useful for cartilage-bone interface tissue engineering [7]. Likewise, the first demonstration of bone and cartilage regeneration was reported by implementing a novel strategy based on a synthetic nanoengineered bio mimicking membrane functionalized with nano-reservoirs of a growth factor (bone morphogenetic protein 2, BMP-2) [117].

On the other hand, electrospun nanofibers such as novel PPDO/ PLLA-b-PEG copolymer [101], composite CS/PEO [99] and PCL grafted with polyNaSS [114] were also reported for cartilage regeneration.

7.4 Skin Tissue Engineering

Skin is the largest body organ, functioning as a barrier to harmful mediums, preventing pathogens from entering into the body. A wound is a result of physical, chemical, mechanical and/or thermal damages. The natural healing process of the skin is complex and continuous.

Alfalfa carries genistein, which is a major phytoestrogen known to accelerate skin repair. In vitro cell culture on PCL/alfalfa nanofiber scaffolds promote cell growth and sustain biocompatibility for epidermal keratinocytes (KCs) and dermal fibroblasts to accelerate skin tissue regeneration in both mouse and human skin, without requiring additional proteins, growth factors or cells [82]. Similarly, in vitro by seeding with human KCs on GT/PCL, nanofibrous membranes were investigated which was further evaluated by transplantation of engineered epidermis into a wound-healing model in the nude mouse showed that the repaired skin at day 14 in the epidermis-treated groups, multiple layers of epithelial cells covered the wound area enhancing a suitable scaffold for tissue engineering [126].

Likewise, the use of G-Rg3/PLLA electrospun fibrous scaffolds rapidly minimizes fibroblast growth and restores the structural and functional properties of wounded skin for patients with deep trauma, severe burn injury and surgical incision [121]. In another study, pbFGF-loaded electrospun fibrous mats showed that the gradual release of pbFGF polyplexes revealed significantly higher wound recovery rate with collagen deposition and maturation, complete reepithelialization and skin appendage regeneration as well as accelerate the healing of skin ulcers for patients with diabetes mellitus [127].

Other electrospun nanofibers such as PCL [38, 91], PVA/γ-Fe$_2$O$_3$ [34], PPDO/ PLLA-b-PEG [101], TA-g-PCL, PLLA [105] and PVA, PVA/HA, PEO [99], also support skin tissue engineering. Moreover, the addition of matrigel or nanoparticles in the nanofibrous scaffolds holds the better potential for skin than only neat electrospun nanofibrous scaffolds [34, 91].

7.5 Clinical Perspective

The creation of biological replacements that can replace or revitalize human is optimistic in regenerative medicine. Although electrospinning is considered to be a simple and efficient process for the preparation of electrospun nanofibers, it is the clinical application on the market not yet fully utilized [128]. Recently, AVflo ™, a vascular access graft, was developed by Nicast using medical grade polycarbonate urethane nanofibers for use as a subcutaneous, arteriovenous conduit for blood access [129]. In the same way, SURGICLOT ®, electrospun-based nanofibers containing protein has been developed by St. Teresa Medical, Inc ®, to promote blood clotting [130]. On the other hand, there are still some economic and technological concerns regarding the safety and effectiveness of electrospun nanofibers in terms of clinical perspective. Since, due to low production yield and requiring highly skilled human resources, it doesn't seem very easy from the economic viewpoint. Furthermore, the large quantities of products in a continuous way are challenging. Since electrospun nanofibers have massive potential in tissue engineering so, the problems should be resolved to have its wider use [131].

7.6 Others

In addition to the above-mentioned issues, the feasibility of using nanofibrous scaffolds to culture stem cells [104, 118], and tissues such as muscles [5, 84, 118], tendon, ligaments [5, 132], and blood vessels [121, 127, 133] have been reported.

8 Challenges and Future Prospects

8.1 Challenges

Scaffold formation for tissue engineering is a delicate and complex process done mainly by electrospinning. While these methods have proved to be successful in the field of tissue engineering, challenges remain for successful regeneration of the tissue. In addition, challenges of deposition and degradation [32], mechanical integrity and porosity [134] and three-dimensional scaffolds formation [134] are faced by many scientists and engineers working on tissue engineering.

The balance between de novo tissue deposition and scaffold degradation is a common challenge of tissue engineering. Quick degradation of scaffolds will affect the mechanical integrity of the scaffolds. So, in order to balance the rate of degradation of scaffolds, optimization in spatial and temporal growth factor needs to be modified for the acceleration of de novo tissue deposition. The ability to withstand physiological loads needs to be ensured. Although regeneration of new bone tissue has been observed, strong mechanical properties were not shown by the human cancellous bone [32, 134, 135].

Fabrication of three-dimensional biomimetic bone tissue scaffolds and alignment of the nanofibers in contact with the liquid environment is a major challenge. Layering and contacting the structure to the liquid environment is difficult, which creates issues of change in size/shape and cell infiltration [134, 135]. Growth of new blood vessels to deliver oxygen and nutrients to the implanted tissue is a long process but implanting these cells to vascular beds of liver, spleen, bone and so on will decrease the growth time, eventually consuming oxygen to engineered tissue at a fast rate.

The diffusion rate of genes and proteins from scaffolds needs to be in an accepted range of physiology to be able to use in human applications. So, a challenge of maintaining this range has also risen. Regulating cell behavior, offering alternatives for enhancing scaffold performance, covalently incorporating growth factors can be done. Specific instruction is required for tissue regeneration in vivo for guided growth of nerve, bone, blood vessels or corneal epithelia for critical injury sites.

Computer simulations for predicting how the cells bind to the extracellular matrix cannot always be used, and the challenge of determining the correct quantity of adhesion is a challenge. If adhesive ligands are less, cells cannot bind effectively for movement, but if there are more adhesive ligands, cells bind firmly to get stuck. So, intermediate adhesion is required for cell movement.

The scale-up process of three-dimensional tissue engineering is a complex challenge as it is limited to laboratory product. Formidable regulatory issues, separate culture for each new patients, high cost are some of the issues which need to be addressed [135].

Spinal injury results in loss of axonal connections and motor functions. Regeneration of injured site is a challenge due to too complex inhibitory environment. Scaffolds promote connections of these axons for regeneration and functional recovery. Finding consistent, quantitative and replicable treatment in preclinical trials for the spinal cord is a challenge.

In the case of the brain, it doesn't promote regeneration capacity, so the fabrication of scaffolds needs to have properties of cell infiltration, degradation and regeneration. Glial scarring should be prevented and thus, reducing the inflammatory response to neuron survival while coating brain implanted devices. Alternative creation of autograft in the peripheral nervous system for the clinical standard is a big issue in tissue engineering. Implantable scaffolds posing as a bridge for long gaps producing results same as autograft without harvesting autologous donor tissue has also become an issue. Biodegradable scaffolds can be the solution for this, capable of fully recovering the damaged tissue.

Thus, the complex task always promotes complex challenges which need to be addressed [136]. These challenges, when addressed accordingly, will give the right solution, thus, helping in modifying the complex structure of tissue engineering basics.

8.2 Future Work

Electrospun nanofibers can be considered a feasible way of generating nano scaffolds for various applications. It is good in every aspect, but still, there is room for improvement. Researchers can give more focus on melt electrospinning. Though solution spinning is easy and convenient, the use of organic solvent can lead to solvent toxicity when implanted in the human body. If more advancement in the scaffold can be done, the work in the future will be easy.

Similarly, cartilage injury, arthritis in bone joints is common nowadays. Repairing of cartilage tissue depends on the collagen orientation. Control tissue growth of damaged joints can be obtained when there is no foreign body involved after the injury. Efforts can be focused to mimic the tissue structure and to fabricate the nano scaffolds exact and biocompatible [137]. When an inert protein gradient is created, it can help to create a masking gradient and conserve the amount of bioactive protein along with the generation of self-assembling peptides [138].

9 Conclusion

This paper presents a review on surface-modified nanofibers produced through electrospinning process and its applications in tissue engineering, particularly on bone tissue regeneration and drug delivery system. The electrospun nanofibers are biocompatible, biodegradable, easily fabricated, and are able to support cell adhesion, proliferation, and differentiation and have emerged as a promising material for constructing replaceable biological components. The first and foremost step of the electrospinning process is the selection of proper polymer solutions with intrinsic functions that can be enhanced by employing several modifications strategies. The properties of electrospun nanofiber such as its morphology, fiber diameter and porosity depend on the flow rate of the polymer solution and the electric voltage applied. The scaffolds prepared through the electrospinning process are usually surface treated before application in the human body to match the bio-properties of the native site and to ensure the safety of the host site. Surface treatment by use of plasma, using chemicals, grafting after plasma treatment, coating and click reactions is mainly performed. Plasma treatment of PCL, PLLA, nylon-6 has increased the hydrophilicity and wetting characteristics of the polymer. The wet chemical method by using NaOH on PCL influenced cellular attachment and proliferation. PAN treated with Fe_2O_3 increased mechanical strength and uniformity. Surface graft polymerization of PCL after air plasma treatment increased cell attachment and growth. CaP coating in PLGA/gelatin increased hydrophilicity and agglomeration. Co-electrospinning PLLA improved fiber morphology, and mechanical properties and coaxial electrospinning of PLLACL increased cell attachment and proliferation. Other physical and chemical functionalization techniques involving the use of various nanofibers were also discussed. The electrospun and surface-modified nanofibers are characterized using SEM, TEM and diffraction before application inside the human body. Scaffolds or nanofiber produced by electrospinning possess characteristics which closely resemble the natural bone (Extra Cellular Matrix) ECM in nanoscale structure and chemical composition; so scaffolding is a good choice for bone tissue engineering. The mechanical properties and cell response of aligned electrospun nanofiber bundles is a promising scaffold for orthopedic tissue engineering applications. Bio-composite scaffolds facilitated hMSC (human mesenchymal stromal cells) colonization and bone formation. The highly porous nanofibrous film with a high surface area is used for the processes of tissue regeneration as it provides more structural space for the accommodation and attachment of cells and enables the efficient exchange of nutrients and metabolic waste. Neural tissues can be designed by incorporating scaffolds with directional cues, bioactive to promote regeneration and repair, neural/progenitor cells for the release of growth factors, and functionalized polymers for better neuro-compatibility. Tissue engineering is mankind's attempt to replicate functioning tissues and organs whose working prototypes are available in nature. Thus, electrospun scaffolds have the potential to provide both a structural and functional mimicry of the native tissue through intelligent biomaterial scaffold design, enhanced time course and the functional outcome of endogenous tissue repair.

References

1. Rim NG Shin CS Shin H (2013) Current approaches to electrospun nanofibers for tissue engineering. Biomed Mater 8. https://doi.org/10.1088/1748-6041/8/1/014102
2. Choi JS, Lee SJ, Christ GJ et al (2008) The influence of electrospun aligned poly(ε-caprolactone)/collagen nanofiber meshes on the formation of self-aligned skeletal muscle myotubes. Biomaterials 29:2899–2906. https://doi.org/10.1016/j.biomaterials.2008.03.031
3. Hu J, Kai D, Ye H et al (2017) Electrospinning of poly(glycerol sebacate)-based nanofibers for nerve tissue engineering. Mater Sci Eng C 70:1089–1094. https://doi.org/10.1016/j.msec.2016.03.035
4. Tonsomboon K, Oyen ML (2013) Composite electrospun gelatin fiber-alginate gel scaffolds for mechanically robust tissue engineered cornea. J Mech Behav Biomed Mater 21:185–194. https://doi.org/10.1016/j.jmbbm.2013.03.001
5. Li WJ, Mauck RL, Cooper JA et al (2007) Engineering controllable anisotropy in electrospun biodegradable nanofibrous scaffolds for musculoskeletal tissue engineering. J Biomech 40:1686–1693. https://doi.org/10.1016/j.jbiomech.2006.09.004
6. Rezk AI, Mousa HM, Lee J et al (2019) Composite PCL/HA/simvastatin electrospun nanofiber coating on biodegradable Mg alloy for orthopedic implant application. J Coat Technol Res 16:477–489. https://doi.org/10.1007/s11998-018-0126-8
7. Fiorani A (2014) Electrospun polymeric scaffolds with enhanced biomimetic properties for tissue engineering applications. 109. https://doi.org/10.6092/unibo/amsdottorato/6483
8. Tanaka T, Ujiie R, Yajima H et al (2011) Ion-beam irradiation into biodegradable nanofibers for tissue engineering scaffolds. Surf Coat Technol 206:889–892. https://doi.org/10.1016/j.surfcoat.2011.04.049
9. Santander J, Fonseca L, Udina S, Marco S (2007) Accepted Musp. https://doi.org/10.1016/j.snb.2007.07.003

10. Abdal-hay A, Vanegas P, Hamdy AS et al (2014) Preparation and characterization of vertically arrayed hydroxyapatite nanoplates on electrospun nanofibers for bone tissue engineering. Chem Eng J 254:612–622. https://doi.org/10.1016/j.cej.2014.05.118
11. Lee JY, Bashur CA, Goldstein AS, Schmidt CE (2009) Polypyrrole-coated electrospun PLGA nanofibers for neural tissue applications. Biomaterials 30:4325–4335. https://doi.org/10.1016/j.biomaterials.2009.04.042
12. Bosworth LA, Hu W, Shi Y, Cartmell SH (2019) Enhancing biocompatibility without compromising material properties: an optimised NaOH treatment for electrospun polycaprolactone fibres. J Nanomater. https://doi.org/10.1155/2019/4605092
13. Cheng Q, Lee BLP, Komvopoulos K et al (2013) Plasma surface chemical treatment of electrospun poly(L-lactide) microfibrous scaffolds for enhanced cell adhesion, growth, and infiltration. Tissue Eng Part A 19:1188–1198. https://doi.org/10.1089/ten.tea.2011.0725
14. Yan S, Xiaoqiang L, Lianjiang T et al (2009) Poly(l-lactide-co-ε-caprolactone) electrospun nanofibers for encapsulating and sustained releasing proteins. Polymer (Guildf) 50:4212–4219. https://doi.org/10.1016/j.polymer.2009.06.058
15. Li Y, Akl TB, Li Y, Akl TB (2016) Electrospinning-material, techniques, and biomedical applications. Electrospinning Mater Tech Biomed Appl. https://doi.org/10.5772/62860
16. Vitchuli N, Shi Q, Nowak J et al (2013) Atmospheric plasma application to improve adhesion of electrospun nanofibers onto protective fabric. J Adhes Sci Technol 27:924–938. https://doi.org/10.1080/01694243.2012.727164
17. Feng C, Liu C, Liu S, et al (2019) Electrospun nanofibers with core-shell structure for treatment of bladder regeneration
18. Senthamizhan A, Balusamy B, Uyar T (2017) Electrospinning: a versatile processing technology for producing nanofibrous materials for biomedical and tissue-engineering applications. Elsevier Ltd.
19. Pordel MA, Maleki A, Khamforosh M, Daraei H (2017) Fabrication , characterization , and microscopic imaging of Fe2O3-modified electrospun nanofibers, pp 146–153. https://doi.org/10.22102/jaehr.2017.80173.1010
20. Kalaoglu-Altan OI, Kirac-Aydin A, Sumer Bolu B et al (2017) Diels-Alder "clickable" biodegradable nanofibers: benign tailoring of scaffolds for biomolecular immobilization and cell growth. Bioconjug Chem 28:2420–2428. https://doi.org/10.1021/acs.bioconjchem.7b00411
21. Rodda AE, Ercole F, Glattauer V et al (2015) Low fouling electrospun scaffolds with clicked bioactive peptides for specific cell attachment. Biomacromol 16:2109–2118. https://doi.org/10.1021/acs.biomac.5b00483
22. Aksoy OE, Ates B, Cerkez I (2017) Antibacterial polyacrylonitrile nanofibers produced by alkaline hydrolysis and chlorination. J Mater Sci 52:10013–10022. https://doi.org/10.1007/s10853-017-1240-1
23. Abdal-hay A, Tijing LD, Lim JK (2013) Characterization of the surface biocompatibility of an electrospun nylon 6/CaP nanofiber scaffold using osteoblasts. Chem Eng J 215–216:57–64. https://doi.org/10.1016/j.cej.2012.10.046
24. Hendrick E, Frey M (2014) Increasing surface hydrophilicity in poly(lactic acid) electrospun fibers by addition of Pla-B-Peg co-polymers. J Eng Fiber Fabr 9:153–164. https://doi.org/10.1177/155892501400900219
25. Lannutti J, Reneker D, Ma T et al (2007) Electrospinning for tissue engineering scaffolds. Mater Sci Eng C 27:504–509. https://doi.org/10.1016/j.msec.2006.05.019
26. Shabani I, Haddadi-Asl V, Seyedjafari E et al (2009) Improved infiltration of stem cells on electrospun nanofibers. Biochem Biophys Res Commun 382:129–133. https://doi.org/10.1016/j.bbrc.2009.02.150
27. Meng ZX, Li HF, Sun ZZ et al (2013) Fabrication of mineralized electrospun PLGA and PLGA/gelatin nanofibers and their potential in bone tissue engineering. Mater Sci Eng C 33:699–706. https://doi.org/10.1016/j.msec.2012.10.021
28. Zhao W, Li J, Jin K et al (2016) Fabrication of functional PLGA-based electrospun scaffolds and their applications in biomedical engineering. Mater Sci Eng C 59:1181–1194. https://doi.org/10.1016/j.msec.2015.11.026

29. Bhaarathy V, Venugopal J, Gandhimathi C et al (2014) Biologically improved nanofibrous scaffolds for cardiac tissue engineering. Mater Sci Eng C 44:268–277. https://doi.org/10. 1016/j.msec.2014.08.018

30. Chow LW (2018) Electrospinning functionalized polymers for use as tissue engineering scaffolds. Methods Mol Biol 1758:27–39. https://doi.org/10.1007/978-1-4939-7741-3_3

31. Al DJT et (2011) No title עטונה וולע. בצמ תנומת :יווייקה ףנעמ 66:37–39

32. Pauly H (2018) Development of a hierarchical electrospun scaffold for ligament replacement

33. Wang X, Gittens RA, Song R et al (2012) Effects of structural properties of electrospun TiO2 nanofiber meshes on their osteogenic potential. Acta Biomater 8:878–885. https://doi.org/10. 1016/j.actbio.2011.10.023

34. Ngadiman NHA, Idris A, Irfan M et al (2015) γ-Fe2O3 nanoparticles filled polyvinyl alcohol as potential biomaterial for tissue engineering scaffold. J Mech Behav Biomed Mater 49:90– 104. https://doi.org/10.1016/j.jmbbm.2015.04.029

35. Abdal-hay A, Lim J, Shamshi Hassan M, Lim JK (2013) Ultrathin conformal coating of apatite nanostructures onto electrospun nylon 6 nanofibers: mimicking the extracellular matrix. Chem Eng J 228:708–716. https://doi.org/10.1016/j.cej.2013.05.022

36. Liao S, Murugan R, Chan CK, Ramakrishna S (2008) Processing nanoengineered scaffolds through electrospinning and mineralization suitable for biomimetic bone tissue engineering. J Mech Behav Biomed Mater 1:252–260. https://doi.org/10.1016/j.jmbbm.2008.01.007

37. Shrestha BK, Mousa HM, Tiwari AP et al (2016) Development of polyamide-6,6/chitosan electrospun hybrid nanofibrous scaffolds for tissue engineering application. Carbohydr Polym 148:107–114. https://doi.org/10.1016/j.carbpol.2016.03.094

38. Li L, Li H, Qian Y et al (2011) Electrospun poly (ε-caprolactone)/silk fibroin core-sheath nanofibers and their potential applications in tissue engineering and drug release. Int J Biol Macromol 49:223–232. https://doi.org/10.1016/j.ijbiomac.2011.04.018

39. Doshi J, Reneker DH (1993) Electrospinning process and applications of electrospun fibers. In: Conference record-IAS annual meeting (IEEE Ind Appl Soc), vol 3, pp 1698–1703. https:// doi.org/10.1109/ias.1993.299067

40. Chen F, Huang P, Mo XM (2010) Electrospinning of heparin encapsulated P(LLA-CL) core/shell nanofibers. Nano Biomed Eng 2:56–60. https://doi.org/10.5101/nbe.v2i1.p56-60

41. Širc J, Hobzová R, Kostina N et al (2012) Morphological characterization of nanofibers: methods and application in practice. J Nanomater. https://doi.org/10.1155/2012/327369

42. Xu X, Zhuang X, Chen X et al (2006) Preparation of core-sheath composite nanofibers by emulsion electrospinning. Macromol Rapid Commun 27:1637–1642. https://doi.org/10.1002/ marc.200600384

43. Xu X, Chen X, Wang X, Jing X (2008) The release behavior of doxorubicin hydrochloride from medicated fibers prepared by emulsion-electrospinning, vol 70, pp 165–170. https://doi. org/10.1016/j.ejpb.2008.03.010

44. Xiaoqiang L, Yan S, Shuiping L et al (2010) Colloids and surfaces B: biointerfaces encapsu- lation of proteins in poly (l-lactide-co-caprolactone) fibers by emulsion electrospinning, vol 75, pp 418–424. https://doi.org/10.1016/j.colsurfb.2009.09.014

45. Hutmacher DW, Dalton PD (2011) Melt electrospinning. Chem Asian J 6:44–56. https://doi. org/10.1002/asia.201000436

46. Zhmayev E, Cho D, Joo YL (2010) Nanofibers from gas-assisted polymer melt electrospin- ning. Polymer (Guildf) 51:4140–4144. https://doi.org/10.1016/j.polymer.2010.06.058

47. Brown TD, Edin F, Detta N et al (2015) Melt electrospinning of poly(ε-caprolactone) scaffolds: Phenomenological observations associated with collection and direct writing. Mater Sci Eng C 45:698–708. https://doi.org/10.1016/j.msec.2014.07.034

48. Ramakrishna S, Fujihara K, Teo WE et al (2005) An introduction to electrospinning and nanofibers

49. Soldate P, Fan J (2019) Controlled deposition of electrospun nanofibers by electrohydrody- namic deflection. J Appl Phys 125. https://doi.org/10.1063/1.5084284

50. Koenig K, Beukenberg K, Langensiepen F, Seide G (2019) A new prototype melt- electrospinning device for the production of biobased thermoplastic sub-microfibers and nanofibers. Biomater Res 23:1–12. https://doi.org/10.1186/s40824-019-0159-9

51. Sill TJ, von Recum HA (2008) Electrospinning: applications in drug delivery and tissue engineering. Biomaterials 29:1989–2006. https://doi.org/10.1016/j.biomaterials.2008.01.011

52. Zulkifli FH, Shahitha F, Yusuff MM et al (2013) Cross-linking effect on electrospun hydroxyethyl cellulose/poly(vinyl alcohol) nanofibrous scaffolds. Procedia Eng 53:689–695. https://doi.org/10.1016/j.proeng.2013.02.089

53. Shao S, Zhou S, Li L et al (2011) Osteoblast function on electrically conductive electrospun PLA/MWCNTs nanofibers. Biomaterials 32:2821–2833. https://doi.org/10.1016/j.biomaterials.2011.01.051

54. Zhang S, Chen L, Jiang Y et al (2013) Bi-layer collagen/microporous electrospun nanofiber scaffold improves the osteochondral regeneration. Acta Biomater 9:7236–7247. https://doi.org/10.1016/j.actbio.2013.04.003

55. Chen MC, Sun YC, Chen YH (2013) Electrically conductive nanofibers with highly oriented structures and their potential application in skeletal muscle tissue engineering. Acta Biomater 9:5562–5572. https://doi.org/10.1016/j.actbio.2012.10.024

56. Yang C, Deng G, Chen W et al (2014) A novel electrospun-aligned nanoyarn-reinforced nanofibrous scaffold for tendon tissue engineering. Colloids Surf B Biointerfaces 122:270–276. https://doi.org/10.1016/j.colsurfb.2014.06.061

57. Ikada Y (1994) Surface modification of polymers for medical applications. Biomaterials 15:725–736. https://doi.org/10.1016/0142-9612(94)90025-6

58. Liao C, Li Y, Tjong SC (2018) Graphene nanomaterials: synthesis, biocompatibility, and cytotoxicity. Int J Mol Sci 19. https://doi.org/10.3390/ijms19113564

59. Singh R, Lillard JW (2009) Nanoparticle-based targeted drug delivery. Exp Mol Pathol 86:215–223. https://doi.org/10.1016/j.yexmp.2008.12.004

60. Sreekumar PA, Thomas SP, Marc SJ et al (2009) Effect of fiber surface modification on the mechanical and water absorption characteristics of sisal/polyester composites fabricated by resin transfer molding. Compos Part A Appl Sci Manuf 40:1777–1784. https://doi.org/10.1016/j.compositesa.2009.08.013

61. Yoo HS, Kim TG, Park TG (2009) Surface-functionalized electrospun nanofibers for tissue engineering and drug delivery. Adv Drug Deliv Rev 61:1033–1042. https://doi.org/10.1016/j.addr.2009.07.007

62. Poulsson AHC, Richards RG (2012) Surface modification techniques of polyetheretherketone, including plasma surface treatment. Elsevier Inc.

63. Rana D, Ramasamy K, Leena M et al (2016) Surface functionalization of nanobiomaterials for application in stem cell culture, tissue engineering, and regenerative medicine. Biotechnol Prog 32:554–567. https://doi.org/10.1002/btpr.2262

64. Kennedy M, Ristau D, Niederwald HS (1998) Ion beam-assisted deposition of MgF2 and YbF3 films. Thin Solid Films 333:191–195. https://doi.org/10.1016/S0040-6090(98)00847-5

65. Chen W, McCarthy TJ (1997) Layer-by-layer deposition: a tool for polymer surface modification. Macromolecules 30:78–86. https://doi.org/10.1021/ma961096d

66. Sarathy KV, Thomas PJ, Kulkarni GU, Rao CNR (1999) Superlattices of metal and metal-semiconductor quantum dots obtained by layer-by-layer deposition of nanoparticle arrays. J Phys Chem B 103:399–401. https://doi.org/10.1021/jp983836l

67. Uğur ŞS, Sariişk M, Hakan Aktaş A (2010) The fabrication of nanocomposite thin films with TiO2 nanoparticles by the layer-by-layer deposition method for multifunctional cotton fabrics. Nanotechnology 21 https://doi.org/10.1088/0957-4484/21/32/325603

68. Gentile P, Carmagnola I, Nardo T, Chiono V (2015) Layer-by-layer assembly for biomedical applications in the last decade. Nanotechnology 26:422001. https://doi.org/10.1088/0957-4484/26/42/422001

69. Kim MS, Ma L, Choudhury S et al (2016) Fabricating multifunctional nanoparticle membranes by a fast layer-by-layer Langmuir-Blodgett process: application in lithium-sulfur batteries. J Mater Chem A 4:14709–14719. https://doi.org/10.1039/c6ta06018h

70. de Souza ID, Cruz MAE, de Faria AN et al (2014) Formation of carbonated hydroxyapatite films on metallic surfaces using dihexadecyl phosphate-LB film as template. Colloids Surf B Biointerfaces 118:31–40. https://doi.org/10.1016/j.colsurfb.2014.03.029

71. Liao WS, Yang T, Castellana ET et al (2006) A rapid prototyping approach to ag nanoparticle fabrication in the 10–100 nm range. Adv Mater 18:2240–2243. https://doi.org/10.1002/adma. 200600589
72. Seif S, Planz V, Windbergs M (2017) Controlling the release of proteins from therapeutic nanofibers: the effect of fabrication modalities on biocompatibility and antimicrobial activity of lysozyme. Planta Med 83:445–452. https://doi.org/10.1055/s-0042-109715
73. Zare S, Kargari A (2018) Membrane properties in membrane distillation. Elsevier Inc.
74. Qin X (2017) Coaxial electrospinning of nanofibers
75. Maria D (2015) Three dimensional scaffolds based on electroactive polymers for tissue engineering applications
76. Mondal K, Ali MA, Agrawal VV et al (2014) Highly sensitive biofunctionalized meso-porous electrospun TiO2 nanofiber based interface for biosensing. ACS Appl Mater Interfaces 6:2516–2527. https://doi.org/10.1021/am404931f
77. Sarkar S (2016) Roles of nanofiber scaffold structure and chemistry in directing human bone marrow stromal cell response. Adv Tissue Eng Regen Med Open Access 1. https://doi.org/ 10.15406/atroa.2016.01.00003
78. Zhang Y, Venugopal JR, El-Turki A et al (2008) Electrospun biomimetic nanocomposite nanofibers of hydroxyapatite/chitosan for bone tissue engineering. Biomaterials 29:4314–4322. https://doi.org/10.1016/j.biomaterials.2008.07.038
79. Luo Y, Wang S, Shen M et al (2013) Carbon nanotube-incorporated multilayered cellulose acetate nanofibers for tissue engineering applications. Carbohydr Polym 91:419–427. https:// doi.org/10.1016/j.carbpol.2012.08.069
80. Huang C-J (2019) Advanced surface modification technologies for biosensors. Elsevier Inc.
81. Ramier J, Grande D, Bouderlique T et al (2014) From design of bio-based biocomposite electrospun scaffolds to osteogenic differentiation of human mesenchymal stromal cells. J Mater Sci Mater Med 25:1563–1575. https://doi.org/10.1007/s10856-014-5174-8
82. Ahn S, Ardoña HAM, Campbell PH et al (2019) Alfalfa nanofibers for dermal wound healing. ACS Appl Mater Interfaces 11:33535–33547. https://doi.org/10.1021/acsami.9b07626
83. Prabhakaran MP, Venugopal J, Ramakrishna S (2009) Electrospun nanostructured scaffolds for bone tissue engineering. Acta Biomater 5:2884–2893. https://doi.org/10.1016/j.actbio. 2009.05.007
84. Ito Y, Hasuda H, Kamitakahara M et al (2005) A composite of hydroxyapatite with electrospun biodegradable nanofibers as a tissue engineering material. J Biosci Bioeng 100:43–49. https:// doi.org/10.1263/jbb.100.43
85. Ma G, Fang D, Liu Y et al (2012) Electrospun sodium alginate/poly(ethylene oxide) core-shell nanofibers scaffolds potential for tissue engineering applications. Carbohydr Polym 87:737–743. https://doi.org/10.1016/j.carbpol.2011.08.055
86. Sharma Y, Tiwari A, Hattori S et al (2012) Fabrication of conducting electrospun nanofibers scaffold for three-dimensional cells culture. Int J Biol Macromol 51:627–631. https://doi.org/ 10.1016/j.ijbiomac.2012.06.014
87. Bhattarai N, Edmondson D, Veiseh O et al (2005) Electrospun chitosan-based nanofibers and their cellular compatibility. Biomaterials 26:6176–6184. https://doi.org/10.1016/j.biomateri als.2005.03.027
88. Frohbergh ME, Katsman A, Botta GP et al (2012) Electrospun hydroxyapatite-containing chitosan nanofibers crosslinked with genipin for bone tissue engineering. Biomaterials 33:9167–9178. https://doi.org/10.1016/j.biomaterials.2012.09.009
89. González E, Shepherd LM, Saunders L, Frey MW (2016) Surface functional poly(lactic acid) electrospun nanofibers for biosensor applications. Mater (Basel) 9:1–11. https://doi.org/10. 3390/ma9010047
90. Kim JH, Choung PH, Kim IY et al (2009) Electrospun nanofibers composed of poly(ε-caprolactone) and polyethylenimine for tissue engineering applications. Mater Sci Eng C 29:1725–1731. https://doi.org/10.1016/j.msec.2009.01.023
91. Jing X, Mi HY, Cordie TM et al (2014) Fabrication of shish-kebab structured poly(ε-caprolactone) electrospun nanofibers that mimic collagen fibrils: Effect of solvents and

matrigel functionalization. Polymer (Guildf) 55:5396–5406. https://doi.org/10.1016/j.pol ymer.2014.08.061

92. Meng ZX, Wang YS, Ma C et al (2010) Electrospinning of PLGA/gelatin randomly-oriented and aligned nanofibers as potential scaffold in tissue engineering. Mater Sci Eng C 30:1204–1210. https://doi.org/10.1016/j.msec.2010.06.018

93. Zulkifli FH, Hussain FSJ, Rasad MSBA, Mohd Yusoff M (2014) Nanostructured materials from hydroxyethyl cellulose for skin tissue engineering. Carbohydr Polym 114:238–245. https://doi.org/10.1016/j.carbpol.2014.08.019

94. Meng ZX, Zheng W, Li L, Zheng YF (2010) Fabrication and characterization of three-dimensional nanofiber membrance of PCL-MWCNTs by electrospinning. Mater Sci Eng C 30:1014–1021. https://doi.org/10.1016/j.msec.2010.05.003

95. Ding Q, Li Z, Yang Y et al (2016) Preparation and therapeutic application of docetaxel-loaded poly(d,l-lactide) nanofibers in preventing breast cancer recurrence. Drug Deliv 23:2677–2685. https://doi.org/10.3109/10717544.2015.1048490

96. Haynie DT, Khadka DB, Cross MC (2012) Physical properties of polypeptide electrospun nanofiber cell culture scaffolds on a wettable substrate. Polym (Basel) 4:1535–1553. https://doi.org/10.3390/polym4031535

97. Su Y, Su Q, Liu W et al (2012) Controlled release of bone morphogenetic protein 2 and dexamethasone loaded in core-shell PLLACL-collagen fibers for use in bone tissue engineering. Acta Biomater 8:763–771. https://doi.org/10.1016/j.actbio.2011.11.002

98. Santillán J, Dwomoh EA, Rodríguez-Avilés YG et al (2019) Fabrication and evaluation of polycaprolactone beads-on-string membranes for applications in bone tissue regeneration. ACS Appl Bio Mater 2:1031–1040. https://doi.org/10.1021/acsabm.8b00628

99. Janković B, Pelipenko J, Škarabot M et al (2013) The design trend in tissue-engineering scaffolds based on nanomechanical properties of individual electrospun nanofibers. Int J Pharm 455:338–347. https://doi.org/10.1016/j.ijpharm.2013.06.083

100. Asheghali D Lee SJ Furchner A et al (2020) Enhanced neuronal differentiation of neural stem cells with mechanically enhanced touch-spun nanofibrous scaffolds. Nanomedicine Nanotechnol Biol Med 24:102152. https://doi.org/10.1016/j.nano.2020.102152

101. Bhattarai SR, Bhattarai N, Yi HK et al (2004) Novel biodegradable electrospun membrane: Scaffold for tissue engineering. Biomaterials 25:2595–2602. https://doi.org/10.1016/j.biomat erials.2003.09.043

102. Ghasemi-Mobarakeh L, Prabhakaran MP, Morshed M et al (2008) Electrospun poly(ε-caprolactone)/gelatin nanofibrous scaffolds for nerve tissue engineering. Biomaterials 29:4532–4539. https://doi.org/10.1016/j.biomaterials.2008.08.007

103. Karuppuswamy P, Venugopal JR, Navaneethan B et al (2014) Functionalized hybrid nanofibers to mimic native ECM for tissue engineering applications. Appl Surf Sci 322:162–168. https://doi.org/10.1016/j.apsusc.2014.10.074

104. Singh RK, Jin GZ, Mahapatra C et al (2015) Mesoporous silica-layered biopolymer hybrid nanofibrous scaffold: a novel nanobiomatrix platform for therapeutics delivery and bone regeneration. ACS Appl Mater Interfaces 7:8088–8098. https://doi.org/10.1021/acsami.5b0 0692

105. Jiang S, Song P, Guo H et al (2017) Blending PLLA/tannin-grafted PCL fiber membrane for skin tissue engineering. J Mater Sci 52:1617–1624. https://doi.org/10.1007/s10853-016-0455-x

106. Chahal S, Hussain FSJ, Kumar A et al (2016) Fabrication, characterization and in vitro biocompatibility of electrospun hydroxyethyl cellulose/poly (vinyl) alcohol nanofibrous composite biomaterial for bone tissue engineering. Elsevier

107. Zhang C, Wang L, Zhai T et al (2016) The surface grafting of graphene oxide with poly(ethylene glycol) as a reinforcement for poly(lactic acid) nanocomposite scaffolds for potential tissue engineering applications. J Mech Behav Biomed Mater 53:403–413. https://doi.org/10.1016/j.jmbbm.2015.08.043

108. Guler Z, Sarac AS (2015) BMP-2 immobilized PCL/P3ANA nanofibers for bone tissue engineering, pp 1–4

109. Azim N, Kundu A, Royse M et al (2019) Fabrication and characterization of a 3D printed, microelectrodes platform with functionalized electrospun nano-scaffolds and spin coated 3D insulation towards multi-functional biosystems. J Microelectromechanical Syst 28:606–618. https://doi.org/10.1109/JMEMS.2019.2913652

110. Ravichandran R, Sundaramurthi D, Gandhi S et al (2014) Bioinspired hybrid mesoporous silica-gelatin sandwich construct for bone tissue engineering. Microporous Mesoporous Mater 187:53–62. https://doi.org/10.1016/j.micromeso.2013.12.018

111. Zeeshan K (2013) Preparation and characterization of electrospun nanofibers for apparel and medical application

112. Wang S, Cao X, Shen M et al (2012) Fabrication and morphology control of electrospun poly(γ-glutamic acid) nanofibers for biomedical applications. Colloids Surf B Biointerfaces 89:254–264. https://doi.org/10.1016/j.colsurfb.2011.09.029

113. Martins A, Duarte ARC, Faria S et al (2010) Osteogenic induction of hBMSCs by electrospun scaffolds with dexamethasone release functionality. Biomaterials 31:5875–5885. https://doi.org/10.1016/j.biomaterials.2010.04.010

114. Amokrane G, Humblot V, Jubeli E et al (2019) Electrospun poly(ε-caprolactone) fiber scaffolds functionalized by the covalent grafting of a bioactive polymer: surface characterization and influence on in vitro biological response. ACS Omega 4:17194–17208. https://doi.org/10.1021/acsomega.9b01647

115. Tang S, Zhao Z, Chen G et al (2016) Fabrication of ampicillin/starch/polymer composite nanofibers with controlled drug release properties by electrospinning. J Sol-Gel Sci Technol 77:594–603. https://doi.org/10.1007/s10971-015-3887-x

116. He X, Cheng L, Zhang X et al (2015) Tissue engineering scaffolds electrospun from cotton cellulose. Carbohydr Polym 115:485–493. https://doi.org/10.1016/j.carbpol.2014.08.114

117. Mendoza-palomares C, Ferrand A, Facca S et al (1977) Infection with contact lens wear. Br J Ophthalmol 61:249. https://doi.org/10.1136/bjo.61.4.249

118. Prabhakaran MP, Ghasemi-Mobarakeh L, Jin G, Ramakrishna S (2011) Electrospun conducting polymer nanofibers and electrical stimulation of nerve stem cells. J Biosci Bioeng 112:501–507. https://doi.org/10.1016/j.jbiosc.2011.07.010

119. Xue J, He M, Niu Y et al (2014) Preparation and in vivo efficient anti-infection property of GTR/GBR implant made by metronidazole loaded electrospun polycaprolactone nanofiber membrane. Int J Pharm 475:566–577. https://doi.org/10.1016/j.ijpharm.2014.09.026

120. He XX, Yu GF, Wang XX et al (2017) Electromagnetic functionalized micro-ribbons and ropes for strain sensors via UV-assisted solvent-free electrospinning. J Phys D Appl Phys 50. https://doi.org/10.1088/1361-6463/aa8101

121. Cui W, Cheng L, Hu C et al (2013) Electrospun Poly(L-Lactide) Fiber with Ginsenoside Rg3 for Inhibiting Scar Hyperplasia of Skin. PLoS ONE 8:1–12. https://doi.org/10.1371/journal.pone.0068771

122. Veras FF, Roggia I, Pranke P et al (2016) Inhibition of filamentous fungi by ketoconazole-functionalized electrospun nanofibers. Eur J Pharm Sci 84:70–76. https://doi.org/10.1016/j.ejps.2016.01.014

123. Abdalkarim SYH, Yu HY, Wang D, Yao J (2017) Electrospun poly(3-hydroxybutyrate-co-3-hydroxy-valerate)/cellulose reinforced nanofibrous membranes with ZnO nanocrystals for antibacterial wound dressings. Cellulose 24:2925–2938. https://doi.org/10.1007/s10570-017-1303-0

124. Qi H, Ye Z, Ren H et al (2016) Bioactivity assessment of PLLA/PCL/HAP electrospun nanofibrous scaffolds for bone tissue engineering. Life Sci 148:139–144. https://doi.org/10.1016/j.lfs.2016.02.040

125. Pournaqi F, Farahmand M, Ardeshirylajimi A (2016) Increase biocompatibility of scaffold made of Polyethersulfone (PES) by combining polyaniline(PANI). J Paramed Sci Winter 7:2008–4978

126. Duan H, Feng B, Guo X et al (2013) Engineering of epidermis skin grafts using electrospun nanofibrous gelatin/polycaprolactone membranes. Int J Nanomedicine 8:2077–2084. https://doi.org/10.2147/IJN.S42384

127. Yang Y, Xia T, Chen F et al (2012) Electrospun fibers with plasmid bFGF polyplex loadings promote skin wound healing in diabetic rats. Mol Pharm 9:48–58. https://doi.org/10.1021/mp200246b
128. Nemati S, Jeong KS, Shin YM, Shin H (2019) Current progress in application of polymeric nanofibers to tissue engineering. Nano Converg 6. https://doi.org/10.1186/s40580-019-0209-y
129. Ryan CNM, Fuller KP, Larrañaga A et al (2015) An academic, clinical and industrial update on electrospun, additive manufactured and imprinted medical devices. Expert Rev Med Devices 12:601–612. https://doi.org/10.1586/17434440.2015.1062364
130. Agarwal S, Wendorff JH, Greiner A (2008) Use of electrospinning technique for biomedical applications. Polymer (Guildf) 49:5603–5621. https://doi.org/10.1016/j.polymer.2008.09.014
131. Greve C, Jorgensen L (2016) Therapeutic delivery. Ther Deliv 7:117–138
132. Pauly HM, Kelly DJ, Popat KC et al (2016) Mechanical properties and cellular response of novel electrospun nanofibers for ligament tissue engineering: Effects of orientation and geometry. J Mech Behav Biomed Mater 61:258–270. https://doi.org/10.1016/j.jmbbm.2016.03.022
133. Gostev AA, Chernonosova VS, Murashov IS, et al (2019) IOPscience. In: Biomedical Materials. https://iopscience.iop.org/article/10.1088/1748-605X/ab550c
134. Katsanevakis E (2013) Engineering a biomimetic structure for human long bone regeneration. ProQuest Dissertations Theses, p 172
135. Griffith LG, Naughton G (2002) Tissue engineering-current challenges and expanding opportunities. Science (80-):295. https://doi.org/10.1126/science.1069210
136. Willerth SM, Sakiyama-Elbert SE (2007) Approaches to neural tissue engineering using scaffolds for drug delivery. Adv Drug Deliv Rev 59:325–338. https://doi.org/10.1016/j.addr.2007.03.014
137. Thevenot PT, Saravia J, Jin N, et al (2012), pp 1–59
138. Tanes ML (2016) Generating a bioactive protein gradient on electrospun nanofiber mats using a bovine serum albumin blocking scheme by COPYRIGHT © 2016 BY Michael Tanes Generating a bioactive protein gradient on electrospun nanofiber mats using a bovine Serum

Functionalized Carbon Nanotubes-Based Electrospun Nano-Fiber Composite and Its Applications for Environmental Remediation

Bharti, Pradeep Kumar, and Pramod Kumar Rai

Abstract Carbon Nanotubes (CNTs) have excellent properties such as high electrical and thermal conductivity and mechanical characteristic owing to their outstanding high specific surface area-to-volume ratio. However, there are restrictions for direct utilization of CNTs for processing and fabrication of devices because of their agglomeration tendency, difficulties in controlling morphology and leaching out problem from the composite material which lead to prevent its objective application with its inherent properties. The purpose of using functionalized CNTs (f-CNTs) to get homogeneous CNTs-based nano-composite which leads to enhancement of mechanical, chemical, electrical and thermal properties of composite materials. The surface area and pore size play an important role for removal and sensing of toxicants for environmental applications. Electrospinning is the most suitable technique for tuning the pore size and surface area of material as per requirement. The f-CNTs having various functional groups (hydroxyl, acetic, phenolic, polymer, etc.) improve their dispersion in matrix, water flux, scavenging of toxicants and attachment to the template for the fabrication of various significant devices such as filtration systems, sensors, high strength conducting fibers, etc. In this chapter, we are focusing on different techniques for the synthesis of f-CNTs, f-CNTs-based polymer nano-composites and fabrication of its nano-fibers using electrospun techniques for environmental remediation. The chapter concludes with the future prospect and challenges.

Keywords Functionalized carbon nanotubes · Electrospun nano-fibers · Polymer nano-composite · Environmental remediation

Bharti (✉) · P. Kumar · P. K. Rai
Centre for Fire, Explosive and Environment Safety (CFEES), Brig. S. K. Mazumdar Road, Timarpur, Delhi 110054, India
e-mail: bharti_2006@rediffmail.com

P. K. Rai
e-mail: pkrai@cfees.drdo.in

© The Author(s), under exclusive license to Springer Nature Switzerland AG 2021
S. K. Tiwari et al. (eds.), *Electrospun Nanofibers*, Springer Series on Polymer and Composite Materials, https://doi.org/10.1007/978-3-030-79979-3_13

1 Introduction

Carbon nanotubes (CNTs) are well-illustrated carbon nano-materials among the carbon family (activated carbon, graphene, carbon nanodiamond, fullerene, etc.) since its discovery in 1991 by Iijima [1]. CNTs are the seamless rolled structure of graphene sheet which may be single, double and multiple layers based on the number of graphitic layers constituting the hollow concentric wall of CNTs. CNTs hollow tube structure having diameter in a range of nanometers makes it special in the field of nano-science and technology [2–5]. It has extraordinary electrical, mechanical, thermal and chemical properties along with high specific surface area [4], but due to the weak interaction with other materials it is difficult to utilize its outstanding properties for fabrication of CNTs-based devices. CNTs remain in agglomerate form due to strong van der Waals forces among themself. Therefore, properties of composite materials do not effectively enhanced by increasing weight percentage of CNTs in composite material [6] due to poor dispersion and weak interaction of CNTs with matrix. The different techniques are applied for uniform dispersion of CNTs in matrix such as mechanical separation (ball milling, ultrasonication, etc.) physical stacking with other molecules (π-π interaction, electrochemical force of attraction, etc.) and chemical functionalization (sigma bond, co-ordination bond formation, etc.) [7–9]. Among the available techniques, functionalization of CNTs (physical/ chemical) is realized as best method for dispersion of CNTs. The mechanical separation leads to temporary dispersion and after prolong duration it gets re-agglomerate [7, 9] in solution or in matrix. Functionalization of CNTs is found to be one of the easiest, stable and effective techniques for loading of CNTs in appropriate quantities in different polymeric matrixes [10]. f-CNTs having various oxygen, nitrogen and sulfur containing groups either physically or chemically attached on the surface of CNTs, which is further helpful for secondary interaction with various solvents and other organic and inorganic molecules [10, 11]. The functionalization of CNTs resolves the agglomeration problem during fabrication of composite and devices [10, 12]. Therefore f-CNTs were directly utilized for composite materials to enhance the mechanical, electrical and thermal properties of composite materials for various applications in different fields. The f-CNTs-based nano-fiber composites have recently gained a lot of attention in various applications such as artificial muscles [12], drug delivery [13], electrodes [14], environment remediation such as toxicants sensors [15] along with fabrication of air and water purification system [4].

The water flux is an important factor for filtration system, the functionalization of CNTs enhances the hydrophilicity of water molecules which increases the water flux for filtration system [16]. Hollow fiber membranes (HFMs) have been extensively used to alleviate the crisis of water resources because of their large contact area per unit volume. The conductive CNT/nano-fiber composite hollow fiber membrane (CNC-HFM) has been successfully fabricated by coating and cross linking CNTs on the hollow fiber support layer made up of electrospun polyacrylonitrile (PAN) nano-fibers. The tensile stress and Young's modulus and water flux of CNC-HFMs composite was enhanced to 3.06, 12.7 and 7.3 times higher than those of commercial

HFMs prepared by phase inversion, which confirms that the mechanical strength along with water flux of CNC-HFMs significantly outperformed that of HFMs [16]. The functionalization of CNTs with chelating ligands leads to scavenging of toxic metals from contaminated medium [17, 18]. f-CNTs provides active sites for holding metal nano-catalyst for degradation of organic pollutants and also for conversion of toxic materials into non-toxic forms [18, 19].

There are several methods available for fabrication CNTs-based template such as self-assembly, vacuum filtration, spin-coating and electrospinning techniques [5, 20–22]. Electrospinning has been explored to a larger extent so as to develop an efficient functionalized CNTs-based nano-fiber composites [20, 22]. Electrospun nano-fibers have been widely used in the fields of electronics, medical and environmental applications due to the advantages of relatively high production rate, simplicity of the setup, uniformity and repeatability of synthesized nano-fibers [23–26].

Nowadays, environmental remediation is a burning issue for the saving of life on earth. Provision of clean water and pure air for living organisms is the major challenge of today's environment. Detection and removal of toxicants from water and air is an important field where researcher should give more devotion to get clean environment. Fabrication of nano-fibers of composite materials using electrospun technique has become increasingly attractive, mainly for removal of toxicants, because of their highly porous structure and large surface-to-volume ratio, which is one of the most important characteristic features required for purification of toxicants from gases and water [27].

Detection of pollutants plays an important role in remediation of environment. Conducting property of CNTs attracted various research groups for the investigation of different types of sensors using CNTs with different type conducting polymers [28].

The release of toxicants from the industries, burning of waste and effluents from households increase the pollution level in environment [29, 30]. They are continuously added into our water reserves and mixed into the fresh air in the form of oxide of sulfur and nitrogen. Particulate matters have been also increased in the atmosphere due to the industrial emissions, vehicle exhaust, household combustion, etc. The particulate matter of various sizes like thoracic particles less than 10 μm, PM 10, fine particles 2.5 μm, PM 2.5, ultrafine particles or nanoparticles; diameter less than 0.1 μm, PM 0.1[33] causes various adverse effect on human beings [31, 32]. The main motivation for choosing the f-CNTs-based electrospun nano-composite is to enhance the removal, degradation and detection efficiency of toxicants from air and water [33, 34]. In this chapter, we are focusing on different techniques for synthesis of functionalized carbon nano tubes (f-CNTs), f-CNTs-based polymer nano-composites and casting of its nano-fibers using electrospun techniques and their application for environmental remediation.

2 Functionalization of Carbon Nanotubes (CNTs)

The main purpose of functionalization of CNTs is to utilize the inherent outstanding properties (electrical, thermal, mechanical and chemical) by breaking the strong van der Waals force of attraction from bundle of CNTs and to make it suitable for effective processing or application in various fields [3–10]. Various organic and inorganic functional groups or molecules were physically adsorbed, chemically attached on the surface of CNTs (Fig. 1) and engulf into hollow concentric tube using mechanical, chemical and electrochemical techniques [10–12]. The surface modification of CNT is feasible due to pentagon and heptagon defects located near the tube end and bending of tubes [2, 3, 5]. The f-CNTs decorated with various types of functional groups and molecules provide accessibility for uniform dispersion CNTs in different solvents and polymeric matrix [35] along with required tailor properties. The spectacular properties of CNTs such as their high strength and stiffness make them ideal candidates for structural applications [2, 3, 5, 36]. Several research groups have reported successful functionalization reactions for single-walled carbon nanotubes (SWCNTs) and multiple-walled carbon nanotubes (MWCNTs) by physical stacking and chemical bond formation methods [34, 37, 38]. In a chemical bond formation method, graphitic surface of CNTs having sp^2 hybridized carbon atom change into sp^3 form by means of formation of new chemical bond with other organic (–OH, –COOH, NH_2, Fulurene, etc.) and inorganic (TiO_2, SnO_2, Au, etc.) moieties [19, 34, 39, 40]. Surface modification with carboxylic group (–COOH) and further functionalization of MWCNTs-COOH were carried by oxidation process using H_2SO_4/HNO_3 (3:1, v/v) at 70 °C and the modified CNTs were successfully utilized for covalent attachment of dodecylamine (DDA) and 3-aminopropyl triethoxysilane (3-APTES) [41].

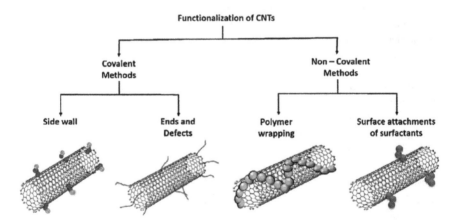

Fig. 1 Functionalization of Carbon nanotubes, reproduced with permission [11]. Copyright 2018 Elsevier

It is observed that pristine MWCNTs, shortly accumulate after finishing the sonication for dispersion of MWCNTs. In contrast, the aqueous solution with MWCNTs–COOH is stable up to 72 h [41]. The stability of MWCNT-COOH is achieved due to the conversion of hydrophobic graphitic CNT to hydrophilic CNT-COOH. The polar functional group attached on the surface of CNTs provides strong inter-facial bonding with polymers and other substrates having suitable functionalities. The well-dispersed CNTs provide active site for further processing of nano-materials and open diversified application of CNTs in different fields [2–5]. Similarly fluorination of CNT provides substitution site for alkyl, hydroxyl and amino group [42]. The fluorinated CNTs have C–F bonds that are weaker than those in alkyl fluorides and thus providing substitution sites for further functionalization [43]. Other methods, including cycloaddition, such as Diels–Alder reaction, carbene and nitrene addition, chlorination, bromination, hydrogenation, azomethine ylides, have also been successfully employed in recent years [37, 42–49]. Three types of functional groups, aminophenyl ($C_6H_4NH_2$), nitrophenyl ($C_6H_4NO_2$) and benzoic acid (C_6H_4COOH) were covalently attached on the surface of MWCNTs in order to make MWCNTs more compatible with the liquid crystalline polymer (LCP) by Sahoo et al. [50]. The effects of the electron donating and withdrawing groups attached to the f-MWCNTs on the dispersion of MWCNTs in the LCP matrix and their interaction with the LCP were investigated. The results showed that the MWCNTs-$C_6H_4NH_2$ demonstrated the highest intermolecular interaction between the f-MWCNTs and LCP, which led to the considerable enhancement in mechanical properties. Hariharasubramanian et al. functionalized SWCNTs with anthracene by urea using green approach [51]. It is observed that the anthracene functionalization demonstrated 300 fold increase in conductivity at room temperature as compare to bare SWCNTs. The π-π interaction between SWCNTs and the anthracene moiety enhances the conductivity of f-SWCNTs. Thus physical and chemical functionalization improves the accessibility of CNTs in different fields and provides active site for interaction with other molecules.

3 f-CNTs-Based Polymer Nano-Composites

At present, polymer nano-composite is one of the biggest application areas in the field of nano-science and technology. f-CNTs have various advantages over pristine CNTs for processing and synthesis of composite materials due to the ability of uniform distribution and loading of sufficient amount of CNTs into polymeric composite materials [52]. f-CNTs have been considered as an ideal nano-fillers for high-performance polymer-based nano-composites [52, 53]. f-CNTs bearing functional group such as carboxylic acid, fluorine, aryldiazonium compounds, nitrenes and azomethine ylides attached by formation of a new chemical bond and the molecules such as sodium dodecyl sulfate (SDS), poly (vinyl pyrrolidone), etc. are stacked by various force of attraction (van der Waals, electrostatic, hydrogen bonding, etc.) are more active toward a variety of monomeric and polymeric materials than bare

CNTs [41]. Gao and co-workers analyzed that in comparison with non-functionalized CNTs, the poly (ethyleneimine) PEI functionalized CNTs shown better dispersion in epoxy matrix and higher thermal conductivity [53]. Also 660% improvement in thermal conductivity was achieved in the CNT/epoxy matrix composite when 8 vol% of the PEI functionalized CNTs was loaded. Similarly, Hu et al. [54] recently prepare MWCNTs covalently functionalized by phenylenediamine (PDA) via an activation-grafting process for enhancement of chemical, mechanical, electrical and thermal properties of nano-composite. Different loading amounts were achieved by changing the activation time and the dosage of activating agent. It was observed that the adsorption capacity of MWCNTs toward CO_2 gas was increased from 0.17mmolg^{-1} to 0.59 mmolg^{-1} after modification with PDA under same condition. The characterization result of materials indicates that crystal structure of MWCNTs remains same whereas defect sites, amorphous structures and pore size increased after functionalization. The effective enhancement of the CO_2 capturing capacity attributed to the increased number of active sites, which offered strong interactions with CO_2 after PDA grafting. Incorporating f-CNTs into polymers for structural improvement is an important means of designing robust polymers for diverse applications; for example, the tensile strength of polystyrene (PS) for structural applications has been improved through the preparation of its composites with fluorinated SWCNT; this increase in mechanical strength was attributed to the formation of strong covalent bonds between SWCNT and the polymer matrix [55]. New functionalization methods, such as ion exchange, chelation and ring opening polymerization, are projected to obtain f-CNTs with improved performance. Rahman's group reported dispersible samples of CNTs which were carried out through immobilizing gold nanoparticles-S-$(CH_2)_{11}$-CH_3 on the CNTs surface [40]. Aromatic compounds with conjugate macrocycle such as pyrene, anthracene and their derivatives could be tightly adsorbed onto CNTs by π- π stacking interactions. Therefore, they were used as linkage to bind metal nanoparticles and other molecules on the surface of CNTs. Detailed scheme for the dispersion of various functionalized CNTs in solvent or polymer is shown in Fig. 2.

4   Electrospun Nano-Fibers Composite of f-CNTs/Polymers

Electrospun (Fig. 3) is the most often used method to fabricate fibers ranging in diameter from the submicron to nanometer level by using an electric force and varying critical parameter such as viscosity of polymeric solution, potential between electrodes and diameter of syringe [56–61]. The electrospinning technique was invented in 1934. The main constituents of the basic setup for electrospinning are polymer solution, a needle, a high-voltage power supply and an electrode collector. The electrospinning needs a high electric field between a droplet of polymer fluid at the needle and a collector. When the voltage reaches a critical value of 5–20 kV, the polymer flows toward the screen and starts whipping around. Keeping the utility of electrospinning in mind, several studies are done for optimizing the process and

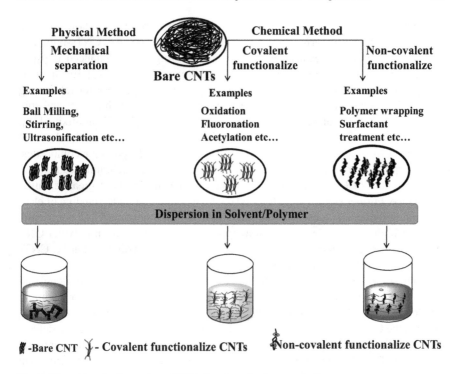

Fig. 2 Illustration for dispersion of CNTs in solvent/ polymer solution

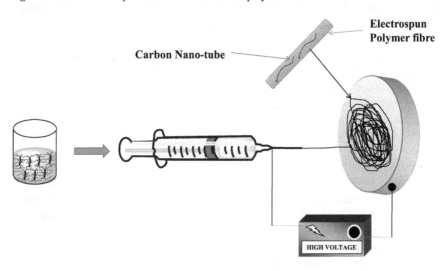

Fig. 3 Fabrication of f-CNTs-based electrospun nano-fiber composite reproduced with permission [61]. Copyright 2020 American chemical society

related parameters [62–65]. Due to unique structural features, CNTs show potential for the development of various nano-composite materials [66–70]. CNTs-polymers being of high interest because of their ultra-light weight and enhanced physical properties make them suitable for various applications [24, 25, 71–73]. Electrospinning provides low-cost, safe and rapid one-step integrated process for fabricating polymer/CNTs composite. Initially host fabrication methods such as film and spin casting [74] combination of solvent casting and melt mixing [75], shear mixing [76], in-situ polymerization [77], matrix-assisted pulsed laser evaporation [78], chemical vapor deposition growth of CNTs [8] were used for CNTs-polymer composites. The incorporation of CNTs into polymer by electrospinning technique leads to the even alignment of CNTs [79, 80] within the polymer matrix is of great scientific importance. Further studies were focused on the fundamental understanding of electrospinning dynamics to escalate the more interest in the polymer/CNT composite. Being so high demanding, various important problems are there in polymer/CNT composite to understand like uniform CNTs dispersion in polymer matrix, CNTs aggregation into bundles and compatibility with polymers [21, 81–83]. In electrospinning, some surfactants like sodium dodecyl sulfate, poly(vinyl pyrrolidone) and natural polysaccharide Gum Arabic were added to CNTs and sonicated for optimum time to obtain stable aqueous CNT dispersions otherwise some defects were introduced in the CNTs. Some other means of dispersion that can overcome these defects are ball milling, electrophoresis and polymer wrapping [84–86]. For the synthesis of CNT-reinforced polymeric nano-fibers, strong inter-facial adhesion and uniform dispersion are more crucial factors for improving the physical and chemical properties of the composite nano-fibers. The covalently and non-covalently functionalized CNTs have proved to be a better option for uniform dispersion of CNTs in polymeric composite as discussed in above section.

The f-MWCNTs/PAN having aromatic amine group on side wall was prepared by Wang et al. [80] compared with p-MWCNTs/PAN composites, f-MWCNTs/PAN displayed distinctly increased dispersion in solvent (DMF). The f-MWCNTs/PAN nano-composite was polymerized by in-situ and ex-situ solution system and composite nano-fibers with a diameter of 400–600 nm have been prepared using electrospinning technique. f-MWCNTs processed excellent compatibility and homogeneous dispersion in the in-situ composite nano-fibers. The mechanical and thermal properties of in-situ composite nano-fibers were more enhanced in comparison to ex-situ. The improved composite nano-fiber crystallinity resulted in the increased modulus and tensile strength. Kaur et al. [79] synthesized electrospun PAN nano-fibers and PAN–CNTs nano-composite nano-fibers and converted the materials into carbon nano-fiber and carbon-CNT composite nano-fiber, respectively. Both types of nano-fibers were undergone through oxidation, stabilization, carbonization and graphitization. It was observed that PAN–CNT composite nano-fiber diameter decreases with increasing concentration of CNT from 1 to 3 wt% in PAN solution. The minimum diameter of fibers in case of 3 wt% of CNT solution is observed due to increase of viscosity which led to reduce the phase separation of PAN and solvent. The degree of graphitization increases on increasing CNT concentration because of additional stress exerting on the nano-fiber surface in the immediate vicinity of CNTs. Similarly, electrospun carbon nano-fibers/carbon nanotubes/Polyaniline nano-composites were prepared by Miao et al. [14] and its shows enhanced electrochemical performance for flexible solid-state super capacitors. A lot of work is carried out for synthesis of CNT-polymer nano-composite using electrospinning techniques to enhance the various properties of materials. f-CNTs play an important role for getting uniform materials in comparison to bare CNTs which enhances the performance of composite materials.

5 Application of CNTs-Based Electrospun Nano-Composite for Environmental Remediation

There is a need of effective, stable and environmentally friendly method for purification and detection of toxicants from environment. A remarkable achievement was accomplished by membrane filtration technologies [23, 87–89] for removal of toxicants from air and water. Comparing with other methods, CNTs-based composite materials open new frontiers in the field of filtration technologies. CNTs-based composites and electrospun CNTs-based nano-fibers composites represent a relatively low cost, highly tunable route for the preparation of efficient material for environmental remediation. The electrospun f-CNTs-based fibers composite are identified as the optimistic materials for removal of toxicants along with sensing of their presence in water and air.

5.1 Removal of Toxic Metals from Water

Various toxic metal ions chromium, cadmium, mercury, lead, zinc, copper, uranium and thorium are present in water due to various activity and they tend to accumulate in living organisms with time. The toxicity of these pollutants prompted the scientific community to develop technologies for their effective removal [89–92]. Water technologies like adsorption, filtration, coagulation sedimentation, ion exchange, flocculation and ozone treatment are available for water detoxification, however, membrane-based water technologies are considered to be the most efficient, economically sustainable and clean process [23, 89, 93, 94].

The f-CNTs-based nano-fiber composites used for metal ion removal are due to high adsorption capacities, high selectivity, tolerance to pH values, high mechanical strength, re-usability, lower pressure drop and high permeation flux [95, 96]. Due to the surface characteristics of electrospun nano-materials, they are being extensively used in wastewater treatment processes for metal ion adsorbent [87, 89, 90, 97, 98]. Interconections, pore size in a range of 0.01–10 mm and maximum porosity up to 90% of electrospun nano-materials make them suitable for the removal of metal ions. Usually polymers used in electrospun are polyvinyl alcohol (PVA), polyacrylic acid (PAA), PAN, polyethyleneimine (PEI), polyesters, and some biopolymers for febrication of fiber.

The polymeric electrospun nano-fibers loaded with f-CNTs are effective nano-composites for removal of toxic metals. As CNTs prefer to functionalize at ends due to bending defect, the end of f-CNTs is hydrophilic in nature which interact with water and polymeric materials. The surface of CNTs can be modified for selective removal toxic metals. The surface of CNTs, functionalized by chelating agent acts as a scavenger for removal of metals from water [17, 18] and non-functionalized part of CNTs favor the flow of water. Recently, a super hydrophobic CNTs coated by spraying method on electrospun nano-fiber used for water desalination [95]. Here presence of CNTs enhances the hydrophobicity that results in anti-wetting and high water flux (28.4 kg/m^{2}h) in desalination process. A CNTs/nano-fiber composite hollow fiber membrane (CNC-HFM) was also designed by electrospun which has the pure water flux and Young's modulus 7.3 and 12.7 times higher than the commercial polyacrylonitrile [96]. Various f-CNTs-based electrospun nano-composites used for treatment of water are listed in Table 1.

Table 1 f-CNTs-based electrospun nano-composites used for treatment of water

S. no	Material	Toxicant	Capacity/% Removal	References
1	CNTs-Laccase	Bisphenol A	96.00%	[99]
	Laccase		70.00%	
2	CNTs-PAN	Pyrene	96.00%	[100]
	PAN		23.00%	
3	CNTs-Polyvinylpyrrolidone-TiO$_2$	Benzene	52.00% (degradation)	[19]
	Polyvinylpyrrolidone-TiO$_2$		18.00% (degradation)	
4	CNTs-Polyvinylpyrrolidone-TiO$_2$	Methylene Blue	58% (degradation)	[19]
	Polyvinylpyrrolidone-TiO$_2$		15% (degradation)	
5	CNTs-ACF	Methyl Orange	90.00%	[101]
	ACF		40.00%	
6	CNTs-PVDF	Water Desalination	28.4 kgM^2/h (Water Flux) > 99.9% (Salt Rejection%)	[102]
	PVDF		22.4 kgM^2/h (Water Flux) 99.9% (Salt Rejection%)	
7	CNT-PAN	Oil (Heptane) absorption	75.00%	[103]
	PAN		0.00%	

5.2   Removal of Organic Pollutants from Water

Water pollution due to organic compounds mainly organic dyes has severe effect on human as well as on our environment due to their toxicity [102]. The prominent producers of organic pollutants are industrial units of textiles, cosmetics, food and paper [102, 104]. Due to the production of various goods in these industries, large amounts of toxic compounds are released into the water, which causes millions of death due to water born diseases. Therefore researchers have huge interest in developing the technologies for their decomposition [105–107] and removal from water [104, 108, 109]. CNTs are the promising candidate for the dye absorption but they exhibit drawbacks related to their environment release [110]. One of the best methods to overcome these drawbacks is to load CNTs via electrospinning with some polymers like PAN [80]. CNTs modified electrospun activated carbon fiber (ACF) were used for electrochemical degradation of methyl orange (MO) dye [111]. The degradation was achieved up to 90% in 60 min at an optimum pH of 3.0 with high applied voltage. Electrospun nano-fibers fabricated using open tips oxidized CNTs were used for the adsorption of pyrene [112]. By using 2% oxidized open tips CNTs by weight with respect to PAN polymer, 96% adsorption of pyrene was obtained in 30 min contact time at room temperature [112]. Laccase (p-diphenol: dioxygen oxidoreductases) belongs to multi-copper polyphenol oxidase modified with MWCNTs by emulsion electrospinning was used for the removal of Bisphenol-A (BPA) from wastewater with the removal efficiency of 90% [113]. Laccase fibrous membrane also shows degradation efficiency of BPA up to 45% and with MWCNTs the degradation efficiency increases up to 80%. Their efficiency is tolerated over wide range of temperature and pH values [113]. The electrospun material of carbon nano-fiber carbon nanotube (CNF—CNT) composite in PAN was used for the absorption of organic micro pollutants like Cotinine, Metoprolol, Caffeine, Acetaminophen, Atrazine, Sulphamethoxazole, Sulphadimethoxine, Naproxen, Gemfibrozil and Bezafribate [114]. The Pd/CNF-MWCNT composite prepared via electrospun technique was reported for the application of nitrite hydrogenation in water purification. The loading of MWCNTs with 1.0–2.5 wt% to PAN significantly enhances the nitrite reduction efficiency compared to the catalyst without MWCNT [101].

5.3   Oil–Water Separation

The water pollution by oil is the mainly found in seawater or rivers by oil spills by shipping accidents, marine vessel leakage, and illegal industrial discharges, etc. [100]. Due to the oil spills in water system there is huge loss to aquatic life. Recently there is a huge oil spillage of about 12 km occurs in Ambarnaya river in Russia which will be cleared in about 10 years. Considering the loss of aquatic systems and the time taken for clearance of oil from water, there is an urgent need to develop [99] efficient technologies to tackle these problems. From the existing technologies like in-situ burning, mechanical extraction, bioremediation used for this purpose, mechanical extraction is most efficient [115] process considering their effectiveness and economy. The large and tunable surface in absorbent materials with the ability of re-usability makes them cost effective and efficient technology in oil–water treatment [116]. In this process various materials [117] are used but electrospun nano-fibrous materials are most studied due to their high oleophilicity/hydrophobicity and the oil selectivity [99, 118]. Base polymer materials like poly (vinylidene fluoride) (PVDF), cellulose acetate (CA), poly (L-lactide) (PLA), polystyrene (PS) are the most commonly studied for electrospinning with inorganic and organic nano-materials [117–123]. Recently, electrospun 3D CNT- PAN nano-fibers nano-composite have been developed for adsorbing oil from water [120]. f-CNTs wrapped in PS polymer were synthesized to create PS-CNTs nano-composite by electrospun technique. The author shows that the fluorine functionalized CNT are more dispersible and better alignment in polymer than bare CNTs. The synthesized material shows two-phase sorption of oil by pores that help in the entering inside the membrane and other by absorption on the surface due to Van der Waals force. The oil sorption capacity of the PS-CNTs was 65% higher than that of the PS sorbent without CNTs, and found to 123, 116, and 112 g/g, for peanut oil, sunflower oil, and motor oils, respectively. Further the re-usability check by oil sorption test shows promising potential in this field [121]. Recently, poly (acrylonitrile-co-vinyl acetate) and SWCNTs composite was synthesized by electrospinning using various proportions of SWCNTs such as 0.1, 0.5 and 1.0% w/w. The oil/water separation rejection shows 97.5% for 0.5 wt% SWCNTs in composite [122]. A superhydrophobic and super oleophilic nano-fibrous membrane was successfully developed using 0.15 wt%. of CNTs by Kai Wang et al. [103] with an improved mechanical strength. The resulting composite membrane was shown to be improved hydrophobicity (water contact angle $= 152°$) and oleophilicity (oil contact angle $\sim 0°$, oil permeation time $=$ 91.63 ms). The composite membrane resulted in a strong acid/alkali resistance (pH $= 1 - 14$) with high thermal stability (160 °C). It enhances oil–water separation flux (max $= 9270 \, L \cdot m^{-2} \cdot h^{-1}$) and a separation efficiency up to 99% under the gravity.

5.4 Removal of Gaseous Pollutants

In today's lifestyle, transportation, manufacturing and construction are essential needs of life but they are the main source of large amounts of dust, gases and particles into the atmosphere leading to the air pollution. Due to the continuous increase in the global air pollution, various technologists came forward to mitigate the problem [124] all around the world. Researchers are looking for coast effective and efficient technologies for the development of toxic gas adsorbents [125]. In this field electrospun materials are showing high performance because wide range tuning of fiber diameters from 40 to 2000 nm [126, 127] and other structural features like large surface area-to-volume ratios [128], controllable morphology [129, 130] and easy to introduce active nano-metal catalyst. For the electrospinning various polymers like PVA [131], polyethylene terephthalate (PET) [129], polyamide [132], Nylon-66 [133], nylon-6 [134], chitosan [135], cellulose acetate [136] were applied for the high performance of materials. The SWCNTs, MWCNTs and vertically aligned CNTs (VACNTs) show SO_2 physi-sorption with heat of adsorption values between 25 and 30 kJ/mol. Among them, the VACNTs have high SO_2 absorption value due to the presence of more absorption sites [137]. The 3-aminopropyltriethoxysilane (APTES) functionalized MWCNT has been synthesized for the adsorption of CO_2/CH_4 and CO_2 /N_2.. A comparative study for removal of CO_2, CH_4 and N_2 by Babaei et al. [138] using MWCNT and 3-aminopropyltriethoxysilane (APTES) functionalized MWCNTs (N-MWCNTs). The removal capacities for CO_2, CH_4 and N_2 using N-MWCNTs were found to be 2.26, 0.71 and 0.2 mmol/gm where for MWCNT 1.19, 0.63 and 0.19 mmol/gm, respectively, at room temperature. Further the author reported zeolite/N-MWCNT composite shows enhancement in gas adsorption capacity of 4.1, 1.86 and 0.27 mmol/gm for CO_2, CH_4 and N_2, respectively, and selectivity because of high micropore volume [139]. The results confirm that the functionalized MWCNTs are promising materials for the separation and purification of gases.

Further the composite materials of these electrospun fibers with CNTs improve the efficiency of filters. A report says an electrospun CNTs composite [140] used to enhance the properties of protective clothing for waterproof and breathable application. A amine functionalized CNTs doped CNFs used for adsorption of CO_2, 6.3 mmol g^{-1} at 1.0 bar and 298 K with an excellent CO_2/N_2 selectivity [141]. The electrospun CNT/TiO_2 nano-fibers were used for adsorption and photochemical degradation of gaseous benzene. The photocatalytic activity of 100 ppm gaseous benzene was performed by using decolorization of methylene blue solution in water in visible light. This system later on checked with simulated air purifier and 52% higher efficiency was observed for 50% $CNT-TiO_2$ composite than other composites [19].

6   Chemo-Sensor for Detection of Toxicants

Chemo-sensors are the important tools in the analytical chemistry and biotechnology for the detection of various contents in the system [142–146]. They detect analytes when comes in their contact by some response like fluorescent, color change, pH change, etc. They are used to detect a very simple to very complex systems. They are used for gas sensing, air quality checks, toxic gas detection, volatile compounds detection and various others [147–151]. Therefore, researchers develop huge interest in the development of cheap and efficient chemo-sensors for the environmental health. Some common methods used for the detection of gases are gas chromatography (GC) and high-performance liquid chromatography (HPLC) [152, 153] but these are expensive and time taking methods. Numerous gas sensing materials are developed by the researchers while the sensor of electrospun materials remains more efficient and cheap. Various types of CNTs-based sensor are being studied for application in different fields. In the electrospun composites, CNTs incorporated materials are highly studied materials (Table 2).

Table 2 f-CNTs-based electrospun nano-composites used for sensing of toxicants

S. no	Material	Polutant	Sensitivity/detection limit	References
1	PU-CNT-AgNP	H_2O_2	117.5 μAmM^-Cm^{-2}	[154]
	PU-AgNP		160.5 μAmM^-Cm^{-2}	
2	MWCNT-filled PANCAA nano-fibrous membrane	Glucose	0.10543 μAmM^{-1}/ 0.557 mM	[155]
	PANCAA nano-fibrous membrane		0.18354 μAmM^{-1}/ 0.668 mM	
3	Electrospun-calcined MWCNTs	Metronidazole	Ip = 3.75×10^{-5} A Ep = 0.902 V	[156]
	functionalized MWCNTs		Ip = 3.86×10^{-5} A Ep = 0.760 V	
4	PA6/PAH_MWCNTs nano-fibers	Dopamine	1.07 $\mu mol\,L^{-1}$	[157]
	MWCNTs		0.15 $\mu mol\,L^{-1}$	

In this context a MWCNTs/polyvinylacetate/titanium oxide (MWCNTs/PVAc/TiO_2) composite material with different loading of MWCNTs was reported for oxygen monitoring and for sensing of ammonia at moderate temperature ranges [158]. Electrospun CNTs composites are also used for metal ion detections. A modified poly m-phenylenediamine/carbon nanotube (PmPDA/CNT) nano-fiber was used to trace the copper from food and water sources with a very low detection limit (nano gram) range [159]. Han et al. [154] have reported, an integrated sensor system using mats formed of electrospun polymer/SWCNTs nano-fibers composite directly printed on the surface to detect volatile organic compounds. The response to different vapors showed a linear relationship between resistance change and vapor concentration. A team of Ouyang et al. [160] demonstrated a H_2O_2 biosensor based on electrospun PU–MWCNT–AgNP hybrid nano-fibers. The electrochemical analysis showed that the electrospun PU–MWCNT–AgNP nano-fiber reveals better electrocatalytic activity toward H_2O_2 than the electrodes modified by traditional casting strategy. Yang et al. [39] studied electrospun nano-fibers of bare SnO_2 and SnO_2 polycrystalline nano-fibers doped with MWCNTs for detection of carbon monoxide (CO) at room temperature. SnO_2/MWCNT nano-fibers are able to detect CO up to 50 ppm at room temperature, while the pure SnO_2 nano-fibers are insensitive up to 500 ppm. Similarly, electrospun MWCNTs/Nylon-6 nano-fiber composite was synthesized by Lala et al. and reported for sensing of various polar and non-polar organic solvent [161]. In recent year electrospun f-MWCNTs poly (acrylonitrile-co-acrylic acid) (PANCAA) nano-fibers membrane and β-phase polyvinylidene difluoride (PVDF) nano-fibrous membrane decorated with MWCNTs and platinum nanoparticles (PtNPs) were used for the sensing of glucose [155, 162]. Similarly Gusmao et al. [156] studied electrospun poly-butylene adipate co-terephthalate (PBAT) and polylactic acid (PLA) containing f-MWCNTs for the sensing of antibiotic pollutant metronidazole from aqueous solution. Electrospun

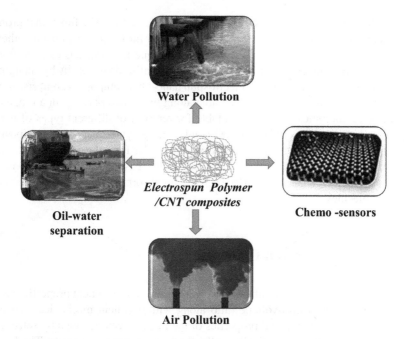

Fig. 4 Illustration for applications of f-CNTs-based electrospun nano-fiber for environmental remediation

f-MWCNTs composite shows higher current intensity (Ip = 3.86 × 10^{-5} A), with lower peak potential (Ep = 0.760 V) as compared to bare MWCNTs (Ip = 3.75 × 10^{-5} A; and Ep = 0.902 V). Mercante et al. [157] reported indium tin oxide (ITO) electrodes coated with CNTs electrospun polyamide 6/poly-(allylamine hydrochloride) (PA6/PAH) nano-fibers composite were synthesized for the detection of dopamine (DA) a neurotransmitter. The detection limit was found to be 0.15 μmol L^{-1} without interference of uric acid and ascorbic acid. By introducing different organic and inorganic functional nano-materials and building blocks into the polymer nano-fibers with electrospinning technique have got a lot of futuristic possibility for application in the field of chemo-sensors (Fig. 4).

7 Conclusions

We have reported the recent trend for synthesis of f-CNTs-based electrospun nano-fiber composite and its application for environmental remediation. CNTs comprise outstanding mechanical, chemical and thermal properties but it is worthless due to strong van der Waals of attraction among themselves, which destruct the uniformity of nano-composites. Various techniques are discussed to overcome from this problem and it is realized that functionalization of CNTs is the suitable method for uniform dispersion in different solvent and polymeric matrix. The functional group attached on the surface of CNTs act as an active site for further interaction with others molecule. Electrospinning is a most suitable technique for synthesis of nano-fiber composite because length, width and diameter can be altered very easily by changing various parameters. f-CNTs-based electrospun nano-fiber composites comprise high specific surface area. The high surface area along with huge porosity in a range of nanometer to micrometer makes it suitable for removal of different types of toxicants from air and water. f-CNTs-based conducting polymer nano-fiber composite is also an ideal material for detection of toxic molecules from the environment. The functional groups attached on the surface of f-CNTs-based electrospun nano-fiber composite material enhance the surface activity for absorptive removal and detection of toxic molecules.

8 Future Prospect and Challenges

The f-CNTs electrospun nano-fibers composites proved its excellent properties such as high specific surface-to-volume ratio along with excellent mechanical properties. The electrical and thermal properties of the material can be tuned by selecting proper polymers and types of CNTs for the synthesis of nano-composite. The functionalization of CNTs leads to uniform distribution of CNTs in polymeric solution for fabrication of electrospun nanofiber composites but strong acids/bases are being used the functionalization, generate defect on the surface of CNTs and also shorten the length of CNTs. Therefore critical analysis is required for optimization of f-CNTs-based electrospun nano-fibers. The orientation of CNTs in the polymeric material is a challenging factor. The proper orientation of CNTs will enhance the electrical and thermal properties along with mechanical properties of composite nano-fibers. Synthesis of vertical/horizontal aligned CNTs over electrospun nano-fiber composite may resolve the problem for fabrication of CNTs-based sensors and other devices but stability of CNTs on the surface of electrospun nanofibers and other properties of the composite materials should be analyzed. For nano-composite fibers, incorporation of inorganic compounds such as nano-metals or metal oxides enhance the various properties such as sensing ability and catalytic activity but maintaining good mechanical properties are a critical issues to be considered. The selection of appropriate functional moiety on the surface of CNTs or of polymers for holding of nano inorganic catalyst may solve the associated problems [160]. The functionalized CNTs-polymer nano-composite prevent the CNTs to leach out but a proper fate and transport study for CNTs and polymer in environment is required. The functionalized CNT nano-composite materials are found to be suitable for the removal of toxicants from air and water but self-toxicity of functionalized CNTs nano-composite materials are unknown. Thus long-term health hazards of the material to be examined so that functionalized CNTs composite nano-fibers can be utilized with its full potential.

Acknowledgements We would like to thank Director, CFEES for his constant support and encouragement during the work.

References

1. Iijima S (1991) Helical microtubules of graphitic carbon. Nature 354:56–58
2. Baughman RH, Zhakidov AA, de Heer WA et al (2002) Carbon nanotubes:the route toward applications. Science 297(5582):787–792
3. Gupta N, Gupta SM, Sharma SK et al (2019) Carbon nanotubes: synthesis, properties and engineering applications. Carbon Lett 29:419–447
4. Kurwadkar S, Hoang TV, Malwade K et al (2019) Application of carbon nanotubes for removal of emerging contaminants of concern in engineered water and wastewater treatment systems. Nanotechnol Environ Eng 4:12
5. Venkataraman A, Amadi EV, Chen Y et al (2019) Carbon Nanotube Assembly and Integration for Applications. Nanoscale Res Lett 14:220
6. Ajayan PM (1999) Nanotubes from carbon. Chem Rev 99(7):1787–1800
7. Manzetti S, Gabriel JP et al (2019) Methods for dispersing carbon nanotubes for nanotechnology applications: liquid nanocrystals, suspensions, polyelectrolytes, colloids and organization control. Int Nano Lett 9:31–49
8. Tucknott R, Yaliraki SN (2002) Aggregation properties of carbon nanotubes at interfaces. Chem Phys 281:455–463
9. Le VT, Ngo CL, Le QT et al (2013) Surface modification and functionalization of carbon nanotube with some organic compounds. Adv Nat Sci Nanosci Nanotechnol 4:1–5
10. Ma P-C, Siddiqui N A, Marom G et al (2010) Dispersion and functionalization of carbon nanotubes for polymer-based nanocomposites: a review. Compos Part A 41:1345–1367.
11. Jun LY, Mubarak NM, Yee MJ et al (2018) An overview of functionalized carbon nanomaterial for organic pollutant removal. J Ind Eng Chem 67(2018):175–186
12. Mottaghitalab V, Xi B, Gordon SGM et al (2006) Polyaniline fibres containing single walled carbon nanotubes: enhanced performance artificial muscles. Synth Met 156(11–13):796–803
13. Bianco A, Kostarelos K, Prato M (2005) Applications of carbon nanotubes in drug delivery. Curr Opin Chem Biol 9(6):674–679
14. Miao F, Shao C, Li X et al (2016) Electrospun carbon nanofibers/carbon nanotubes/polyaniline ternary composites with enhanced electrochemical performance for flexible solid-state supercapacitors. ACS Sustain Chem Eng 4(3):1689–1696
15. Kumar S, Pavelyev V, Mishra P et al (2018) A review on chemooresistive gas sensors based on carbon nanotubes: device and technology transformations. Sens Actuators A 2831:174–186
16. Du L, Quan X, Fan X et al (2019) Conductive CNT/nanofiber composite hollow fiber membranes with electrospun support layer for water purification. J Membr Sci. https://doi.org/10.1016/j.memsci.2019.117613
17. Sankararamakrishanan N, GuptaA VSR et al (2014) Enhanced arsenic removal at neutral pH using functionalized multiwalled carbon nanotubes. J Environ Chem Eng 2(2):802–810
18. Salehi E, MadaeniS S, Rajabi L et al (2012) Novel chitosan/poly(vinyl) alcohol thin adsorptive membranes modified with amino functionalized multi-walled carbon nanotubes for Cu(II) removal from water: preparation, characterization, adsorption kinetics and thermodynamics. Sep Purif Technol 89:309–319
19. Wongaree M, Chiarakorn S, Chuangchote S et al (2016) Photocatalytic performance of electrospun CNT/TiO$_2$ nanofibers in a simulated air purifier under visible light irradiation. Environ Sci Pollut Res 23:21395–21406
20. Dror Y, Salalha W, Khalfin RL et al (2003) Carbon nanotubes embedded in oriented polymer nanofibers by electrospinning. Langmuir 19(17):7012–7020

21. Ge JJ, Hou H, Li Q et al (2004) Assembly of well-aligned multiwalled carbon nanotubes in confined polyacrylonitrile environments: electrospun composite nanofiber sheets. J Am Chem Soc 126(48):15754–15761
22. Sobolciak P, Tanvir A, Popelka A et al (2018) Electrospun copolyamide mats modified by functionalized multiwall carbon nanotubes. Polym Compos 40(S2):E1451–E1460
23. Suja PS, Reshmi CR, Sagitha P et al (2017) Electrospun nanofibrous membranes for water purification. Polym Rev 57(3):467–504
24. Wang J, Lin Y (2008) Functionalized carbon nanotubes and nanofibers for biosensing Applications. Trends Analyt Chem 27(7):619–626
25. Tahhan M, Truong VT, Spinks GM et al (2003) Carbon nanotube and polyaniline composite actuators. Smart Mater Struc 12:26–31
26. Zong X, Kim K, Fang D et al (2002) Structure and process relationship of electrospun bioabsorbable nanofiber membranes. Polymer 43:4403–4412
27. Yoon K, Hsiao BS, Chu B et al (2008) Functional nanofibers for environmental applications. J Mater Chem 18:5326–5334
28. Zhou Y, Fang Y, Ramasamy RP et al (2019) Non-covalent functionalized of carbon nanotubes for electrochemical biosensor development. Sensors 19(2):392
29. Kjellstrom T, Lodh M, McMichael T et al (2006) Air and water pollution: burden and strategies for control in disease control priorities in developing countries, 2nd ed, pp 817–832
30. Miller KA, Siscovick DS, Sheppard L et al (2007) Long-term exposure to air pollution and incidence of cardiovascular events in women. N Engl J Med 356:447–458
31. Brook RD, Rajagopalan S, Pope CA et al (2010) Particulate matter air pollution and cardiovascular disease. Circulation 121:2331–2378
32. Mannucci PM, Harari S, Martinelli I et al (2015) Effects on health of air pollution: a narrative review. Intern Emerg Med 10:657–662
33. Rahimpour A, Jahanshahi M, Khalili S et al (2012) Novel functionalized carbon nanotubes for improving the surface properties and performance of polyethersulfone (PES) membrane. Desalination 286:99–107
34. Dyke CA, James MT (2004) Covalent functionalization of single-walled carbon nanotubes for materials applications. J Phys Chem A 108(51):11151–11159
35. Liu L, Barber AH, Nuriel S et al (2005) Mechanical properties of functionalized single-walled carbon-nanotube/poly(vinyl alcohol) nanocomposites. Adv Func Mater 15:975–980
36. Wang C, Zhou G, Liu H et al (2006) Chemical functionalization of carbon nanotubes by carboxyl groups on stone-wales defects: a density functional theory study. J Phys Chem B 110(21):10266–10271
37. Li H, Cheng F, Duft AM et al (2005) Functionalization of single-walled carbon nanotubes with well-defined polystyrene by click coupling. J Am Chem Soc 127(41):14518–14524
38. Mickelson ET, Huffman CB, Rinzler AG et al (1998) Fluorination of single-wall carbon nanotubes. Chem Phys Lett 296:188–194
39. Yang A, X Wang T R, Lee S, et al (2007) Room temperature gas sensing properties of SnO2/multiwallcarbon-nanotube composite nanofibers. Appl Phys Lett 91:133110–133113
40. Liu L, Wang TX, Li JX et al (2003) Self-assembly of gold nanoparticles to carbon nanotubes using a thiol-terminated pyrene as interlinker. Chem Phys Lett 367(5–6):747–752
41. Khabashesku VN, Billups WE, Margrave JL et al (2002) Fluorination of single-wall carbon nanotubes and subsequent derivatization reactions. Acc Chem Res 35:1087–1095
42. Touhara H, Inahara J, Mizuno T et al (2002) Fluorination of cup-stacked carbon nanotubes, structure and properties. Fluorine Chem 114:181–188
43. Tagmatarchis N, Prato MJ (2004) Functionalization of carbon nanotubes via 1,3- dipolar cycloadditions. J Mater Chem 14:437–439
44. Dyachkova TP, Rukhov AV, Tkachev AG et al (2018) Functionalization of carbon nanotubes: methods, mechanisms and technological realization. Adv Mater Technol 02:018–041
45. Balasubramanian K, Burghard M (2005) Chemically functionalized carbon nanotubes. Small 1:180–192

46. Hu H, Zhao B, Hamon MA et al (2003) Sidewall functionalization of single-walled carbon nanotubes by addition of dichlorocarbene. J Am Chem Soc 125:14893–14900
47. Holzinger M, Steinmetz J, Samaille D et al (2004) [2+1] Cycloaddition for cross linking SWCNTs. Carbon 42:941–947
48. Kim KS, Bae DJ, Kim JR et al (2002) Modification of electronic structures of a carbon nanotube by hydrogen functionalization. Adv Mater 14:1818–1821
49. Holzinger M, Abraham J, Whelan P et al (2003) Functionalization of single walled carbon nanotubes with (R-) oxycarbonyl nitrenes. J Am Chem Soc 125:8566–8580
50. Sahoo NG, Cheng HKF, Li L (2011) Covalent functionalization of carbon nanotubes for ultimate interfacial adhesion to liquid crystalline polymer.Soft Matter 7:9505–9514.
51. Hariharasubramanian A, Ravichandran YD, Rajesh R et al (2014) Covalent functionalization of single-walled carbon nanotubes with anthracene by green chemical approach and their temperature dependent magnetic and electrical conductivity studies. Mater Chem Phys 143:838–844
52. Wang S, Richard L, Ben W et al (2008) Load-transfer in functionalized carbon nanotubes/polymer composites. Chem Phys Lett 457:371–375
53. Huang J, Gao M, Pan T et al (2014) Effective thermal conductivity of epoxy matrix filled with poly(ethyleneimine) functionalized carbon nanotubes. Compos Sci Technol 95:16–20
54. Hu A, Zhang T, Yuan S et al (2017) Functionalization of multi-walled carbon nanotubes with phenylenediamine for enhanced CO_2 adsorption. Adsorption 23:3–85
55. Qian D, Dickeya EC, Andrews R et al (2000) Load transfer and deformation mechanisms in carbon nanotube-polystyrene composite. Appl Phys Lett 76:2868–2870
56. Feng JJ (2002) The stretching of an electrified non-Newtonian jet: a model for electrospinning. Phys Fluids 14:3912–3926
57. Yarin AL, Koombhongse S, Reneker DH et al (2001) Bending instability in electrospinning of nanofibers. J Appl Phys 89:3018–3026
58. Bognitzki M, Czado W, Frese T et al (2001) Nanostructured fibers via electrospinning. Adv Mater 13:70–72
59. Theron A, Zussman E, Yarin AL et al (2001) Electrostatic field-assisted alignment of electruspun nanofibres. Nanotechnology 12:384–390
60. Megelski S, Stephens JS, Rabolt JF et al (2002) Micro and nanostructured surface morphology on electrospun polymer fibre. Macromolecules 35:8456–8466
61. Tian X, He Y, Song Y et al (2020) Flexible cross-linked electrospun carbon nanofiber mats derived from pitch as dual-functional materials for supercapacitors. Energy Fuels. https://doi.org/10.1021/acs.energyfuels.0c02847
62. Kim J-S, Reneker DH (1999) Mechanical properties of composites using ultrafine electrospun fibers. Polym Compos 20:124–131
63. Koombhongse S, Liu WX, Reneker DH et al (2001) Flat polymer ribbons and other shapes by electrospinning. J Polym Sci Part B Polym Phys 39:2598–2606
64. Matthews J, Wnek GE, Simpson DG et al (2002) Electrospinning of collagen nanofibers. Biomacromol 3(2):232–238
65. Fong H, Chun I, Reneker DH et al (1999) Beaded nanofibers formed during electrospinning. Polymer 40:4585–4592
66. Breuer O, Sundararaj U (2004) Big returns from small fibers: a review of polymer/carbon nanotube composites. Polym Comp 25:630–645
67. Dersch R, Steinhart M, Boudriot U et al (2005) Nanoprocessing of polymers: applications in medicine, sensors, catalysis, photonics. Polym Adv Technol 16:276–282
68. Curran SA, Ajayan PA, Blau WJ et al (1998) A composite from poly(m-phenylenevinylene-co-2,5-dioctoxyp-phenylenevinylene) and carbon nanotubes: a novel material for molecular optoelectronics. Adv Mater 10:1091–1093
69. Bradley K, Gabriel JCP, Gruner G et al (2003) Flexible nanotube electronics. Nano Lett 3:1353–1355
70. Kymakis E, Amaratunga GAJ (2004) Optical properties of polymer-nanotube composites. Synth Met 142:161–167

71. Smith JG Jr, Connell JW, Delozier DM et al (2004) Space durable polymer/carbon nanotube films for electrostatic charge mitigation. Polymer 45:825–836
72. Wnek G, Carr ME, Simpson DG et al (2003) Electrospinning of nanofiber fibrinogen structures. Nano Lett 3:213–216
73. Joshi PP, Merchant SA, Wang Y et al (2005) Amperometric biosensors based on redox polymer–carbon nanotube–enzyme composites. Anal Chem 77:3183–3188
74. Safadi B, Andrews R, Grulke EA et al (2002) Multiwalled carbon nanotube polymer composites: synthesis and characterization of thin films. J Appl Polym Sci 84:2660–2669
75. Haggenmueller R, Gommans HH, Rinzler AG et al (2000) Aligned single-wall carbon nanotubes in composites by melt processing methods. Chem Phys Lett 330:219–225
76. Andrews R, Jacques D, Minot M et al (2002) Fabrication of carbon multiwall nanotube/polymer composites by shear mixing. Macromol Mater Eng 287:395–403
77. Jia Z, Wang Z, Xu C et al (1999) Study on poly(methyl methacrylate)/carbon nanotube composites. Mater Sci Eng A 271:395–400
78. Wu P K, Fitz-Gerald J, Pique A et al Deposition of nanotubes and nanotube composites using matrix-assisted pulsed laser evaporation. Mater Res Soc Proc 617:J2.3.1–6.
79. Kaur N, Kumar V, Dhakate SR et al (2016) Synthesis and characterization of multiwalled CNT–PAN based composite carbon nanofibers via electrospinning. Springerplus 5:483
80. Kaitao W, Mingbo G, Jian-jun W et al (2012) Functionalized carbon nanotube/polyacrylonitrile composite nanofibers: fabrication and properties. Polym Adv Technol 23. https://doi.org/10.1002/pat.1866
81. Ko F, Gogotsi Y, Ali A et al (2003) Electrospinning of continuous carbon nanotube-filled nanofiber yarns. Adv Mater 15:1161–1165
82. Ye H, Lam H, Titchenal N et al (2004) Reinforcement and rupture behavior of carbon nanotubes–polymer nanofibers. Appl Phys Lett 85:1775–1777
83. Liu J, Wang T, Uchida T et al (2005) Carbon nanotube core–polymer shell nanofibers. J Appl Polym Sci 96:1992–1995
84. Kim YA, Hayashi T, Fukai Y et al (2002) Effect of ball milling on morphology of cup-stacked carbon nanotubes. Chem Phys Lett 355:279–284
85. Yamamoto K, Akita S, Nakayama Y et al (1998) Orientation and purification of carbon nanotubes using ac electrophoresis. J Phys D Appl Phys 31:L34–L36
86. O'Connell MJ, Boul P, Ericson LM et al (2001) Reversible water-solubilization of single-walled nanotubes by polymer wrapping. Chem Phys Lett 342:265–271
87. Fane AG, Wang R, Hu MX et al (2015) Synthetic membranes for water purification: status and future. Angew Chem Int Ed 54:3368–3386
88. Xunda F, Tousley ME, Cowan MG et al (2014) Scalable fabrication of polymer membranes with vertically aligned 1 nm pores by magnetic field directed self assembly. ACS Nano 8:11977–11986
89. Ma H, Hsiao BS, Chu B et al (2013) Electrospun nanofibrous membrane for heavy metal ion adsorption. Curr Org Chem 17:1361–1370
90. Pereao OK, Bode-Aluko C, Ndayambaje G et al (2017) Electrospinning: polymer nanofibre adsorbent applications for metal ion removal. J Polym Environ 25(4):1175–1189
91. Nouri L, Ghodbane I, Hamdaoui O et al Batch sorption dynamics and equilibrium for the removal of cadmium ions from aqueous phase using wheat bran. J Hazard Mater 149(1):115–125.
92. Ogata T, Narita H, Tanaka M et al (2015) Adsorption behavior of rare earth elements on silica gel modified with diglycol amic acid. Hydrometallurgy 152:178–182
93. Shannon MA, Bohn PW, Elimelech M et al (2008) Science and technology for water purification in the coming decades. Nature 452:301–310
94. Wang EN, Karnik R (2012) Water desalination: graphene cleans up water. Nat Nanotechnol 7:552–554
95. Yan K-K, Jiao L, Lin S et al (2018) Superhydrophobic electrospun nanofiber membrane coated by carbon nanotubes network for membrane distillation. Desalination 437:26–33

96. Du L, Quan X, Fan X et al (2020) Conductive CNT/nanofiber composite hollow fiber membranes with electrospun support layer for water purification.J Membr Sci 596:117613. https://doi.org/10.1016/j.memsci.2019.117613.
97. Ahmed FE, Lalia BS, Hashaikeh R et al (2015) A review on electrospinning for membrane fabrication: challenges and applications. Desalination 356:15–30
98. Wang X, Hsiao BS (2016) Electrospun nanofiber membranes. Curr Opin Chem Eng 12:62–81
99. Wu J, Wang N, Wang L et al (2012) Electrospun porous structure fibrous film with high oil adsorption capacity. ACS Appl Mater Interfaces 4(6):3207–3212
100. Jiang Z, Tijing LD, Amarjargal A et al (2015) Removal of oil from water using magnetic bicomponent composite nanofibers fabricated by electrospinning. Compos B Eng 77:311–318
101. Ye T, Durkin DP, Maocong Hu et al (2016) Enhancement of nitrite reduction kinetics on electrospun Pd-carbon nanomaterial catalysts for water purification. ACS Appl Mater Interf 8:17739–17744
102. Rasheed T, Bilal M, Nabeel F et al (2019) Environmentally-related contaminants of high concern: potential sources and analytical modalities for detection, quantification, and treatment. Environ Int 122:52–66
103. Almasian A, Jalali M, Chizari L et al (2017) Surfactant grafted PDA-PAN nanofiber: optimization of synthesis, characterization and oil absorption property. Chem Eng J 326:1232–1241
104. Katheresan V, Kansedo J, Lau SY et al (2018) Efficiency of various recent waste water dye removal methods: a review. J Environ Chem Eng 6(4):4676–4697
105. Malwal D, Gopinath P (2016) Fabrication and applications of ceramic nanofibers in water remediation: a review. Crit Rev Environ Sci Technol 46(5):500–534
106. Gopal P, Mira P, Hak-Yong K et al (2015) Electrospun ZnO hybrid nanofibers for photo degradation of wastewater containing organic dyes: a review. J Ind Eng Chem 21:26–35
107. Peng C, Zhang J, Xiong Z et al (2015) Fabrication of porous hollow g-Al2O3 nanofibers by facile electrospinning and its application for water remediation. Microporous Mesoporous Mater 215:133–142
108. Gupta VK, Suhas (2009) Application of low-cost adsorbents for dye removal : a review. J Environ Manag 90(8):2313–2342
109. Wei L, Bingnan M, Yiqi Y et al (2019) Feasibility of industrial-scale treatment of dye wastewater via bio-adsorption technology. Bioresour Technol 277:157–170
110. Simate GS, Iyuke SE, Ndlovu S et al (2012) Human Health Effects of Residual Carbon Nanotubes and Traditional Water Treatment Chemicals in Drinking Water. Environ Int 39(1):38–49
111. Sun Y, Wang Y, Dong Q et al (2014) Electrolysis removal of methyl orange dye from water by electrospun activated carbon fibers modified with carbon nanotubes. Chem Eng J 253(1):73–77
112. Jadhav AH (2015) Preparation, characterization, and kinetic study of end opened carbon nanotubes incorporated polyacrylonitrile electrospun nanofibers for the adsorption of pyrene from aqueous solution. Chem Eng J 259:348–356
113. Dai Y (2016) enhanced performance of immobilized laccase in electrospun fibrous membranes by carbon nanotubes modification and its application for bisphenol A removal from water. J Hazard Mater 317:485–493
114. Peter AT, Vargo JD, Rupasinghe TP et al (2016) Synthesis, optimization, and performance demonstration of electrospun carbon nanofiber-carbon nanotubes composite sorbents for point of use water treatment. ACS Appl Mater Interfaces 8(18):11431–11440
115. Zhu H, Qiu S, Jiang W et al (2011) Evaluation of electrospun polyvinyl chloride/polystyrene fibers as sorbent materials for oil spill cleanup. Environ Sci Technol 45(10):4527–4531
116. Xianfeng W, Jianyong Y, Sun G et al (2016) Electrospun nanofibrous materials: a versatile medium for effective oil/water separation. Mater Today 19(7):403–414
117. Jianliang X, Weiyang L, Yihu S et al (2018) Graphene/nanofiber aerogels: performance regulation towards multiple applications in dye adsorption and oil/water separation. Chem Eng J 338:202–210

118. Dorneanu PP, Cojocaru C, Olaru N et al (2017) Electrospun PVDF fibers and a novel PVDF/CoFe$_2$O$_4$ fibrous composite as nanostructured sorbent materials for oil spill cleanup. Appl Surf Sci 424:389–396

119. Dorneanu PP, Cojocaru C, Samoila P et al (2018) Novel fibrous composites based on electrospun PSF and PVDF ultrathin fibers reinforced with inorganic nanoparticles: evaluation as oil spill sorbents. Polym Adv Technol 29(5):1435–1446

120. Jin L, Hu B, Kuddannaya S et al (2018) A three-dimensional carbon nanotube nanofiber composite foam for selective adsorption of oils and organic liquids. Polym Compos 39(S1):E271–E277

121. Bandegi A, Moghbeli MR (2018) Effect of solvent quality and humidity on the porous formation and oil absorbency of SAN electrospun nanofibers. J Appl Polym Sci 135(1):45586

122. Khalaf DM, Elkatlawy SM, Sakr A-HA et al (2020) Enhanced oil/water separation via electrospun poly(acrylonitrile-co-vinyl acetate)/single-wall carbon nanotubes fibrous nanocomposite membrane. J Appl Polym Sci 137:49033

123. Tian L, Zhang C, He X et al (2017) Novel reusable porous polyimide fibers for hot-oil adsorption. J Hazard Mater 340:67–76

124. Wang K, Zhang T C, Wei B et al (2020) Durable CNTs reinforced porous electrospun superhydrophobic membrane for efficient gravity driven oil/water separation. Colloids Surf A Physicochem Eng Aspects. https://doi.org/10.1016/j.colsurfa.2020.125342

125. Li P, Wang C, Zhang Y et al (2014) Air filtration in the free molecular flow regime: a review of high-efficiency particulate air filters based on carbon nanotubes. Small 10:4543–4561

126. Givehchi R, Tan Z (2015) The effect of capillary force on airborne nanoparticle filtration. J Aerosol Sci 83:12–24

127. Park H-S, Park YO (2005) Filtration properties of electrospun ultrafine fiber webs. Korean J Chem Eng 22:165–172

128. Kosmider K, Scott J (2002) Polymeric nanofibres exhibit an enhanced air filtration performance. Filtr Sep 39:20–22.

129. Sridhar R, Lakshminarayanan R, Madhaiyan K et al (2015) Electrosprayed nanoparticles and electrospun nanofibers based on natural materials: applications in tissue regeneration, drug delivery and pharmaceuticals. Chem Soc Rev 44:790–814

130. Lu P, Ding B (2008) Applications of electrospun fibers. Recent Pat Nanotechnol 2:169–182

131. Bhardwaj N, Kundu SC (2010) Electrospinning: a fascinating fiber fabrication technique. Biotechnol Adv 28:325–347

132. Qin XH, Wang SY (2006) Filtration properties of electrospinning nanofibers. J Appl Polym Sci 102:1285–1290

133. Matulevicius J, Kliucininkas L, Martuzevicius D et al (2014) Design and characterization of electrospun polyamide nanofiber media for air filtration applications. J Nanomater. https://doi.org/10.1155/2014/859656

134. Vitchuli N, Shi Q, Nowak J et al (2010) Electrospun ultrathin nylon fibers for protective applications. J Appl Polym Sci 116:2181–2187

135. Li L, Frey MW, GreenT B et al (2006) Modification of air filter media with nylon-6 nanofibers. J Eng Fibers Fabr 1:1–22

136. Desai K, Kit K, Li J et al (2009) Nanofibrous chitosan non-wovens for filtration applications. Polymer 50:3661–3669

137. Chattopadhyay S, Hatton TA, Rutledge GC et al (2016) Aerosol filtration using electrospun cellulose acetate fibers. J Mater Sci 51:204–217

138. Babu DJ, Puthusseri D, Kühl FG et al (2018) SO$_2$ gas adsorption on carbon nanomaterials: a comparative study. Beilstein J Nanotechnol 9:1782–1792

139. Babaei M, Anbia M, Kazemipour M (2019) study of the effect of functionalization of carbon naotubes on gas separation. Braz J Chem Eng 36:1613–1620

140. Babaei M, Anbia M, Kazemipour M (2016) Synthesis of zeolite/carbon nanotube composite for gas separation. Can J Chem. https://doi.org/10.1139/cjc-2016-0305

141. Li Y, Zhu Z, Yu J et al (2015) Carbon nanotubes enhanced fluorinated polyurethane macroporous membranes for waterproof and breathable application. ACS Appl Mater Interf 7:13538–13546

142. Iqbal N, Wang X, Yu J et al (2017) Robust and flexible carbon nanofibers doped with amine functionalized carbon nanotubes for efficient CO2 capture. Adv Sustain Syst 1:1600028
143. Zhang X, Yin J, Yoon J et al (2014) Recent advances in development of chiral fluorescent and colorimetric sensors. Chem Rev 114:4918–4959
144. Zhou Y, Zhang JF, Yoon J et al (2014) Fluorescence and colorimetric chemosensors for fluoride-ion detection. Chem Rev 114:5511–5571
145. Chen X, Zhou G, Peng X et al (2012) Biosensors and chemosensors based on the optical responses of polydiacetylenes. Chem Soc Rev 41:4610–4630
146. Kim HN, Ren WX, Kim JS et al (2012) Fluorescent and colorimetric sensors for detection of lead, cadmium, and mercury ions. Chem Soc Rev 41:3210–3244
147. Ko SK, Chen X, Yoon J et al (2011) Zebrafish as a good vertebrate model for molecular imaging using fluorescent probes. Chem Soc Rev 40:2120–2130
148. Bencic-Nagale S, Sternfeld T, Walt DR et al (2006) Microbead chemical switches: an approach to detection of reactive organophosphate chemical warfare agent vapors. J Am Chem Soc 128:5041–5048
149. Diaz de Grenu B, Moreno D, Torroba T et al (2014) Fluorescent discrimination between traces of chemical warfare agents and their mimics. J Am Chem Soc 136:4125–4128
150. Lei Z, Yang Y (2014) A concise colorimetric and fluorimetric probe for sarin related threats designed via the covalent-assembly approach. J Am Chem Soc 136:6594–6597
151. Ishida M, Kim P, Choi J et al (2013) Benzimidazole-embedded N-fused aza-indacenes: synthesis and deprotonation-assisted optical detection of carbon dioxide. Chem Commun 49:6950–6952
152. Tomas-Barbera FA, Gil MI, Cremin P et al (2001) HPLC–DAD–ESIMS analysis of phenolic compounds in nectarines, peaches, and plums. J Agric Food Chem 49:4748–4760
153. Ashley DL, Bonin MA, Cardinali FL et al (1992) Determining volatile organic compounds in human blood from a large sample population by using purge and trap gas chromatography/mass spectrometry. Anal Chem 64:1021–1029
154. Han L, Andrady AL, Ensor DS et al (2013) Chemical sensing using electrospun polymer/carbon nanotube composite nanofibers with printed-on electrodes. Sensors Actuators B 186:52–55
155. Zhang P, Zhao X, Zhang X et al (2014) Electrospun doping of carbon nanotubes and platinum nanoparticles into the β-phase polyvinylidene difluoride nanofibrous membrane for biosensor and catalysis applications. ACS Appl Mater Interf 6:7563–7571
156. Gusmao AP, Rosenberger AG, Muniz EC et al (2021) Characterization of microfibers of carbon nanotubes obtained by electrospinning for use in electrochemical sensor. J Polym Environ 29:1551–1565
157. Mercante LA, Pavinatto A, Iwaki Le EO et al (2015) Electrospun polyamide 6/poly(allylamine hydrochloride) nanofibers functionalized with carbon nanotubes for electrochemical detection of dopamine. ACS Appl Mater Interf https://doi.org/10.1021/am508709c
158. Khuspe GD, Navale ST, Bandgar DK et al (2014) SnO2 nanoparticles-modified polyaniline films as highly selective, sensitive, reproducible and stable ammonia sensors. Electron Mater Lett 10:191–197
159. Mehrani Z, Ebrahimzadeh H, Asgharinezhad AA et al (2019) Determination of copper in food and water sources using poly m-phenylenediamine/CNT electrospun nanofiber. Microchem J 149:103975. https://doi.org/10.1016/j.microc.2019.103975
160. Ouyang Z, Li J, Wang J et al (2013) Fabrication, characterization and sensor application of electrospun polyurethane nanofibers filled with carbon nanotubes and silver nanoparticles. J Mater Chem B 1:2415–2424
161. Lala N, Thavasi V, Ramakrishna S et al (2009) Preparation of surface adsorbed and impregnated multiwalled carbon nanotube/nylon-6 nanofiber composites and investigation of their gas sensing ability. Sensors 9:86–101
162. Wang Z-G, Wang Y, Xu H et al (2009) Carbon nanotube-filled nanofibrous membranes electrospun from poly(acrylonitrile-co-acrylic acid for glucose biosensor. J Phys Chem C 113, 2955–2960

163. Patil PT, Anwane RS, Kondawar SB et al (2015) Development of electrospun polyaniline/ZnO composite nanofibers for LPG sensing. Proc Mater Sci 10:195–204

CPSIA information can be obtained
at www.ICGtesting.com
Printed in the USA
LVHW080704210922
728862LV00007B/311

9 783030 799816